Serono Symposia USA
Norwell, Massachusetts

W0105711

Springer
New York
Berlin
Heidelberg
Barcelona
Hong Kong
London
Milan
Paris
Singapore
Tokyo

PROCEEDINGS IN THE SERONO SYMPOSIA USA SERIES

Continued after Index

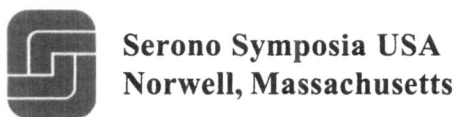

Serono Symposia USA
Norwell, Massachusetts

Francis L. Bellino

Editor

Biology of Menopause

With 58 Figures

Springer

Francis L. Bellino, Ph.D.
Biology of Aging Program
National Institute on Aging
Gateway Building, Suite 2C231
Bethesda, MD 20892-9205
USA

Proceedings of the International Symposium on Biology of Menopause, sponsored by Serono Symposia USA, Inc., held September 10 to 13, 1998, in Newport Beach, California.

For information on previous volumes, please contact Serono Symposia USA, Inc.

Library of Congress Cataloging-in-Publication Data
Biology of menopause / edited by Francis L. Bellino.
    p. cm.
  "Serono symposia USA."
  Includes bibliographical references and index.

    1. Menopause—Congresses. I. Bellino, Francis L.
  RG186 .B55 2000
  612.6'65—dc21   00-020064

Printed on acid-free paper.

Production coordinated by Chernow Editorial Services, Inc., and managed by Francine McNeill; manufacturing supervised by Jacqui Ashri.
Typeset by KP Company, Brooklyn, NY.

9 8 7 6 5 4 3 2 1

ISBN 978-1-4684-9530-0   ISBN 978-0-387-21628-7 (eBook)
DOI 10.1007/978-0-387-21628-7

SYMPOSIUM ON BIOLOGY OF MENOPAUSE

Scientific Committee

Francis L. Bellino, Ph.D., Chair
National Institute on Aging
Bethesda, Maryland

Sue P. Duckles, Ph.D.
University of California at Irvine
Irvine, California

William B. Ershler, M.D.
Institute for Advanced Studies in Immunology and Aging
Washington, D.C.

Andrew A. Monjan, Ph.D.
National Institute on Aging
Bethesda, Maryland

James F. Nelson, Ph.D.
University of Texas Health Sciences Center
San Antonio, Texas

Sherry Sherman, Ph.D.
National Institute on Aging
Bethesda, Maryland

William Sonntag, Ph.D.
Wake Forest University School of Medicine
Winston-Salem, North Carolina

Phyllis M. Wise, Ph.D.
University of Kentucky
Lexington, Kentucky

Organizing Secretary

Leslie Nies
Serono Symposia USA, Inc.
100 Longwater Circle
Norwell, Massachusetts

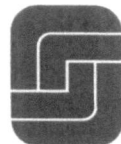

Preface

While it is not officially recognized as inevitable like death and taxes, menopause is a fact of life for women who live long enough. Until the 1990s, substantive research efforts into the biological and physiological mechanisms underlying the menopausal process were sparse. With the increased attention to women's health in the early 1990s came renewed interest in supporting and conducting studies intended to reveal the physiological manifestations and underlying mechanisms associated with the menopause, not only in Caucasian women in which most of the previous studies had been conducted, but in racially and ethnically diverse populations of women as well.

The International Symposium on the Biology of Menopause, held in September 1998, at Newport Beach, California, was cosponsored by Serono Symposia USA, Inc., and the National Institute on Aging (NIA). The purpose of this symposium was first to review the current state of the science of female reproductive aging across the hypothalamic–pituitary–ovarian (H-P-O) axis. The intention to subsequently review physiological connections between the menopausal process and the development of pathophysiology associated with the menopause, such as osteoporosis, cardiovascular disease, hot flashes, and excessive uterine bleeding, was compromised to some extent by the dearth of studies addressing those connections. Thus, to explore the role of estrogen in connecting menopause with pathophysiological processes in nonreproductive tissues, several of the presentations approached the question from the extensively studied "estrogen replacement therapy" perspective. Clearly, estrogen is a major player, perhaps even the major player, but is it the only H-P-O axis factor that impacts nonreproductive physiological systems in middle-aged and older women? The final session of the symposium addressed new techniques and experimental approaches that may help in resolving many of the long-standing questions related to the biology of the menopausal process and connections to other physiological processes. The ultimate goal of this symposium was to highlight gaps in knowledge that can be addressed by further focused research studies: basic, clinical, and behavioral. To improve the health, productivity, and enjoyment of life of older women, the NIA, along with several other institutes of the National Institutes of Health, continues to encourage and support studies designed to elucidate underlying molecular and cellular mechanisms of the menopausal process and its association with pathophysiological processes associated with the menopause (for further information, see program announcement PA-95-006, "Biology of the Menopause: Change of Ovarian Function," available through the internet at http://grants.nih.gov/grants/guide/pa-files/PA-95-006.html).

The organizing committee for this symposium, Sue Duckles, Bill Ershler, Andy Monjan, Jim Nelson, Sherry Sherman, Bill Sonntag, and Phyllis Wise, worked long and hard with me for over a year in planning this symposium. Their help is gratefully and enthusiastically acknowledged, and reflected in the extraordinarily high quality of the presentations in this volume. All of the speakers and participants are also recognized as contributing to a highly successful symposium. Finally, but not less importantly, I would like to recognize the extremely helpful and professional attitude of the Serono Symposia USA staff who contributed substantively to the success of this symposium, both by handling the details of the symposium itself and providing valuable and extensive assistance in editing this volume.

FRANCIS L. BELLINO

Contents

Contributors

TY W. ABEL, Department of Pathology, University of Arizona School of Medicine, Tucson, Arizona, USA.

ROLAND BARON, Department of Orthopedics and Rehabilitation, Yale University School of Medicine, New Haven, Connecticut, USA.

DAVID E. BATTAGLIA, Department of Obstetrics and Gynecology, University of Washington, Seattle, Washington, USA.

AYKUT BAYRAKCEKEN, Cornell University Medical College and New York Methodist Hospital, New York, New York, USA.

ADRIAN BONEV, Department of Pharmacology, The University of Vermont College of Medicine, Burlington, Vermont, USA.

JOSEPH E. BRAYDEN, Department of Pharmacology, The University of Vermont College of Medicine, Burlington, Vermont, USA.

HENRY BURGER, Prince Henry's Institute of Medical Research, Monash Medical Centre, Clayton, Victoria, Australia.

STEVE C. DANZER, Department of Pathology, University of Arizona School of Medicine, Tucson, Arizona, USA.

LORRAINE DENNERSTEIN, The Melbourne Women's Midlife Health Project, Office for Gender and Health, Department of Psychiatry, University of Melbourne, Carlton, Victoria, Australia.

R. REX DENTON, Department of Orthopedics and Rehabilitation, Yale University School of Medicine, New Haven, Connecticut, USA.

GIANLUCA D'IPPOLITO, Geriatric Research, Education, and Clinical Center, Veterans Affairs Medical Center, and Department of Medicine, University of Miami School of Medicine, Miami, Florida, USA.

EMMA DUDLEY, The Melbourne Women's Midlife Health Project, Office for Gender and Health, Department of Psychiatry, University of Melbourne, Carlton, Victoria, Australia.

JULIA A. ELVIN, Departments of Pathology and Molecular and Human Genetics, Baylor College of Medicine, Houston, Texas, USA.

WILLIAM S. EVANS, Department of Internal Medicine, University of Virginia, Charlottesville, Virginia, USA.

JODI ANNE FLAWS, Epidemiology and Preventive Medicine, University of Maryland, Baltimore, Maryland, USA.

SAMIR K. GHOSH, Department of Physiology, Vidyasagar College, Calcutta, West Bengal, India.

KATHRYN A. GREAVES, Department of Pathology, Section on Comparative Medicine, Wake Forest University School of Medicine, Winston-Salem, North Carolina, USA.

PATTIE S. GREEN, Department of Pharmacodynamics, Center for the Neurobiology of Aging, University of Florida College of Pharmacy, Gainesville, Florida, USA.

KELLY E. GRIDLEY, Department of Pharmacodynamics, Center for the Neurobiology of Aging, University of Florida College of Pharmacy, Gainesville, Florida, USA.

ANNE NEWMAN HIRSHFIELD, Office for Research Subjects, University of Maryland, Baltimore, Maryland, USA.

GUY A. HOWARD, Geriatric Research, Education, and Clinical Center, Veterans Affairs Medical Center, and Departments of Medicine and Biochemistry and Molecular Biology, University of Miami School of Medicine, Miami, Florida, USA.

MERCEDES JIMENEZ-LINAN, Department of Anatomy and Cellular Biology, Tufts University School of Medicine, Boston, Massachusetts, USA.

GUVENC KARLIKAYA, Cornell University Medical College and New York Methodist Hospital, New York, New York, USA.

KATALIN KAUSER, Cardiovascular Department, Berlex Biosciences, Richmond, California, USA.

MARIE C. KERBESHIAN, Department of Biology, University of Virginia, Charlottesville, Virginia, USA.

NANCY A. KLEIN, Department of Obstetrics and Gynecology, University of Washington. Seattle, Washington, USA.

HARM KNOT, Department of Medicine, The University of Vermont College of Medicine, Burlington, Vermont, USA.

LEWIS H. KULLER, Department of Epidemiology, University of Pittsburgh Graduate School of Public Health, Pittsburgh, Pennsylvania, USA.

PHILIPPE LEHERT, Sciences Economics Appliquees, University of Mons-Hainaut, Mons, Belgium.

ELLIS R. LEVIN, Department of Medicine, University of California at Irvine, Irvine, California, USA.

KAREN LOUNSBURY, Department of Pharmacology, The University of Vermont College of Medicine. Burlington. Vermont, USA.

NOBUYO MAEDA, Pathology and Laboratory Medicine, University of North Carolina at Chapel Hill, Chapel Hill, North Carolina, USA.

STAVROS C. MANOLAGAS, Division of Endocrinology and Metabolism, Center for Osteoporosis and Metabolic Bone Diseases, and the McClellan VA Medical Center, University of Arkansas for Medical Sciences, Little Rock, Arkansas, USA.

DENNIS W. MATT, Department of Obstetrics and Gynecology, Medical College of Virginia, Richmond, Virginia, USA.

KAREN A. MATTHEWS, Department of Psychiatry, University of Pittsburgh School of Medicine, Pittsburgh, Pennsylvania, USA.

MARTIN M. MATZUK, Departments of Pathology, Molecular and Human Genetics, and Cell Biology, Baylor College of Medicine, Houston, Texas, USA.

ISTVAN MERCHENTHALER, Women's Health Research Institute, Wyeth-Ayerst Research, Radnor, Pennsylvania, USA.

EILEEN K. MONCK, Department of Pharmacodynamics, Center for the Neurobiology of Aging, University of Florida College of Pharmacy, Gainesville, Florida, USA.

FREDERICK NAFTOLIN, Department of Obstetrics and Gynecology, Center for Research in Reproductive Biology, Yale University School of Medicine, New Haven, Connecticut, USA.

MARK T. NELSON, Department of Pharmacology, The University of Vermont College of Medicine, Burlington, Vermont, USA.

KUTLUK OKTAY, Cornell University Medical College and New York Methodist Hospital, New York, New York, USA.

SUZANNE OPARIL, Department of Medicine, Division of Cardiovascular Disease, Vascular Biology and Hypertension Program, University of Alabama at Birmingham, Birmingham, Alabama, USA.

STEVEN M. PINCUS, Guilford, Connecticut, USA.

NAOMI E. RANCE, Departments of Pathology, Cell Biology and Anatomy, and Neurology, University of Arizona School of Medicine, Tucson, Arizona, USA.

ANURADHA RAY, Department of Internal Medicine, Pulmonary and Critical Care Section, Yale University School of Medicine, New Haven, Connecticut, USA.

CAMILLO RICORDI, Department of Surgery and Diabetes Research Institute, University of Miami School of Medicine, Miami, Florida, USA.

BERNARD A. ROOS, Geriatric Research, Education, and Clinical Center, Veterans Affairs Medical Center, and Departments of Medicine and Neurology, University of Miami School of Medicine, Miami, Florida, USA.

GABOR M. RUBANYI, Cardiovascular Department, Berlex Biosciences, Richmond, California, USA.

BEVERLY S. RUBIN, Department of Anatomy and Cellular Biology, Tufts University School of Medicine, Boston, Massachusetts, USA.

PAUL C. SCHILLER, Geriatric Research, Education, and Clinical Center, Veterans Affairs Medical Center, and Departments of Medicine and Biochemistry and Molecular Biology, University of Miami School of Medicine, Miami, Florida, USA.

DAWN C. SCHWENKE, Department of Pathology, Wake Forest University School of Medicine, Winston-Salem, North Carolina, USA.

BARBARA B. SHERWIN, Departments of Psychology and Obstetrics and Gynecology, McGill University, Montreal, Quebec, Canada.

JIONG SHI, Department of Pharmacodynamics, Center for the Neurobiology of Aging, University of Florida College of Pharmacy, Gainesville, Florida, USA.

PAUL J. SHUGHRUE, Women's Health Research Institute, Wyeth-Ayerst Research, Radnor, Pennsylvania, USA.

JAMES W. SIMPKINS, Department of Pharmacodynamics, Center for the Neurobiology of Aging, University of Florida College of Pharmacy, Gainesville, Florida, USA.

OLIVER SMITHIES, Pathology and Laboratory Medicine, University of North Carolina at Chapel Hill, Chapel Hill, North Carolina, USA.

MICHAEL R. SOULES, Department of Obstetrics and Gynecology, University of Washington, Seattle, Washington, USA.

KIM SUTTON-TYRRELL, Department of Epidemiology, University of Pittsburgh Graduate School of Public Health, Pittsburgh, Pennsylvania, USA.

C. DOMINIQUE TORAN-ALLERAND, Departments of Anatomy and Cell Biology and Neurology, Columbia University College of Physicians and Surgeons, New York, New York, USA.

JOHANNES D. VELDHUIS, Department of Internal Medicine, University of Virginia, Charlottesville, Virginia, USA.

JANICE D. WAGNER, Department of Pathology, Section on Comparative Medicine, Wake Forest University School of Medicine, Winston-Salem, North Carolina, USA.

GERSON WEISS, Department of Obstetrics and Gynecology, University of Medicine and Dentistry of New Jersey–New Jersey Medical School, Newark, New Jersey, USA.

GEORGE C. WELLMAN, Department of Pharmacology, The University of Vermont College of Medicine, Burlington, Vermont, USA.

C. ROGER WHITE, Department of Medicine, Division of Cardiovascular Disease, Vascular Biology and Hypertension Program, University of Alabama at Birmingham, Birmingham, Alabama, USA.

PHYLLIS M. WISE, Department of Physiology, University of Kentucky, Lexington, Kentucky, USA.

LI ZHANG, Department of Pathology, Section on Comparative Medicine, Wake Forest University School of Medicine, Winston-Salem, North Carolina, USA.

1

New Understanding of the Complexity of the Menopause and Challenges for the Future

PHYLLIS M. WISE

Introduction

The menopause signals the permanent end of the reproductive period in a woman's life. Throughout medical history, humans have sought to understand this "change of life." Until about 1900, however, most women died before they ever experienced the menopause, so it never was the focus of attention of biomedical researchers, clinicians, epidemiologists, and social scientists as it is today. During this century, the average life span of humans has increased dramatically. Today, in the United States, the average life span of women is 83 years. There are more than 35 million postmenopausal women, and more than 1 million will reach this stage each year. Thus, an ever-increasing number and proportion of women will live a larger fraction of their lives in the postmenopausal state than ever before. It becomes critical, therefore, that we gather extensive and in-depth information about the ensemble of events that regulate the menopausal transition and the biomedical and societal repercussions of this dramatic transition in women from a reproductive to nonreproductive status. Our goals are to (1) summarize our current understanding of the natural course of the menopause and its repercussions, (2) discuss the mechanisms by which ovarian hormones influence reproductive and "nonreproductive" functions, (3) identify gaps in our knowledge, and (4) determine potential avenues and directions for research in reproductive aging. This chapter will highlight some of the issues that will be addressed in this volume.

Interactions Among the Hypothalamus, Pituitary, and Ovary Leads to the Menopause

At one time it was thought that the exhaustion of ovarian follicles was the sole reason for the menopausal transition. We are beginning to appreciate that it may be more complex. Long before the follicular reserve is exhausted, fertility and fecundity decrease, reproductive cycles become increasingly irregular in length and patterns of gonadotropin, and ovarian peptides and steroid secretion are altered (1). What are the critical variables that control these changes? Is there a single culprit? Is there a more complex interacting series of virtually simultaneous changes that play on and with each other that ultimately lead to the postmenopausal state. Several speakers will address this question from different perspectives. To understand the menopausal state, investigators have begun to focus on the period prior to the establishment of permanent acyclicity (i.e., the events that occur during the fourth decade of life in women or the equivalent stage in experimental animal models). Other chapters will discuss the important changes that occur during middle age as women approach the menopause and how they may interact.

For many years, it was thought that a decrease in the ovarian steroid, estradiol, was one of the first signals to herald the imminent transition to acyclicity in women. Data from Santoro and her colleagues (2) clearly demonstrate that estrogen levels in urine are maintained or elevated during the entire menstrual cycle during the pre- and perimenopausal period (Fig. 1.1, left panel). These levels of estrogen clearly indicate that the ovary remains capable of producing estrogen during these initial stages. The patterns of gonadotropin secretion also appear to change early during the middle-age. Matt and colleagues (3) reported that the interpulse interval of LH pulses

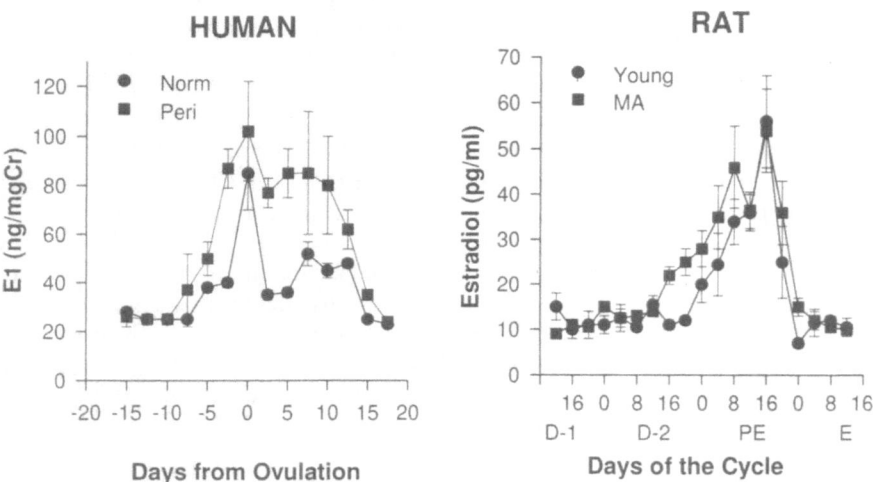

FIGURE 1.1. Left panel: Mean ± standard error daily urinary estrone conjugates excretion patterns in perimenopausal women compared to midreproductive-aged women. Data are standardized to day 0, the putative day of ovulation. (Modified with permission from Santoro N et al. Characterization of reproductive hormonal dynamics in the perimenopause. J Clin Endocrinol Metab 1996;81:1495–501. © The Endocrine Society.) Right panel: Mean ± standard error of plasma estradiol in 4- and 11-month-old regularly cycling female rats on various days of the estrous cycle. (Modified with permission from Ref. 9.)

and their duration increase on day 10 of the menstrual cycle in women who are still cycling regularly prior to any changes in estradiol or FSH (Fig. 1.2, left panel). In addition, one of the earliest changes during the peri-menopausal period is an increase in FSH secretion, which is accompanied by a decrease in inhibin-B (4) (Fig. 1.3, left panel). Because gonado-tropin-releasing hormone (GnRH) cannot be measured in the peripheral plasma, measurement of gonadotropins is used as a surrogate marker of GnRH secretory patterns. Thus, changes in the frequency of luteinizing hor-mone (LH) pulses and selective increases in FSH secretion may result from changes in the pattern of hypothalamic GnRH secretory patterns. Changes in the interpulse interval and duration of LH pulses may indicate a slow-ing of GnRH pulses and a desynchronization in pulses, such that the du-ration of increased secretion is greater. Knobil et al. (5) showed that a slowing of exogenously administered GnRH pulses can stimulate the selec-tive secretion of FSH. Thus, these early changes in patterns of gonado-

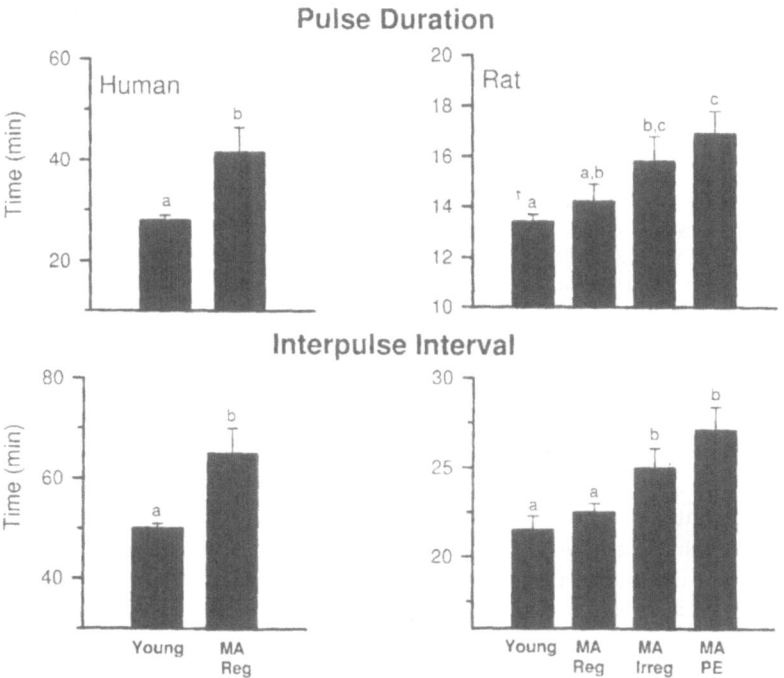

FIGURE 1.2. Left panel: Mean ± standard error of LH pulse duration and interpulse interval obtained from 8 hours of sampling during the mid- to late-follicular phase of menstrual cycle in young and middle-aged women. (Modified with permission from Matt DW et al. Characteristics of LH secretion in younger versus older premenopausal women. Am J Obstet Gynecol 1998;178:504–10.) Right panel: Mean ± standard error of LH pulse duration and interpulse interval of ovariectomized young and middle-aged rats at various stages of reproductive senescence obtained from 3 hours of sampling. (Modified with permission from Ref. 11.)

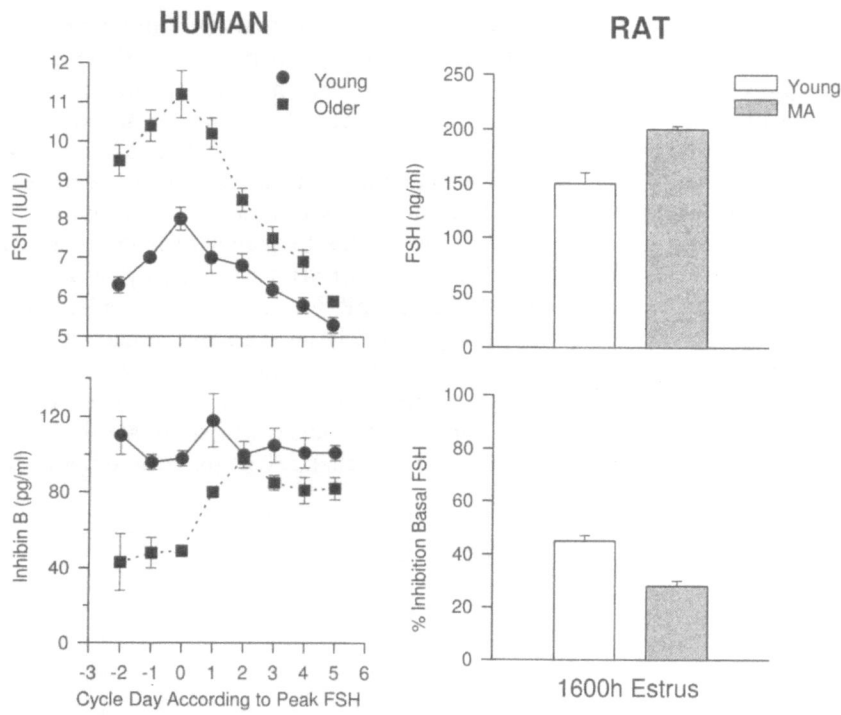

FIGURE 1.3. Left panel: Mean ± standard error daily FSH (top) and inhibin B (bottom) in older and younger women. (Modified with permission from Klein NA et al. Decreased inhibin B secretion is associated with the monotropic FSH rise in older, ovulatory women. J Clin Endocrinol Metab 1996;81:2742–45. © The Endocrine Society.) Right panel: Mean ± standard error of FSH and FSH suppressing activity in ovarian vein serum in 3- to 7-month-old rats during the periovulatory period. (Modified with permission from Ref. 12.)

tropin secretion may indicate changes in hypothalamic function. This is more difficult to assess in humans. In fact, I am unaware of any study that examines the status of GnRH neurons or the activity of any neurotransmitter that regulates GnRH during the perimenopausal period. For this information, we have relied on studies performed using animal models (see later). New findings on the changes in ovarian, pituitary, and hypothalamic function that occur during the perimenopausal period will be reviewed later in this volume.

The Rodent as a Model of the Human Menopause

There has been an ongoing and lively debate as to whether rodents serve as a good model for human reproductive aging. In some sense, because rodents do not undergo a true menstrual cycle, (i.e., there in no sloughing of the uterine wall at the end of each cycle) they cannot undergo a true "meno-pause." In addition, in rats and mice, there is no true luteal phase of the estrous cycle because the corpus luteum regresses rapidly after ovulation and progesterone is not secreted for a prolonged period of time. Arguments that rodents are not good models center primarily around two findings. First, in women, the loss of primordial follicles is log-linear during the initial stages of life and accelerates dramatically around the time women are 37 years old. This accelerated loss leads to the total absence of follicles when women are between 50 and 55 years old, when they are postmenopausal (6). In contrast, in rodents there is striking variation in the rate of follicular loss across different strains. Although no studies have followed the rate of follicular loss across the entire life span, Faddy and colleagues (7) showed that during the first 100 days of life, the rate of follicular loss, plotted on a log-linear scale to enable a fair comparison with the work of Richardson et al. (6), does not change with age in mice. These data would suggest that exhaustion of the follicular pool may not be a limiting factor. Second, in postmenopausal women, gonadotropin levels are clearly elevated in response to lowered estradiol. Gonadotropin concentrations decrease in response to estrogen replacement, demonstrating that the pathways that regulate negative feedback remain intact. In contrast, gonadotropin concentrations remain relatively normal in aged acyclic, repeatedly pseudopregnant rats (8) (Fig. 1.4). Thus, despite decreases in estradiol, gonadotropins do not appear to exhibit the dramatic increases observed in women. This suggests that decreased hypothalamic influences are paramount to the postreproductive state in rats.

Despite these differences, I am struck by the commonalities in endocrine changes that occur in female rats and in women. Changes in estradiol, LH, and FSH during the perimenopausal period in women and the middle-age transition to acyclicity in rats are strikingly parallel. Lu and colleagues (9) and Butcher and colleagues (10) reported that periovulatory estradiol concentrations are maintained or elevated in middle-aged rats (Fig. 1.1, right panel). In particular, during the irregularly cycling period of time, estradiol may be elevated for 2–3 days instead of a single day prior to ovulation. Thus, in both women and rats, diminished estradiol concentrations cannot explain the initial transition to irregular cycles. Likewise, changes in the patterns of pulsatile LH during the menstrual cycle of middle-aged women are remarkably parallel to those reported in ovariectomized middle-aged rats which were displaying different estrous cycle status at the time of ovariectomy (11) (Fig. 1.2, right panel). In both rats and women, interpulse interval and the duration of LH pulses increases during middle-age. Finally a rise in FSH has been observed (12) in middle-aged rats, particularly during the secondary FSH surge (Fig. 1.3, right panel). Together, these findings lead me to the conclusion that rodents are an excellent experimental model of female reproductive aging and have provided key insights into the mechanisms and factors that regulate reproductive aging. We assume that information gained from these species can be extrapolated, with care, to humans and will allow us to glean concepts that can be generalized to human reproductive aging.

6   P.M. Wise

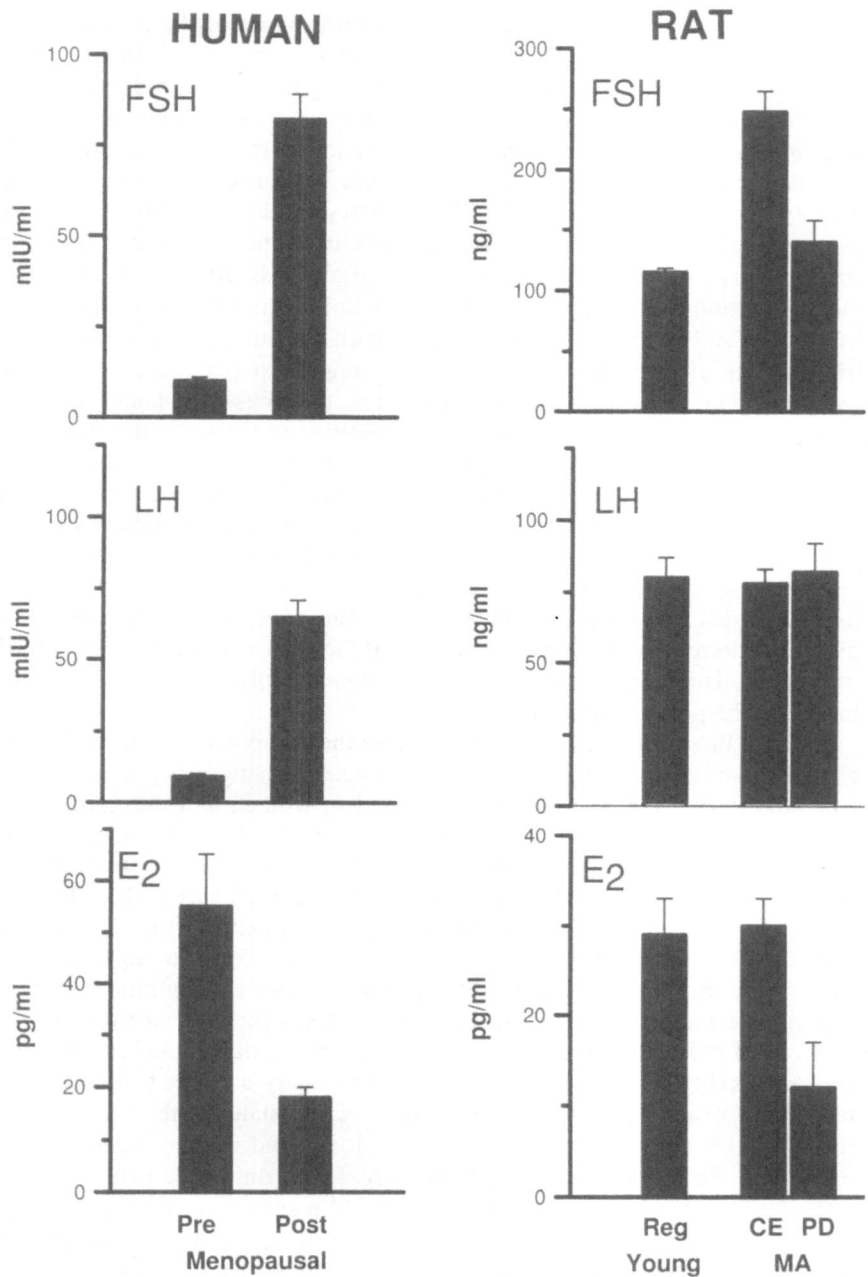

FIGURE 1.4. Left panel: Mean ± S.E. of FSH (top panel), LH (middle panel), and
estradiol (bottom panel) in young premenopausal and old postmenopausal women.
(Modified from Ref. 13.) Right panel: Mean ± S.E. of FSH (top panel), LH (middle
panel), and estradiol (bottom panel) in young regularly cycling rats compared to middle-
aged constant estrous or pseudopregnant rats. (Modified from Ref. 8.)

The Menopause Brings About
Many Physiological Changes

In some senses our growing realization that the menopause signals the beginning of a multitude of other biological changes in systems that are traditionally considered nonreproductive is one of the most exciting new areas of research. These repercussions of lowered levels of estrogen will be the focus of many chapters. During the 1990s, we began to understand that estrogen is not merely a female reproductive hormone that influences only traditional reproductive tissues (e.g., hypothalamus, anterior pituitary, mammary glands, uterus, and vagina). We are now beginning to appreciate that estrogen also affects a plethora of functions from bone and mineral metabolism to cardiovascular function. In addition, it affects memory and cognition, organization and expression of daily biological rhythms, and modulates the incidence of age-related diseases. In addition, we are beginning to appreciate novel and complex mechanisms of the action of estradiol that involve more than just the traditional estradiol receptor. Chapters will provide evidence of the repertoire of repercussion of the lack of estrogen and why it is so essential to multiple physiological systems.

Accumulating data clearly demonstrates that estrogens play a neuroprotective role in adults, during normal aging, in neurodegenerative diseases, and after brain injury. It modulates brain structure and function in areas that regulate memory. It may therefore also be an important factor in maintaining cognition and memory. Animal models and in vitro methods have been used to further explore the neuroprotective effects and to begin to determine the mechanisms of action of estrogen in this context. In a parallel manner, estrogens protect against bone loss by regulating the production of bone cell progenitors in the bone marrow. Thus, the postmenopausal hypoestrogenic state leads to stimulation of increased bone remodeling and ultimately increased osteoclastogenesis and bone loss. Finally increasing evidence demonstrates that estrogen protects against cardiovascular disease via multiple mechanisms, including modulation of systolic blood pressure, pulse pressure, total high density lipoprotein (HDL) and low density lipoprotein (LDL) cholesterol, and triglycerides. Estrogen appears to decrease vascular injury in models of atherogenesis through direct antiproliferative effects and indirectly via promotion of endothelial cell integrity and nitric oxide function.

There are many indications that estrogen's protective effects are not mediated exclusively by a classical estrogen receptor. It is very possible that the trophic and protective effects of estrogen are mediated genomically by estrogen receptors and/or by cross-talk with other second messenger systems. There is accumulating evidence that estrogen exerts effects by novel mechanisms that are different from the well-characterized mechanism involving transactivation, nuclear receptor dimerization, and binding to consensus estrogen response elements. Many different second messenger pathways, including cAMP or MAP kinases, may be involved. Some of these novel mechanisms of estrogen action may require cross-talk with the estradiol re-

8   P.M. Wise

ceptor; which receptor subtype (a or b) is involved is unclear. Whether ligand-bound receptors must bind to classic estrogen response elements to evoke protection and growth is less clear. Other trophic and protective effects of estrogen occur in the absence of any known estrogen receptor.

In summary, these are exciting times because of new findings and challenges that face us. These proceedings are timely and should help to bring us up to date in terms of the advances that we have made and the progress that should be forthcoming.

References

1. Prior JC. Perimenopause: the complex endocrinology of the menopausal transition. Endocrine Rev 1998;19:397–28.
2. Santoro N, Brown JR, Adel T, Skurnick JH. Characterization of reproductive hormonal dynamics in the perimenopause. J Clin Endocrinol Metab 1996;81:1495–501.
3. Matt DW, Kauma SW, Pincus SM, Veldhuis JD, Evans WS. Characteristics of LH secretion in younger versus older premenopausal women. Am J Obstet Gynecol 1998;178:504–10.
4. Klein NA, Illingworth PJ, Groome NP, McNeilly AS, Battaglia DE, Soules MR. Decreased inhibin B secretion is associated with the monotropic FSH rise in older, ovulatory women: a study of serum and follicular fluid levels of dimeric inhibin A and B in spontaneous menstrual cycles. J Clin Endocrinol Metab 1996;81:2742–45.
5. Knobil E. The neuroendocrine control of the menstrual cycle. Rec Prog Hormone Res 1980;36:53–88.
6. Crowley WF, Jr., Filicori M, Spratt DI, Santoro NF. The physiology of gonadotropin-releasing hormone (GnRH) secretion in men and women. Rec Prog Hormone Res 1985;41:473–31.
7. Faddy MJ, Telfer E, Gosden RG. The kinetics of pre-antral follicle development in ovaries of CBA/Ca mice during the first 14 weeks of life. Cell Tissue Kin 1987;20:551–60.
8. Lu KH, Hopper BR, Vargo TM, Yen SSC. Chronological changes in sex steroid, gonadotropin, and prolactin secretion in aging female rats displaying different reproductive states. Biol Reprod 1979;21:193–203.
9. Lu JKH. Changes in ovarian function and gonadotropin and prolactin secretion in aging female rats. In: Neuroendocrinology of aging. Meites J, ed. New York: Plenum Press, 1983:103–22.
10. Butcher RL, Page RD. Role of the aging ovary in cessation of reproduction. In: Dynamics of ovarian function. Schwartz NB, Hunzicker-Dunn M, eds. New York: Raven Press, 1981:253–71.
11. Scarbrough K, Wise PM. Age-related changes in pulsatile luteinizing hormone release precede the transition to estrous acyclicity and depend upon estrous cycle history. Endocrinology 1990;126:884–90.
12. DePaolo LV. Age-associated increases in serum follicle-stimulating hormone levels on estrus are accompanied by a reduction in the ovarian secretion of inhibin. Exp Aging Res 1987;13:3–7.
13. Yen SSC. The biology of menopause. J Reprod Med 1977;18:287–93.

Part I

Hypothalamic–Pituitary–Ovarian Axis During Reproductive Aging

2

Luteinizing Hormone Releasing Hormone (LHRH) Neuronal Function in Middle-Aged Female Rats

BEVERLY S. RUBIN AND MERCEDES JIMENEZ-LINAN

Regular reproductive cycles cease relatively early in the life span of many female mammals. The age-associated loss of reproductive fertility has been well characterized in the female rat (1–5). Female rates typically cease exhibiting regular 4–5-day estrous cycles by 10–12 months of age (1,2). As female rats age, regular estrous cycles may be replaced by lengthened cycles, irregular cycles, and, finally, by an acyclic state. Evidence of alterations in the function of the reproductive axis, particularly in patterns of gonadotropin secretion, have been documented in female rats prior to detectable disturbances in regular estrous cycles. These subtle but measurable changes effectively herald the beginning of age-related alterations within the hypothalamic–pituitary–ovarian axis. Although age-related deficits have been identified at all levels of the reproductive axis (1), there is considerable evidence to suggest that alterations at the hypothalamic level are particularly important to the reproductive decline with age in this species (2–5).

Alterations in Circulating Hormone Levels in Aging Female Rats

Numerous studies have documented altered gonadotropin secretion with age (for review, see Refs. 1–4,6). Before the cessation of estrous cyclicity in rats, the preovulatory luteinizing hormone (LH) surge on proestrus is typically attenuated in magnitude and may also be delayed in time. In addition, the elevation in LH levels in response to ovariectomy is significantly reduced, and the steroid-induced LH surge is also diminished in ovariectomized middle-aged relative to young females. The data currently available suggest that aging rats have a diminished capacity for LH release that may ultimately be responsible for the loss of estrous cyclicity in this species.

Alterations in circulating steroid levels have also been observed in middle-aged cycling females. Estradiol levels are increased on the day of proestrus in middle-aged rats (6). In contrast, age-related deficits in progesterone secretion have been observed in female rats and mice, and contribute to the increase in the estradiol–progesterone ratio that characterizes aging in females of both species (1). This elevated estrogen–progesterone ratio may promote the loss of estrous cycles by interfering with normal feedback mechanisms or by desensitizing the reproductive axis to the actions of estradiol (1,2). The intricate interrelationships between gonadotropin secretion and gonadal steroid production make it difficult to determine the degree to which the well-documented age-related changes in gonadotropin secretion contribute to the altered steroid hormone levels noted in middle-aged females prior to the loss of estrous cyclicity.

LHRH Regulates Pituitary Gonadotropin Synthesis and Secretion

LHRH (also known as gonadotropin-releasing hormone, GnRH) is the primary hypothalamic signal known to regulate pituitary gonadotropin synthesis and secretion. It is likely, therefore, that the well-documented alterations in gonadotropin secretion in middle-aged female rats reflect, to some degree, age-related changes in the parameters of LHRH secretion from the hypothalamus. For this reason, much research in our lab has focused on the examination of LHRH neuronal function in middle-aged female rats.

The parameters of LHRH secretion reflect the exquisite orchestration of a large number of signals that originate from within the hypothalamus and from other regions of the brain and brainstem (7). The changes in the pattern of LHRH release in conjunction with the preovulatory or steroid-induced LH surge in this species is dependent upon the precise coordination of a vast array of signals extrinsic to the LHRH neurons as well as increased activity of the subset of LHRH neurons that participate in LH surge induction. Parameters of LHRH secretion provide a barometer of hypothalamic function and may also reflect the functional integrity of the population of LHRH neurons in aging females.

LHRH Neurons Are Dispersed Throughout the Basal Hypothalamus

LHRH neurons are diffusely distributed within the basal forebrain (8). The majority of the widely distributed population of LHRH cell bodies project to the median eminence (9). There, LHRH is released into the pituitary portal vasculature for transport to the anterior pituitary for the modulation of gonadotropin synthesis and secretion. The dispersed localization of LHRH neurons allows the population of LHRH neurons to receive widespread input from the internal and external environments. This arrangement probably helps to ensure that reproduction occurs at the appropriate time to maximize reproductive success. The diffuse distribution, however, makes the population of LHRH neurons difficult to study. Protocols typically employed to examine peptidergic neurons localized within discrete nuclei in the brain cannot be effectively utilized to examine the population of LHRH neurons. Also adding to the complexity of the study of these neurons is the heterogeneity within the population of LHRH neurons (8). One cannot presume that examining a few LHRH neurons or studying a subset of the population of LHRH neurons will provide accurate information about function within the entire population of neurons.

LHRH Is Secreted in a Pulsatile Manner

Many studies have demonstrated the importance of the pulsatile pattern of LHRH secretion for the regulation of gonadotropin synthesis and secretion. In addition to regulating the levels of LH and FSH secretion, the parameters of pulsatile LHRH release modulate the levels of pituitary LHRH receptors (10) and pituitary gonadotropin subunit gene expression (11). In contrast to the stimulatory effects of pulsatile LHRH release on gonadotropin secretion, constant stimulation with LHRH down-regulates the hypothalamic pituitary gonadal axis.

At present, the nature of the LHRH pulse generator remains poorly understood. Electrical recordings in monkeys (12) and data from rat hypothalamic explant preparations (13) have suggested that the mechanism for LHRH pulse generation is located within the mediobasal hypothalamus. The ability of immortalized LHRH neuronal cell lines and enzymatically dispersed hypothalamic explants to demonstrate pulsatile LHRH release is of considerable interest (14,15). These findings suggest that mechanisms intrinsic to LHRH neurons themselves may be responsible for the basic pulsatile pattern of LHRH release which is then subject to modulation by extrinsic influences.

LHRH Neurons Exhibit High Levels of Neuronal Activity

LHRH neurons reportedly maintain a high level of transcriptional activity (16) and exhibit a high rate of LHRH mRNA turnover (17). In fact, the rate of LHRH mRNA turnover estimated in hypothalamic slice preparations is higher than any other peptidergic mRNA studied to date (17). Maurer and Wray (17) have proposed that the rapid rate of LHRH mRNA turnover may play a role in driving pulsatile LHRH release. Increases in LHRH primary transcript and LHRH mRNA levels have been reported in association with the preovulatory LH surge on proestrus (18,19), and with the steroid-induced LH surge (20–22). The subset of LHRH neurons that participate in the LH surge also express detectable levels of the immediate early genes Fos and Jun (23,24) as well as the peptide galanin (25) in conjunction with the LH surge. Synthesis of these additional gene products may further heighten the level of transcriptional activity and enhance the energy demands required of LHRH neurons during this time of dynamic activity.

LHRH Neurons and LHRH Secretion in Middle-Aged Females

LHRH Immunoreactivity (IR)

Middle-aged females do not exhibit obvious reductions in LHRH immunoreactivity when examined in stagnant conditions relative to LHRH neuronal activity, such as after the cessation of estrous cyclicity (26). In contrast, when LHRH neurons are examined during a dynamic time with regard to LHRH neuronal activity such as in conjunction with the spontaneous LH surge on proestrus, age-related differences in LHRH immunoreactivity do become apparent (27–29). The characteristic accumulation of LHRH in the median eminence of young female rats prior to the start of the LH surge on proestrus (7) is not observed in middle-aged females (28,29), and this deficit may ultimately influence the levels of LHRH released during the LH surge. Young females also reveal a reduction in the level of IR-LHRH in the median eminence at time points examined after the peak of the LH surge relative to time points examined prior to the start of the LH surge. This is an observation that is consistent with evidence of LHRH release (28). This change in IR-LHRH over time on the day of proestrus is not noted in the brains of middle-aged females; therefore, it provides little indication of accumulation and release of the peptide in aging females. Dynamic changes are also observed in the number of LHRH neurons detected in the brains of young females, but not middle-aged females on the day of proestrus (27). These data together provide evidence of significantly enhanced LHRH synthesis, transport, and release in young but not middle-aged females in association with the preovulatory LH surge.

LHRH Secretion

In vivo studies in our laboratory have documented diminished levels of LHRH output from middle-aged relative to young females in conjunction with a steroid-induced LH surge (30). Data from push–pull perfusion studies revealed lower overall mean levels of LHRH output from the median eminence of middle-aged females. They also suggested an age-related decrease in LHRH pulse frequency (see Fig. 2.1). These alterations in LHRH secre-

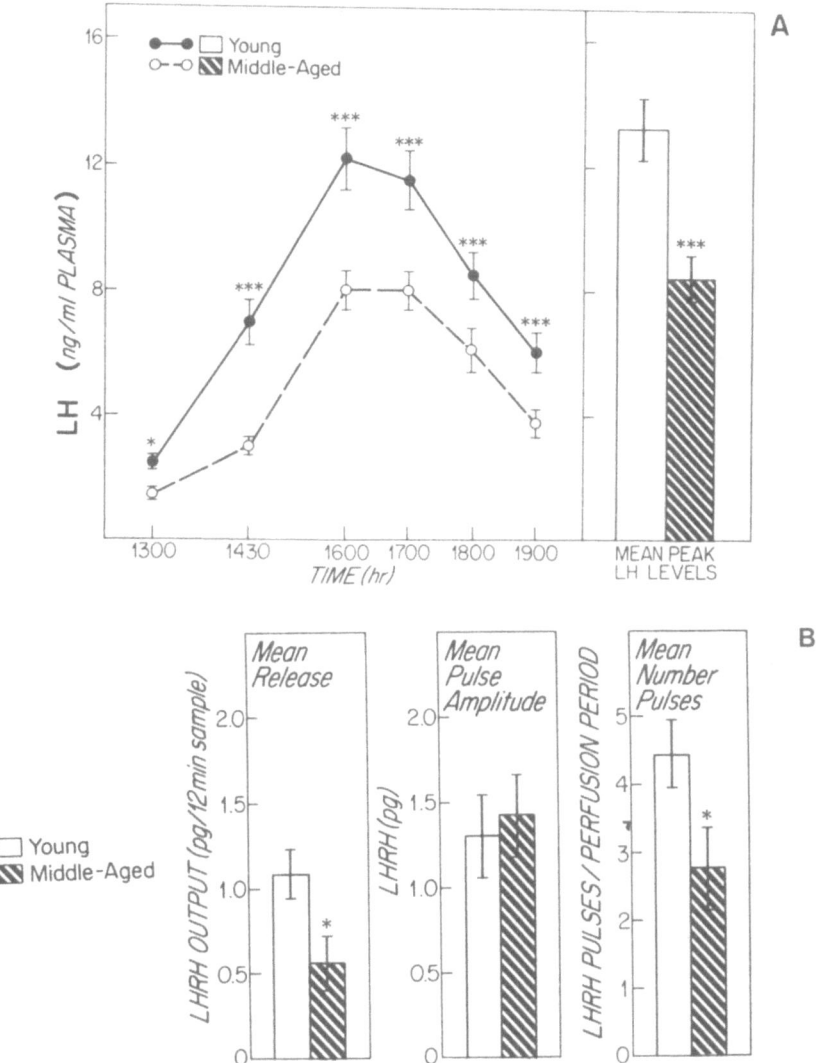

FIGURE 2.1. Comparisons of serum LH levels (1A) and parameters of in vivo LHRH output (1B) in young and middle-aged female rats on the day of a steroid-induced LH surge. As depicted, circulating levels of LH (1A) and mean levels of LHRH release (1B) are diminished in middle-aged relative to young females. An age-related decrease in the number of LHRH pulses detected is also noted (1B). LHRH levels were measured in effluents collected from the medial basal hypothalamus by push–pull perfusion. Blood samples for LH measurements were collected via a surgically implanted jugular catheter. $*p < 0.05$; $***p < 0.001$: middle-aged versus young. Reprinted from Neuroendocrinology 1989;49: 225–32 with permission from Karger, Basel (Ref. 30).

tion would be expected to effect gonadotropin secretion associated with the LH surge.

Other available data provide indirect evidence of altered LHRH secretion. The decrease in LH pulse frequency observed in middle-aged relative to young ovariectomized females (31) is consistent with an age-related increase in the interpulse interval of LHRH release. The reduction in pituitary LHRH receptor mRNA levels and altered gonadotropin subunit gene expression documented in middle-aged females (32,33, and unpublished observation) provide additional indications of an age-related decline in LHRH output. As mentioned previously, pituitary LHRH receptor levels and gonadotropin subunit gene expression are regulated by parameters of pulsatile LHRH secretion (10,11). These endpoints, therefore, afford an index of LHRH action at the pituitary although the possibility that alterations at the pituitary level might influence this measure must be considered.

Immediate Early Gene Expression Within LHRH Neurons

The precise role of immediate early gene expression within LHRH neurons has not yet been elucidated; however, it is clear that a proportion of LHRH neurons express detectable levels of Fos and Jun proteins within their nuclei in conjunction with an LH surge and not at other times of the estrous cycle (23,24). The expression of immediate early gene products within LHRH neurons has been suggested to connote transynaptic activation of LHRH neurons. As such it may provide one estimate of the level of excitatory stimulation received by a proportion of the population of LHRH neurons during the time surrounding LH surge induction.

A smaller proportion of LHRH neurons in the brains of middle-aged compared with young females exhibit nuclear Fos and/or Jun expression in association with the preovulatory LH surge on proestrus (34–36, see Fig. 2.2) or the steroid-induced LH surge (36). The decline and delay in Fos and/or Jun expression noted in middle-aged females may reflect age-related changes in the excitation of the subset of LHRH neurons required for induction and maintenance of the LH surge.

LHRH Gene Expression

The age-related alterations in LHRH immunoreactivity and LHRH output observed in conjunction with an LH surge may result from a decline in LHRH gene expression with age. As discussed, increases in LHRH primary transcript and LHRH mRNA levels are observed within the population of LHRH neurons on the day of an LH surge (18–22). Studies in our laboratory have confirmed dynamic changes in LHRH mRNA levels in the brains of young females in conjunction with the LH surge that are not apparent in the brains of middle-aged females (22). Mean levels of LHRH mRNA/neuron and the total hybridization area measured in tissue sections through the rostral preoptic area, a

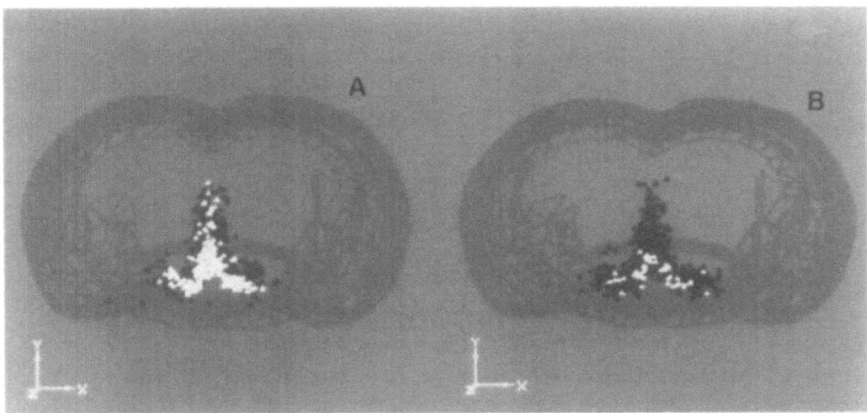

FIGURE 2.2. Coronal views of three-dimensional reconstructions of the population of LHRH neurons (black triangles) and LHRH/Fos neurons (white triangles) in young (A) and middle-aged (B) females on the afternoon of proestrus during the ascending phase of the LH surge. Black triangles indicate LHRH-labeled neurons; white triangles represent LHRH/Fos-labeled neurons. As depicted, significantly fewer LHRH neurons in middle-aged females exhibit evidence of Fos expression at this time. A mean of 28% of the LHRH neurons detected in young and 10% of the LHRH neurons detected in middle-aged females express Fos protein at this time point (see Refs. 36,37).

region containing a large number of LHRH neurons, are significantly diminished in the brains of middle-aged relative to young females (see Fig. 2.3). Whereas the precise relationship between LHRH gene expression and LHRH release is not known, enhanced levels of gene expression would be expected to promote LHRH synthesis. If newly synthesized LHRH is released prior to stores of LHRH, as has been suggested (17), then increased levels of LHRH mRNA may be important to facilitate the enhanced levels of LHRH release required for induction and maintenance of an LH surge.

The Role of Alterations Extrinsic and/or Intrinsic to LHRH Neurons to the Age-Associated Reproductive Decline

The cause of the progressive disruption of LHRH neuronal function in middle-aged animals remains inadequately defined. The complex sequence of events involved in triggering the LH surge is known to include restraint of inhibitory influences, increased excitatory signals, and appropriate input from the neural clock or suprachiasmatic nucleus (SCN; for review, see 7). Alterations in excitatory and inhibitory influences on LHRH secretion have been documented to occur with age (for review, see Refs. 1,3,4). Moreover, the

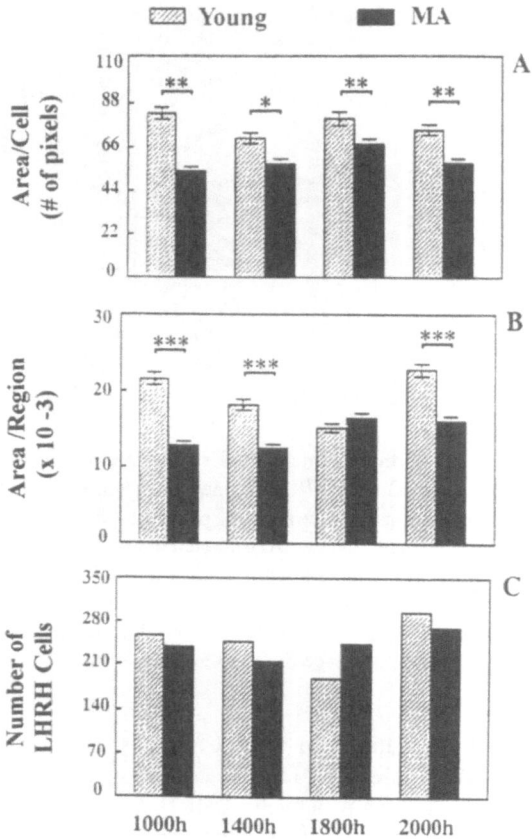

FIGURE 2.3. Detailed cellular analysis of in situ hybridization data from a subset of LHRH mRNA positive neurons present in four anatomically matched tissue sections through the rostral preoptic area on the day of a steroid-induced LH surge in young and middle-aged females (A). The mean hybridization area/cell is depicted in relative units. (B) The mean hybridization area/region provides an estimate of the summed hybridization area within the region of the rostral preoptic area analyzed. (C) The number of LHRH mRNA positive neurons present and analyzed at each of the time points. As shown, LHRH mRNA levels are decreased in middle-aged relative to young females. $*p < 0.05$; $**p < 0.01$; $***p < 0.001$: middle-aged vs. young. Reprinted from Brain Res 1997;770:267–76 with permission from Elsevier Science (see Ref. 22).

age-associated changes observed in SCN function would also be expected to influence the LH surge (5,37,38). The importance of the SCN to the estrogen-dependent increase in LHRH/LH release on the day of the LH surge has been well documented (39), and data have demonstrated direct contact between SCN neurons and LHRH neurons in rats (40). The data currently available, therefore, indicate that alterations extrinsic to LHRH neurons undoubtedly play an important role in the decline in LHRH neuronal function with age.

Alterations intrinsic to LHRH neurons, however, may also contribute to the decline in LHRH neuronal activity. The ability of LHRH neurons in aging animals to respond to relevant signals with significant enhancement of transcription, biosynthesis, and secretion has not yet been evaluated. For example, an overall decline in the transcriptional activity of LHRH neurons may contribute to the decrease in LHRH mRNA levels, LHRH accumulation, LHRH secretion, and Fos and Jun expression in LHRH neurons of aging females on the day of the preovulatory LH surge. The possibility that age-related alterations in membrane receptor levels, signaling pathways, or energy availability may interfere with the ability of LHRH neurons in middle-aged females to maintain high levels of normal activity associated with the LH surge on proestrus has not been explored.

The hypothesis that pulsatile LHRH release is an intrinsic property of the LHRH neuronal network is of particular relevance with regard to the potential importance of age-associated alterations intrinsic to LHRH neurons (41). If some component of the basic pattern of pulsatile LHRH release is endogenous to LHRH neurons, then it is feasible that this essential aspect of LHRH neuronal function may be susceptible to changes with age. Age-related alterations in rhythmic gene expression have been reported in neurons of the SCN, where levels of some peptides are reduced and rhythms are blunted (37).

Do Hypothalamic Changes Contribute to Reproductive Decline in Women?

There are numerous differences in the regulation of estrous and menstrual cycles, and the data available implicate very different mechanisms in the aging of the reproductive axis in women relative to the rat model. As mentioned previously, the hypothalamus is considered a primary site of the deficit responsible for the age-associated reproductive decline in rats (1–5). In contrast, the ovary is considered the primary site of the deficit leading to menopause, and the depletion of the functional ovarian follicular reserve is clearly the immediate cause of the hypoestrogenicity associated with menopause (42); however, studies of the premenopausal period provide evidence of potential alterations in hypothalamic-pituitary function prior to menopause (43).

Alterations in LH pulse frequency (43) and in circulating FSH levels have been observed in women during the years preceding menopause (42). In addition, the increase in estradiol and the decrease in progesterone production reported in premenopausal women (44) are reminiscent of the circulating ovarian steroid milieu described in middle-aged rats prior to cessation of regular estrous cycles. To date, speculation regarding the cause of elevated FSH levels in the face of elevated estradiol levels during the premenopausal period has centered on a decline in inhibin production (42); however, the decrease in LHRH pulse frequency implied from LH measurements in pre-menopausal women (43) may also be expected to influence circulating go-nadotropin levels. Studies in rats have revealed that slower LHRH pulse frequency favors FSH release over LH release, and there is some evidence to suggest that a similar relationship may exist in humans and primates (11).

LHRH is obligatory for fertility in women, and changes in LHRH pulse fre-quency and pulse amplitude during the menstrual cycle have been deduced from measurements of serum LH (42,44). Changes in parameters of LHRH output, if they occur with age in women, could therefore contribute to the elevation in FSH levels that has been hypothesized to hasten the depletion of the ovarian follicular reserve (42). Assessment of this possibility awaits further study.

Summary

A significant body of evidence implicates the hypothalamus as a primary site of deficit in the age-related reproductive decline in female rats. LHRH is the primary signal leaving the hypothalamus to influence pituitary go-nadotropin release. As such, parameters of LHRH output are influenced by the summation of hypothalamic activity as well as the functional integrity of the population of LHRH neurons themselves. Several lines of investigation reveal alterations in LHRH neuronal function in middle-aged female rats. The alterations noted might be expected to influence LHRH synthesis, trans-port, and secretion; therefore, they may be responsible for the well-docu-mented, age-associated alterations in gonadotropin secretion in this species. Whether the alterations in LHRH neuronal function can be explained solely by alterations extrinsic to LHRH neurons or whether alterations intrinsic to LHRH neurons may also contribute to the age-related deficit in LHRH output remains to be determined. Critical evaluation of the potential importance of age-related changes at the hypothalamic level to the cascade of events that lead to reproductive decline in women awaits further investigation.

References

1. vom Saal FS, Finch CE, Nelson JF. Natural history and mechanisms of reproductive aging in humans, laboratory rodents, and other selected vertebrates. In: Knobil E, Neill JD, eds. The physiology of reproduction, second edition. New York: Raven Press, 1994:1213–314.
2. Lu JKH, Anzalone CR, LaPolt PS. Relation of neuroendocrine function to re-productive decline during aging in the female rat. Neurobiol Aging 1994;15(4): 541–44.

3. Wise PM. Neuroendocrine ageing: its impact on the reproductive system of the female rat [review]. J Reprod Fertil 1993;46:(suppl)35–46.
4. Wise PM. The role of the hypothalamus in aging of the female reproductive system. J Steroid Biochem 1987;27:713–19.
5. Wise PM, Weiland NG, Scarbrough K, Lloyd JM. Changing diurnal and pulsatile rhythms during aging. Neurobiol Aging 1994;15:503–7.
6. Nass TE, LaPolt PS, Judd HL, Lu JKH. Alterations in ovarian steroid and gonadotrophin secretion preceding the cessation of regular oestrus cycles in ageing. J Endocrinol 1984;100:43–50.
7. Kalra SP. Mandatory neuropeptide-steroid signaling for the preovulatory luteinizing hormone-releasing hormone discharge. Endocrinol Rev 1993;14:507–38.
8. King JC, Rubin BS. GnRH subgroups: a microarchitecture. In: Conn PM, Crowley WF, Jr., eds. Modes of action of GnRH and GnRH analogs. New York: Springer-Verlag, 1992:161–78.
9. Silverman A-J, Witkin JW, Silverman RC, Gibson MJ. Modulation of GnRH neuronal activity as evidenced by uptake of fluorogold from the vasculature. Synapse 1990;6:154–60.
10. Yasin M, Dalkin AC, Haisenleder DJ, Kerrigan JR, Marshall JC. Gonadotropin-releasing hormone (GnRH) pulse pattern regulates GnRH receptor gene expression: augmentation by estradiol. Endocrinology 1995;136:1559–64.
11. Marshall JC, Dalkin AC, Haisenleder DJ. Regulation of gonadotropin gene expression by gonadotropin releasing hormone. In: Crowley WFJ, Conn PM, eds. Modes of action of GnRH and GnRH analogs. New York: Springer-Verlag, 1992:55–68.
12. Knobil E. The electrophysiology of the GnRH pulse generator in the rhesus monkey. J Steroid Biochem 1989;33:669–71.
13. Purnelle G, Gérard A, Czajkowski V, Bourguignon JP. Pulsatile secretion of gonadotropin-releasing hormone by rat hypothalamic explants of GnRH neurons without cell bodies. Neuroendocrinology 1997;66:305–12.
14. Wetsel WC, Valença MM, Merchenthaler I, Liposits Z, López FJ, Weiner RI, et al. Intrinsic pulsatile secretory activity of immortalized luteinizing hormone-releasing hormone-secreting neurons. Proc Natl Acad Sci USA 1992;89:4149–53.
15. Woller M, Nichols E, Herdendorf T, Tutton D. Release of LHRH from enzymatically dispersed hypothalamic explants is pulsatile. Biol Reprod 1998;59:(3) 587–90.
16. Yeo TT, Gore AC, Jakubowski M, Dong KW, Blum M, Roberts JL. Characterization of gonadotropin-releasing hormone gene transcripts in a mouse hypothalamic neuronal GT1 cell line. Brain Res Mol Brain Res 1996;42:255–62.
17. Maurer JA, Wray S. Luteinizing hormone-releasing hormone (LHRH) neurons maintained in hypothalamic slice explant cultures exhibit a rapid LHRH mRNA turnover rate. J Neurosci 1997;17:9481–91.
18. Gore AC, Roberts JL. Regulation of gonadotropin-releasing hormone gene expression in the rat during the luteinizing hormone surge. Endocrinology 1995; 136:889–96.
19. Porkka-Heiskanen T, Urban JH, Turek FW, Levine JE. Gene expression in a subpopulation of luteinizing hormone-releasing hormone (LHRH) neurons prior to the preovulatory gonadotropin surge. J Neurosci 1994;14(9):5548–58.
20. Petersen SL, McCrone S, Keller M, Shores S. Effects of estrogen and progesterone on luteinizing hormone-releasing hormone messenger ribonucleic acid levels: con-

sideration of temporal and neuroanatomical variables. Endocrinology 1995; 136:3604–10.

21. Petersen SL, Gardner E, Adelman J, McCrone S. Examination of steroid-induced changes in LHRH gene transcription using 33P- and 35S-labeled probes specific for intron 2. Endocrinology 1996;137:234–39.

22. Rubin BS, Lee CE, Ohtomo M, King JC. Luteinizing hormone-releasing hormone gene expression differs in young and middle-aged females on the day of a steroid-induced LH surge. Brain Res 1998;770:267–76.

23. Lee W-S, Smith MS, Hoffman GE. LHRH neurons express Fos protein during the proestrous surge of luteinizing hormone. Proc Natl Acad Sci USA 1990;87: 5163–67.

24. Lee W-S, Abbud R, Smith MS, Hoffman GE. LHRH neurons express cJun protein during the proestrous surge of luteinizing hormone. Endocrinology 1992;130: 3101–3.

25. Marks DL, Smith MS, Vrontakis M, Clifton DK, Steiner RA. Regulation of galanin gene expression in gonadotropin-releasing hormone neurons during the estrous cycle of the rat. Endocrinology 1993;132:1836–44.

26. Rubin BS, King JC, Bridges RS. Immunoreactive forms of LHRH in the brains of aging rats exhibiting persistent vaginal estrus. Biol Reprod 1984;31:343–51.

27. Rubin BS, King JC. The number and distribution of detectable luteinizing hormone (LH)-releasing hormone cell bodies changes in association with the preovulatory LH surge in the brains of young but not middle-aged female rats. Endocrinology 1994;134:467–74.

28. Rubin BS, King JC. A relative depletion of luteinizing hormone-releasing hormone was observed in the median eminence of young but not middle-aged rats on the evening of proestrus. Neuroendocrinology 1995;62:259–69.

29. Wise PM. Alterations in the proestrous pattern of median eminence LHRH, serum LH, FSH, estradiol and progesterone concentrations in middle-aged rats. Life Sci 1982;31:165–73.

30. Rubin BS, Bridges RS. Alterations in luteinizing hormone-releasing hormone release from the mediobasal hypothalamus of ovariectomized, steroid-primed middle-aged rats as measured by push-pull perfusion. Neuroendocrinology 1989;49:225–32.

31. Scarbrough K, Wise PM. Age-related changes in pulsatile luteinizing hormone release precede the transition to estrous acyclicity and depend upon estrous cycle history. Endocrinology 1990;126:884–90.

32. Hogg BB, Matt DW, Sayles TE. Attenuated proestrous luteinizing hormone surges in middle-aged rats are associated with decreased pituitary luteinizing hormone-beta messenger ribonucleic acid expression. Am J Obstet Gynecol 1992;167:303–7.

33. Matt DW, Sayles TE, Jih MH, Kauma SW, Lu JKH. Alterations in the postovariectomy increases in gonadotropin secretion in middle-aged persistent-estrous rats: correlation with pituitary gonadotropin subunit gene expression. Neuroendocrinology 1993;57:351–58.

34. Lloyd JM, Hoffman GE, Wise PM. Decline in immediate early gene expression in gonadotropin-releasing hormone neurons during proestrus in regularly cycling, middle-aged rats. Endocrinology 1994;134:1800–5.

35. Rubin BS, Mitchell S, Lee CE, King JC. Reconstructions of populations of LHRH neurons in young and middle-aged rats reveal progressive increases in subgroups

expressing Fos protein on proestrus and age-related deficits. Endocrinology 1995;136:3823–30.

36. Rubin BS, Lee CE, King JC. A reduced proportion of luteinizing hormone (LH)-releasing hormone neurons express Fos protein during the preovulatory or steroid-induced LH surge in middle-aged rats. Biol Reprod 1994;51:1264–72.

37. Harney JP, Scarbrough K, Rosewell KL, Wise PM. *In vivo* antisense antagonism of vasoactive intestinal peptide in the suprachiasmatic nuclei causes aging-like changes in the estradiol-induced luteinizing hormone and prolactin surges. Endocrinology 1996;137:3696–701.

38. Wise PM, Krajnak KM, Kashon ML. Menopause: The aging of multiple pacemakers. Science 1996;273:67–70.

39. Wiegand SJ, Terasawa E. Discrete lesions reveal functional heterogeneity of suprachiasmatic structures in regulation of gonadotropin secretion in the female rat. Neuroendocrinology 1982;34:395–404.

40. Van der Beek EM, Horvath TL, Wiegant VM, Van den Hurk R, Buijs RM. Evidence for a direct neuronal pathway from the suprachiasmatic nucleus to the gonadotropin-releasing hormone system: combined tracing and light and electron microscopic immunocytochemical studies. J Comp Neurol 1997;384:569–79.

41. López FJ, Merchenthaler IJ, Moretto M, Negro-Vilar A. Modulating mechanisms of neuroendocrine cell-activity: the LHRH pulse generator. Cell Mol Neurobiol 1998;18:125–46.

42. Prior JC. Perimenopause: the complex endocrinology of the menopausal transition. Endocrine Rev 1998;19(4):397–428.

43. Matt DW, Kauma SW, Pincus SM, Veldhuis JD, Evans WS. Characteristics of luteinizing hormone secretion in younger versus older premenopausal women. Am J Obstet Gynecol 1998;178:504–10.

44. Santoro N, Brown JR, Adel T, Skurnick JH. Characterization of reproductive hormonal dynamics in the perimenopause. J Clin Endocrinol Metab 1996; 81:1495–501.

3

Reproductive Aging and the Human Hypothalamus

Naomi E. Rance, Ty W. Abel, and Steve C. Danzer

The goal of our laboratory in the 1990s has been to characterize the molecular events that occur in the human hypothalamus in response to menopause. Despite the major clinical significance of menopause on a growing proportion of our society, there have been few studies of the aging human hypothalamus. In addition, from a basic science perspective, menopause provides a unique opportunity to study hypothalamic control mechanisms in the human. Decades of animal research have also provided a rationale for focusing on specific hypothalamic systems and neuronal populations. Finally, the development of in situ hybridization technology has allowed the measurements of cellular levels of mRNAs in human postmortem material with the precision achieved in laboratory animals. As a result, we have been able to provide some of the first descriptions of changes in neuronal morphology and neuropeptide gene expression in the hypothalamus of postmenopausal women.

Alterations in peripheral plasma hormone levels in postmenopausal women are consistent and dramatic. The loss of ovarian follicles (1) results in castrate levels of ovarian steroids (2). Because of removal of steroid negative feedback on the pituitary and hypothalamus, gonadotropins are significantly increased in peripheral plasma (3–5). This rise in plasma gonadotropins is similar in magnitude to that seen after ovariectomy in young women (6–8). Indeed, ovariectomy in young women is commonly referred to as "surgical menopause" (9).

More than 75% of postmenopausal women experience hot flushes (10), a symptom of estrogen withdrawal. Flushes also occur in young women after ovariectomy and are treated effectively by estrogen replacement (9,10). Flushes are characterized by peripheral vasodilatation, perspiration, and an intense sensation of heat (11,12). Although these are physiological mechanisms that normally reduce body temperature, the core temperature paradoxically is normal at the onset of a flush. Because heat loss mechanisms are activated in the presence of normal core temperature, flushing is considered a disorder of central (hypothalamic) thermoregulation (10). A second line of evidence that the flushes originate within the hypothalamus is the finding that each flush episode coincides with a pulse of LH secretion into the systemic circulation (13,14).

The starting point of the present series of investigations was the seminal report by Sheehan and Kovács on neuronal hypertrophy in the infundibular (arcuate) nucleus of postmenopausal women (15). The phenomenon of postmenopausal neuronal hypertrophy was intriguing because of its association with estrogen withdrawal and the location of the neurons in the putative control center for reproduction (16). The enlargement of neurons more importantly did not appear to be secondary to a degenerative event. On the contrary, it was accompanied by morphologic signs of increased activity, such as enlarged nuclei and nucleoli, and increased Nissl substance (rough endoplasmic reticulum) (15,17).

Postmenopausal Neuronal Hypertropy Occurs in a Subpopulation of Neurons Expressing Estrogen Receptor Gene Transcripts

Because the initial observations of neuronal hypertrophy were qualitative, our first task was to confirm and quantify these changes (17). Formalin-fixed, paraffin-embedded, cresyl violet stained sections were prepared from the hypothalami of pre- and postmenopausal women. An image-combining computer microscope was used to digitize neuronal profiles (18). This study clearly demonstrated that neurons in the infundibular nucleus of postmenopausal women are hypertrophied. We next hypothesized that the hypertrophy was due to the removal of the inhibitory feedback of ovarian steroids, analogous to the enlargement of pituitary gonadotrophs that occurs after castration in laboratory animals (19). We reasoned that if the cellular changes were secondary to alterations in levels of estrogen, then the hypertrophied cells would express the gene for the estrogen receptor. This was indeed the case (17). It is interesting to note that the hypertrophied neurons did not express gonadotropin releasing hormone (GnRH) mRNA (Fig. 3.1), a finding that was consistent with previous studies showing that the estrogen receptor and GnRH are localized within different subpopulations of hypothalamic neurons (20).

Neuronal Hypertrophy Also Occurs in the Infundibular Nucleus of Older Men

In contrast to the complete loss of ovarian steroids in older women (2), the reduction of gonadal steroid levels in older men is mild (21). Based on the modest reduction in testosterone, we predicted no change in the infundibular nucleus of older men, or that hypertrophy, if present, would be small in

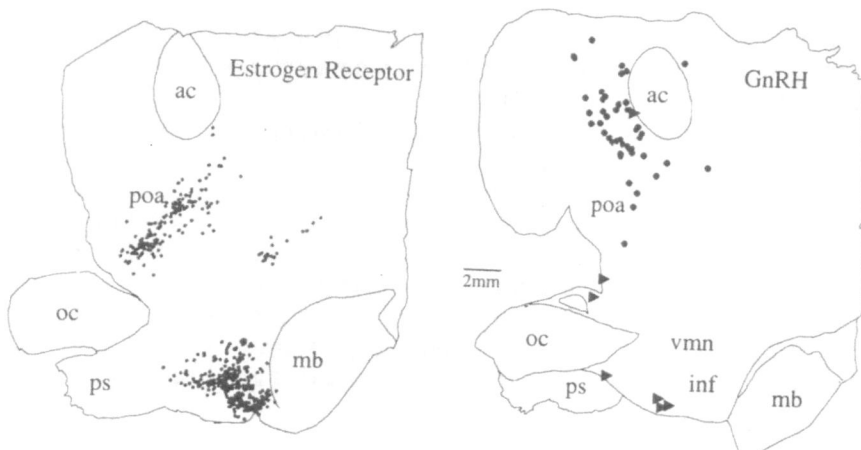

FIGURE 3.1. Neurons containing estrogen receptor alpha mRNA (left) and GnRH mRNA (right) display different distributions in the human hypothalamus. These are computer-assisted maps of the sagittal human hypothalamus with each symbol marking the location of one neuron. In the diagram on the right, the Type I GnRH neurons are represented by triangles and Type II GnRH neurons are represented by filled circles. The hypertrophied neurons containing estrogen receptor alpha mRNA are located in the infundibular nucleus. Increased GnRH gene expression occurs in a separate subpopulation of neurons (Type I) in the medial basal hypothalamus. Ac, anterior commissure; inf, infundibular nucleus; mb, mammillary bodies; oc, optic chiasm; ps, pituitary stalk; poa, preoptic area. Illustration on left based on data from Rance et al. 1990 (Ref. 17); illustration on right modified with permission from Rance NE, Uswandi SV. Gonadotropin-releasing hormone gene expression is increased in the medial basal hypothalamus of postmenopausal women. J Clin Endocrinol Metab 1996;81:3540–46. © The Endocrine Society.

magnitude. Hypothalami were collected from young and older men in the same age range as the previous studies of pre- and postmenopausal women. The methods of tissue processing and data collection were identical to our previous studies (17). We were surprised to find a significant increase in the mean cell area (ca. 8%) of infundibular neurons from older men relative to that of young men, but the magnitude of the changes is reduced relative to older women. The degree of hypertrophy correlates well with the extent of gonadal steroid reduction between the two groups (22).

The Hypertrophy of Neurons in Postmenopausal Women Occurs in a Subpopulation of Neurons Expressing Substance P, Neurokin B, and Estrogen Receptor Gene Transcripts

The next goal was to determine the neuropeptide content of the hypertrophied neurons. Sections from the hypothalami of postmenopausal women were hybridized with synthetic ^{35}S-labeled 48-base cDNA probes complementary to estrogen receptor, corticotropin releasing hormone, dynorphin, neuropeptide Y, growth hormone releasing hormone, cholecystokinin, enkephalin, proopiomelanocortin (POMC), substance P (SP), neurokinin B (NKB), tyrosine hydroxylase, or galanin cDNAs. This study showed that the hypertrophied infundibular neurons in postmenopausal women contained the mRNA of two tachykinins, NKB and SP. In addition, these data provided evidence that these mRNAs are colocalized in the human infundibular nucleus. The remainder of the probes labeled neurons that were either too small or too few in number in the infundibular nucleus to contribute significantly to the hypertrophied population (23).

Postmenopausal Neuronal Hypertrophy Is Accompanied by Increased Tachykinin Gene Expression

Because hypertrophy is accompanied by morphologic signs of increased activity (enlarged nuclei and increased rough endoplasmic reticulum), we predicted that elevations in tachykinin mRNA levels would be found in the hypertrophied cells. A comparison of the size and gene expression of NKB and SP neurons in the hypothalami of premenopausal and postmenopausal women revealed neuronal hypertrophy, as well as marked increases in the numbers of cells expressing tachykinin mRNAs and the autoradiographic grain density of each neuron (Fig. 3.2). These studies were the first to dem-

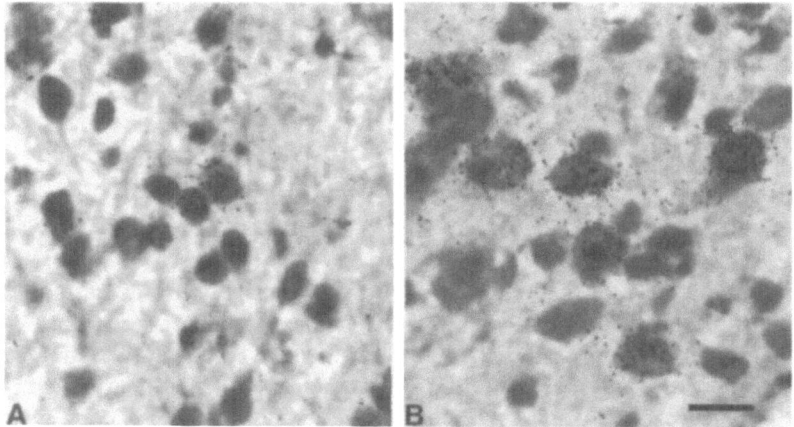

FIGURE 3.2. Photomicrographs of neurons in the infundibular nucleus of a premenopausal (A) and postmenopausal (B) woman labeled by in situ hybridization using a probe complementary to substance P mRNA. The silver grains mark the location of substance P mRNA, and the sections are counterstained with toluidine blue. In postmenopausal women, substance P neurons are larger, more numerous, and display increased autoradiographic grain densities. With permission from Rance NE, Young WS, III. Hypertrophy and increased gene expression of neurons containing neurokinin-B and substance-P messenger ribonucleic acids in the hypothalami of postmenopausal women. Endocrinology 1991;128:2239–47. © The Endocrine Society.

onstrate that menopause is accompanied by marked alterations in neuropeptide gene expression in the human hypothalamus. We have proposed that these tachykinin neurons participate in the hypothalamic circuitry mediating estrogen negative feedback (23,24). This hypothesis is based upon previous experimental studies indicating a role for hypothalamic tachykinins in the regulation of LH secretion (25–28).

Neurokinin B Gene Expression Is Increased in the Arcuate Nucleus of Ovariectomized Rats

Our working hypothesis is that the anatomical changes and increases in gene expression in the hypothalamus of postmenopausal women are secondary to ovarian failure. In addition to the differences in ovarian status, however, postmenopausal women differ from premenopausal women in age. An animal model was therefore used to determine if similar changes in neuropeptide gene expression could be induced in the rat arcuate nucleus in response to long-term ovariectomy. Four groups were examined: proestrus; diestrous day 1; ovariectomized; and constant estrus induced by an injection of estradiol valerate. Rats were sacrificed 2 months after treatment. We found that the number of neurons containing NKB gene transcripts was significantly greater in ovariectomized rats relative to all other groups. In addition, the autoradiographic grain density of NKB neurons was doubled in the ovariectomized group compared with intact animals. Thus, marked changes in NKB gene expression occurred in ovariectomized rats in a direction predicted by our previous studies of postmenopausal women. These data provide strong support for the hypothesis that increased tachykinin gene expression in the hypothalamus of postmenopausal women is due to ovarian failure (29).

Dendritic Growth Occurs in the Arcuate Nucleus of Adult Male Rats Following Orchidectomy

It is interesting that the neuronal hypertrophy observed in the infundibular nucleus of older men was accompanied by an expansion of the volume of the infundibular nucleus associated with each neuron (22). These findings suggested that the neuronal hypertrophy was not confined to neuronal somata, but could involve changes in dendritic architecture as well. To address this question, retrograde labeling was combined with intracellular injection of neurons in a fixed slice preparation to examine the morphology of neuroendocrine neurons in the arcuate nucleus of intact and gonadectomized rats (30). Arcuate neurons were subsequently reconstructed in three dimensions using computer-assisted microscopy. Consistent with our previous studies, the arcuate neuroendocrine neurons in the orchidectomized group had significantly larger somatic areas than did the intact group. These neurons also exhibited significant increases in dendrite length, dendrite volume, terminal branch number, and spines per unit length of dendrite. Furthermore, growth conelike processes were present on many dendrites in the castrate group. These data show that in addition to somatic hypertrophy, gonadectomy results in striking changes in the dendritic morphology of arcuate neuroendocrine neurons. Thus, the changes in neuronal morphology in older men and women are probably much more extensive than previously recognized and may involve a substantial "rewiring" of hypothalamic circuitry (30).

GnRH Gene Expression Is Increased in the Medial Basal Hypothalamus of Postmenopausal Women

Gonadotropin-releasing hormone is the hypothalamic peptide that regulates LH release from the anterior pituitary gland (31,32). Because LH secretion is markedly elevated in postmenopausal women, we decided to find out if corresponding changes in hypothalamic GnRH gene expression could be detected in this group of women. As a preliminary step, the distribution of neurons expressing GnRH mRNA was mapped in the human hypothalamus and basal forebrain (33).

A much more extensive distribution of GnRH neurons was observed using hybridization histochemistry than previously described using immunocytochemical methods (34). In addition, three morphological subtypes of GnRH neurons were identified (Fig. 3.1). Type I GnRH neurons were small, heavily labeled, oval or fusiform neurons, located primarily in the medial basal hypothalamus, ventral preoptic area, and periventricular zone. Type II neurons were small, oval, sparsely labeled neurons located in the septum and dorsal preoptic region. Finally, Type III neurons were large, round neurons scattered within the magnocellular basal forebrain complex, extended amygdala, ventral pallidum, and putamen. The pronounced differences in morphology, labeling density, and location of the three subtypes suggest that distinct functional subgroups of GnRH neurons exist in the human brain.

In the next study, cellular levels of GnRH mRNAs were measured in the two subtypes of hypothalamic GnRH neurons in premenopausal and postmenopausal women (35). This study revealed that menopause is associated with a 50% increase in the number of autoradiographic grains in the Type I GnRH neurons in the medial basal hypothalamus. In contrast, there was no change in the number of autoradiographic grains associated with Type II GnRH neurons in the septal–dorsal preoptic region (Fig. 3.3). The differential response of the two types of hypothalamic neurons provided additional evidence that functional subgroups of GnRH neurons exist in the human brain. Previous studies have established the central role of the medial basal hypothalamus in the regulation of gonado-

Type I GnRH Neurons

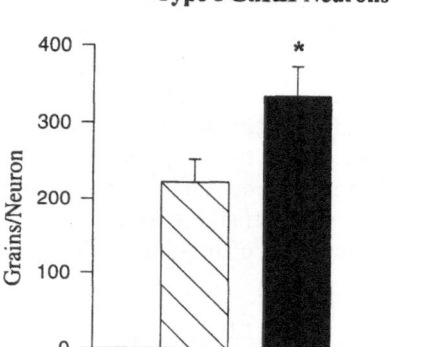

Type II GnRH Neurons

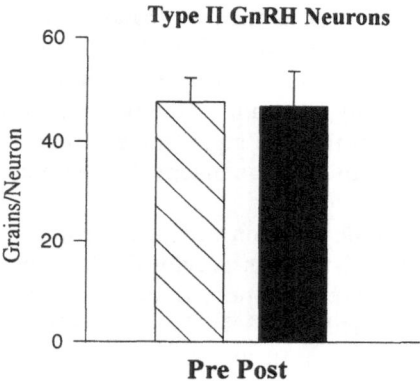

Pre Post

FIGURE 3.3. Number of grains per neuron associated with Type I (top) and Type II (bottom) GnRH neurons in the hypothalamus of premenopausal and postmenopausal women. *, $p <$ 0.05. GnRH gene expression is increased in the heavily labeled neurons (Type I) in the medial basal hypothalamus but not the lightly labeled (Type II) neurons in the septal-preoptic area. Modified with permission from Rance NE, Uswandi SV. Gonadotropin-releasing hormone gene expression is increased in the medial basal hypothalamus of postmenopausal women. J Clin Endocrinol Metab 1996;81:3540–46. © The Endocrine Society.

tropin secretion in the monkey (36,37). Our finding of a selective increase in GnRH gene expression in the medial basal hypothalamus of postmenopausal women provides evidence that this region is also involved in the regulation of gonadotropin secretion in the human.

Because mRNA levels correlate with changes in neuronal activity, the finding of increased GnRH gene expression suggests that the activity of these neurons is increased in postmenopausal women. Previous studies have shown that the hypothalamic content of the GnRH peptide is decreased in post-menopausal women (38); unfortunately, these data do not distinguish among decreased biosynthesis of GnRH, increased hypothalamic degradation, or increased secretion of GnRH into the portal circulation. The finding of increased hypothalamic GnRH gene expression (35) and previous demonstrations of increased LH secretion (3–5) is consistent with the interpretation of increased secretion of GnRH into the portal circulation. Taken together, these studies provide evidence that hypothalamic GnRH secretion contributes to the LH hypersecretion of menopause. This conclusion is controversial, however, because of the ongoing debate over the relative contributions of hypothalamic versus pituitary mechanisms underlying steroid negative feedback (16,39–42).

Proopiomelanocortin Gene Expression Is Decreased in the Infundibular Nucleus of Postmenopausal Women

Our most recent studies have focused on the POMC system of neurons in the medial basal hypothalamus (43). Considerable evidence suggests that hypothalamic β-endorphin neurons have an important inhibitory role in the regulation of GnRH secretion in the primate (39,44). We determined, therefore, that if menopause is associated with a change in the levels of POMC mRNA, then the precursor mRNA for β-endorphin. In contrast to the other neuropeptide systems, the expression of POMC mRNA was decreased in the infundibular nucleus of postmenopausal women (43). These data are consistent with studies that document a decrease in the inhibitory influence of the endogenous opioids on gonadotropin secretion in postmenopausal women (45). A drop in the activity of inhibitory POMC neurons could contribute to the elevation in GnRH gene expression observed in postmenopausal women (35).

Summary and Conclusions

The ovarian failure of menopause is associated with marked changes in morphology and neuropeptide gene expression in the human hypothalamus. Neuronal hypertrophy occurs in a subpopulation of neurons expressing estrogen receptor, neurokinin B, and substance P gene transcripts in the infundibular nucleus (17,23). The neuronal hypertrophy is accompanied by increased tachykinin gene expression. GnRH gene expression is also elevated in a separate subpopulation of neurons in the medial basal hypothalamus (35). In contrast, the number of neurons expressing POMC gene transcripts is decreased within the infundibular nucleus of postmenopausal women (43). We hypothesize that the increased GnRH gene expression in postmenopausal women could be a consequence of increased excitatory input from tachykinin neurons, decreased input of opioid peptides, or both. Contributions from other hypothalamic neurons are also likely given the number of neuropeptides that have been shown to regulate GnRH secretion.

Animal models have provided important insight into the relative contribution of two confounding variables: age and gonadal status. For example, we have found that ovariectomy increases NKB gene expression in the rat arcuate nucleus (29) and, conversely, that hormone replacement decreases NKB gene expression in ovariectomized cynomolgus monkeys (46). These findings lend strong support to the hypothesis that the increase in NKB gene expression in postmenopausal women is secondary to gonadal failure. Animal models have also allowed us to demonstrate that neuronal hypertrophy, induced by steroid withdrawal, is accompanied by dramatic changes in dendritic architecture (30). Because menopause is such a consistent and predictable event and is characterized by marked changes in ovarian hormones, we have had the rare opportunity to use the human brain to address basic biological questions.

Acknowledgments. This work was supported by NIH grant AG-09214. Ty W. Abel and Steve C. Danzer were recipients of Predoctoral Fellowships from the Robert S. Flinn Biomedical Research Initiative.

References

1. Block E. Quantitative morphological investigations of the follicular system in women. Acta Anat 1952;14:108–23.
2. Baird DT, Guevara A. Concentration of unconjugated estrone and estradiol in peripheral plasma in nonpregnant women throughout the menstrual cycle, castrate and postmenopausal women and in men. J Clin Endocrinol Metab 1969;29:149–56.
3. Chakravarti S, Collins WP, Forecast JD, Newton JR, Oram DH, Studd JWW. Hormonal profiles after the menopause. Br Med J 1976;2:784–86.
4. Gambacciani M, Melis GB, Paoletti AM, Cagnacci A, Mais V, Petacchi FD, et al. Pulsatile luteinizing hormone release in postmenopausal women: effect of chronic bromocriptine administration. J Clin Endocrinol Metab 1987;65:465–68.
5. Kazer RR, Liu CH, Yen SSC. Dependence of mean levels of circulating luteinizing hormone upon pulsatile amplitude and frequency. J Clin Endocrinol Metab 1987;65:796–800.
6. Wallach EE, Root AW, Garcia C-R. Serum gonadotropin responses to estrogen and progestogen in recently castrated human females. J Clin Endocrinol Metab 1970;31:376–81.
7. Monroe SE, Jaffe RB, Midgley AR, Jr. Regulation of human gonadotropins. XIII. Changes in serum gonadotropins in menstruating women in response to oophorectomy. J Clin Endocrinol Metab 1972;34:420–22.
8. Chakravarti S, Collins WP, Newton JR, Oram DH, Studd JWW. Endocrine changes and symptomatology after oophorectomy in premenopausal women. Br J Obstet Gynaecol 1977;84:769–75.
9. Sherwin BB, Gelfand MM. Effects of parenteral administration of estrogen and androgen on plasma hormone levels and hot flushes in the surgical menopause. Am J Obstet Gynecol 1984;148:552–57.
10. Casper RF, Yen SSC. Neuroendocrinology of menopausal flushes: an hypothesis of flush mechanism. Clin Endocrinol (Oxf) 1985;22:293–312.
11. Molnar GW. Body temperatures during menopausal hot flashes. J Appl Physiol 1975;38:499–503.

12. Ginsburg J, Swinhoe J, O'Reilly B. Cardiovascular responses during the menopausal hot flush. Br J Obstet Gynaecol 1981;88:925–30.
13. Tataryn IV, Meldrum DR, Lu KH, Frumar AM, Judd HL. LH, FSH and skin temperature during the menopausal hot flash. J Clin Endocrinol Metab 1979;49:152–54.
14. Casper RF, Yen SSC, Wilkes MM. Menopausal flushes: a neuroendocrine link with pulsatile luteinizing hormone secretion. Science 1979;205:823–25.
15. Sheehan HL, Kovács K. The subventricular nucleus of the human hypothalamus. Brain 1966;89:589–614.
16. Knobil E. The neuroendocrine control of the menstrual cycle. Recent Prog Horm Res 1980;36:53–88.
17. Rance NE, McMullen NT, Smialek JE, Price DL, Young WS, III. Postmenopausal hypertrophy of neurons expressing the estrogen receptor gene in the human hypothalamus. J Clin Endocrinol Metab 1990;71:79–85.
18. Glaser EM, Tagamets M, McMullen NT, Van der Loos H. The image-combining computer microscope—an interactive instrument for morphometry of the nervous system. J Neurosci Methods 1983;8:17–32.
19. Addison WHF. The cell-changes in the hypophysis of the albino rat, after castration. J Comp Neurol 1917;28:441–61.
20. Shivers BD, Harlan RE, Morrell JI, Pfaff DW. Absence of oestradiol concentration in cell nuclei of LHRH-immunoreactive neurones. Nature 1983;304:345–47.
21. Neaves WB, Johnson L, Porter JC, Parker CR, Jr., Petty CS. Leydig cell numbers, daily sperm production, and serum gonadotropin levels in aging men. J Clin Endocrinol Metab 1984;59:756–63.
22. Rance NE, Uswandi SV, McMullen NT. Neuronal hypertrophy in the hypothalamus of older men. Neurobiol Aging 1993;14:337–42.
23. Rance NE, Young WS, III. Hypertrophy and increased gene expression of neurons containing neurokinin-B and substance-P messenger ribonucleic acids in the hypothalami of postmenopausal women. Endocrinology 1991;128:2239–47.
24. Rance NE. Hormonal influences on morphology and neuropeptide gene expression in the infundibular nucleus of postmenopausal women. In Swaab DF, Hofman MA, Mirmiran M, et al., eds. The human hypothalamus in health and disease, progress in brain research. Amsterdam: Elsevier, 1992:221–36.
25. Vijayan E, McCann SM. In vivo and in vitro effects of substance P and neurotensin on gonadotropin and prolactin release. Endocrinology 1979;105:64–68.
26. Dees WL, Skelley CW, Kozlowski GP. Central effects of an antagonist and an antiserum to substance P on serum gonadotropin and prolactin secretion. Life Sci 1985;37:1627–31.
27. Ohtsuka S, Miyake A, Nishizaki T, Tasaka K, Aono T, Tanizawa O. Substance P stimulates gonadotropin-releasing hormone release from rat hypothalamus in vitro with involvement of oestrogen. Acta Endocrinol (Copenh) 1987;115:247–52.
28. Arisawa M, De Palatis L, Ho R, Snyder GD, Yu WH, Pan G, et al. Stimulatory role of substance P on gonadotropin release in ovariectomized rats. Neuroendocrinology 1990;51:523–29.
29. Rance NE, Bruce TR. Neurokinin B gene expression is increased in the arcuate nucleus of ovariectomized rats. Neuroendocrinology 1994;60:337–45.
30. Danzer SC, McMullen NT, Rance NE. Dendritic growth of arcuate neuroendocrine neurons following orchidectomy in adult rats. J Comp Neurol 1998;390:234–46.
31. Amoss M, Burgus R, Blackwell R, Vale W, Fellows R, Guillemin R. Purifica-

tion, amino acid composition and N-terminus of the hypothalamic luteinizing hormone releasing factor (LRF) of ovine origin. Biochem Biophys Res Commun 1971;44:205–10.

32. Matsuo H, Baba Y, Nair RMG, Arimura A, Schally AV. Structure of the porcine LH- and FSH-releasing hormone. I. The proposed amino acid sequence. Biochem Biophys Res Commun 1971;43:1334–39.

33. Rance NE, Young WS, III, McMullen NT. Topography of neurons expressing luteinizing hormone-releasing hormone gene transcripts in the human hypothalamus and basal forebrain. J Comp Neurol 1994;339:573–86.

34. Stopa EG, Koh ET, Svendsen CN, Rogers WT, Schwaber JS, King JC. Computer-assisted mapping of immunoreactive mammalian gonadotropin-releasing hormone in adult human basal forebrain and amygdala. Endocrinology 1991;128:3199–207.

35. Rance NE, Uswandi SV. Gonadotropin-releasing hormone gene expression is increased in the medial basal hypothalamus of postmenopausal women. J Clin Endocrinol Metab 1996;81:3540–46.

36. Krey LC, Butler WR, Knobil E. Surgical disconnection of the medial basal hypothalamus and pituitary function in the rhesus monkey. I. Gonadotropin secretion. Endocrinology 1975;96:1073–87.

37. Knobil E. The GnRH pulse generator. Am J Obstet Gynecol 1990;163:1721–27.

38. Parker CR, Jr., Porter JC. Luteinizing hormone-releasing hormone and thyrotropin-releasing hormone in the hypothalamus of women: effects of age and reproductive status. J Clin Endocrinol Metab 1984;58:488–91.

39. Ferin M, Van Vugt D, Wardlaw S. The hypothalamic control of the menstrual cycle and the role of endogenous opioid peptides. Recent Prog Horm Res 1984;40:441–85.

40. Kalra SP. Neural circuits involved in the control of LHRH secretion: a model for estrous cycle regulation. J Steroid Biochem 1985;23:733–42.

41. Kalra SP, Kalra PS. Do testosterone and estradiol-17 beta enforce inhibition or stimulation of luteinizing hormone-releasing hormone secretion? Biol Reprod 1989;41:559–70.

42. Sagrillo CA, Grattan DR, McCarthy MM, Selmanoff MK. Hormonal and neurotransmitter regulation of GnRH gene expression and related reproductive behaviors. Behav Genet 1996;26:241–77.

43. Abel TW, Rance NE. Proopiomelanocortin (POMC) gene expression is decreased in the hypothalamus of postmenopausal women. Mol Brain Res 1999;66:202–8.

44. Gindoff PR, Ferin M. Brain opioid peptides and menstrual cyclicity. Semin Reprod Endocrinol 1987;5:125–33.

45. Reid RL, Quigley ME, Yen SSC. The disappearance of opioidergic regulation of gonadotropin secretion in postmenopausal women. J Clin Endocrinol Metab 1983;57:1107–10.

46. Abel TW, Voytko ML, Rance NE. Effects of hormone replacement therapy on neuropeptide gene expression in a primate model of menopause. J Clin Endocrinol Metab 1999;84:2111–18.

4

Alterations in the Pattern of Gonadotropin Secretion During the Menopausal Transition

DENNIS W. MATT, MARIE C. KERBESHIAN, STEVEN M. PINCUS, JOHANNES D. VELDHUIS, AND WILLIAM S. EVANS

Introduction

It is well recognized that ultradian rhythms of circulating gonadotropins, luteinizing hormone (LH) and follicle stimulating hormone (FSH), play a central role in mammalian reproduction. The pulsatile secretion of these gonadotropins, particularly LH, is generally believed to be directly correlated to the episodic discharge of hypothalamic gonadotropin-releasing hormone (GnRH) into the pituitary portal circulation. Numerous experimental studies, as well as investigations into pathophysiological conditions in females, have unequivocally concluded that alterations in the periodicity of episodic gonadotropin secretion have a profound impact on reproductive cyclicity. The aim of this chapter is to delineate our current understanding of the changes in gonadotropin secretion that occur as a consequence of mammalian female aging, and to address the possibility that such changes may contribute to the cessation of human menstrual cyclicity.

Age-Related Changes in Gonadotropin Secretion in Rodents

Numerous investigations have compared the frequency and amplitude of LH pulsatility in young and aging females under intact and ovariectomized conditions (1,2). In all instances, an age-related deficit in LH pulsatility has been identified. To determine if these changes are intrinsic to the hypothalamic/pituitary axis or secondary to age-related changes in ovarian hormone secretion, previous approaches have removed the ovarian influence (by ovariectomy) and examined LH pulsatility (1). Such studies have convincingly

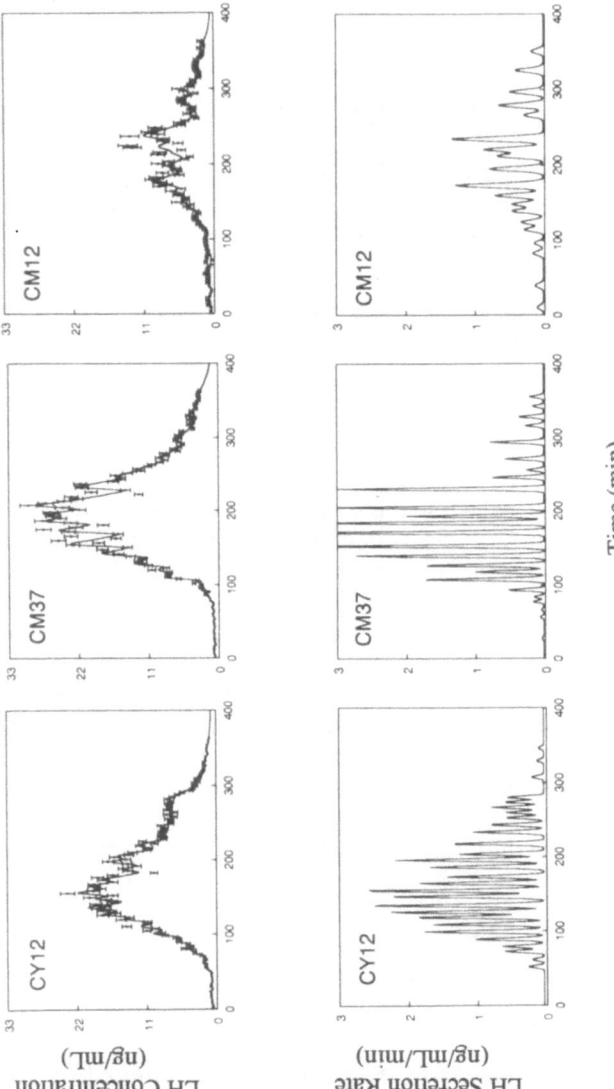

FIGURE 4.1. Representative pulsatile serum immunoreactive LH profiles (top panels) and calculated LH secretion plots (bottom panels) on proestrus in a young female rat (left panels), a middle-aged female rat with a normal LH surge (middle panels), and a middle-aged female rat with an attenuated LH surge (right panels). Blood was sampled at 3-minute intervals from 1600 to 2200 hours, and immunoreactive LH determinations were performed in triplicate for each blood sample. Reproduced with permission from Matt et al. Characteristics of attenuated proestrus luteinizing hormone surges in middle-aged rats by deconvolution analysis. Biol Reprod 1998;59:1477–82.

demonstrated that age-related alterations in LH pulsatility are in part attributable to changes in the hypothalamic/pituitary release mechanism.

Our laboratory, as well as others, have concentrated our studies on the well-documented attenuation of the preovulatory (proestrous) LH surge that occurs in middle-aged female rats prior to the loss of reproductive cyclicity. Previous studies have shown that middle-aged females with attenuated proestrous LH surges will change from regular to irregular estrous cyclicity within 2 months, whereas females of the same age demonstrating LH surges of similar amplitude to young proestrous females will continue to cycle beyond 2 months (3). This has been a particularly attractive model to investigate because the attenuated surges appear to be the first apparent neuroendocrine change observed prior to the loss of estrous cyclicity. One recent study from our laboratory has performed detailed analyses of the proestrous LH surge of young and middle-aged regularly cyclic rats. Figure 4.1 shows representative LH pulsatile and deconvolution profiles on proestrus in a young female, a middle-aged female with a normal LH surge, and a middle-aged female with an attenuated LH surge. As presented in Table 4.1, the attenuated LH surges in middle-aged females are associated with no changes in LH secretory burst frequency, but a profound decrease in LH burst mass (amplitude). These findings suggest that the primary deficit associated with attenuated LH surges in middle-aged rats is a decrease in the amount of GnRH released into the pituitary–portal system rather than an alteration in the timing of the GnRH signal during the preovulatory LH surge. These findings are in accord with investigations that demonstrate decreased GnRH neuronal activation on proestrus in middle-aged rats, whereby fewer GnRH neurons synthesized Fos protein throughout the LH surge as compared with young proestrous rats (4,5).

TABLE 4.1. LH secretory activity in young and middle-aged rats on proestrus.*

Parameter	Young ($n = 5$)	Middle-aged normal surge ($n = 7$)	Middle-aged attenuated surge ($n = 6$)
LH secretory bursts/hour	3.6 ± 0.9	3.4 ± 0.5	3.4 ± 0.7
Interburst interval, minute	14.8 ± 1.4	15.7 ± 1.1	17.0 ± 1.2
LH secretory burst half-duration, minute	6.4 ± 2.4	5.4 ± 0.9	5.6 ± 0.9
LH half-life, minute	13.3 ± 1.8	16.5 ± 1.1	15.0 ± 1.6
Mass of LH secreted/burst, ng/ml	9.1 ± 3.5[a]	8.2 ± 1.9[a]	2.4 ± 0.9[b]
Maximal rate of LH secretion/burst (amplitude), ng • ml^{-1} • min^{-1}	1.6 ± 0.3[a]	2.1 ± 1.0[a]	0.4 ± 0.1[b]
Total burstlike release, ng • ml^{-1} • h^{-1}	190 ± 76[a]	163 ± 35[a]	45 ± 15[b]
Basal secretion rate, ng • ml^{-1} • h^{-1}	0.05 ± 0.01	0.05 ± 0.01	0.03 ± 0.01

*Data are expressed as mean ± SEM. For each mean deconvolution parameter, values with different superscripts differ significantly ($p < 0.05$).
(Reproduced with permission from Matt et al., Characteristics of attenuated proteus luteinizing hormone surges in middle-aged rats by deconvolution analysis. Biol Reprod 1998;59:1477–82.

Age-Related Changes in Gonadotropin Secretion in Humans

Although the prevailing evidence on aging female rodents indicates a slowing of GnRH pulse generator, few studies have been performed in postmenopausal women. It is clear that in the absence of ovarian feedback, postmenopausal women become hypergonadotropic with robust episodic release of LH and FSH; however, a definitive decline in gonadotropin secretion occurs in aging postmenopausal women (6,7). A previous study has convincingly demonstrated that the decline in gonadotropin secretion in older postmenopausal women is associated with a decrease in LH pulse frequency and amplitude as well as a blunted response of gonadotropin secretion to exogenous GnRH administration (8). Thus, as has been well demonstrated in aging rodents, aging in postmenopausal women appears to be associated with a slowing of the GnRH pulse generator.

It is unequivocal that a major event associated with the onset of menopause is the rapid exhaustion of the ovarian follicular reserve. Such decline is heralded by increased circulating levels of FSH (but not LH) during the perimenopausal period, presumably as a result of declining circulating levels of ovarian inhibin or a deceased negative feedback to estrogen. It is still under debate whether intrinsic alterations in the hypothalamic–pituitary axis participate in the events of this transitional period. Relatively few detailed investigations have been conducted to compare the episodic release of LH between young and premenopausal women prior to the onset of the perimenopausal period.

With the prevailing evidence that the loss of reproductive cyclicity in rodents is associated with intrinsic deficits in the hypothalamus and pituitary, our laboratory has questioned whether analogous changes also occur in humans. To investigate this, we designed studies to examine the pulsatile LH secretion in regularly cycling middle-aged women. Unlike regularly cycling rats, it is not possible prospectively to determine the day of ovulation in the human menstrual cycle. Thus, our ideal goal to examine the preovulatory LH surge in middle-aged women was not feasible. We therefore decided to investigate pulsatile LH secretion during the mid- to late-follicular phase, which is a time when previous studies have shown that pulsatile LH frequency is the greatest (9). Following documentation of normal cyclicity, 10-minute blood samples were withdrawn for 8 hours during the mid- to late-follicular phase in young and middle-aged women.

To assess episodic LH secretion, we used the pulse detection algorithm, *Cluster*, which has been previously well characterized and validated on in vivo biologic data as well as computer simulations (10). Because blood concentrations of LH reflect the multiple parameters of pituitary secretory events and metabolic clearance, we also assessed LH secretory events with the well-characterized deconvolution analysis procedure, *Deconv* (11). Figure 4.2 illustrates representative profiles of serum LH concentration-time series in a young and a middle-aged premenopausal woman assessed by *Cluster* and deconvolution analysis. Comparison of the young and middle-aged women revealed a significant decrease in the LH interpulse interval with increases in pulse width and circulating LH half-life in middle-aged women (Fig. 4.3; 12). These findings strongly suggest that a slowing in the GnRH pulse generator occurs in aging premenopausal women. Other investigators have undertaken similar experimental approaches with different findings (13–15). Klein et al. (13) found no differences in LH pulse frequency during the early

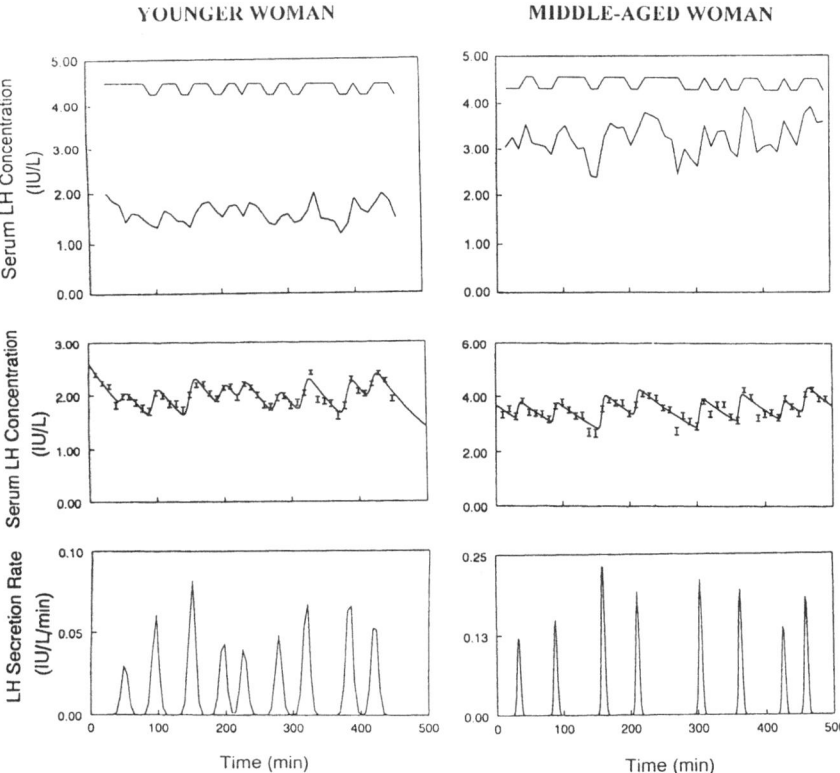

FIGURE 4.2. Representative profiles of serum LH concentration-time series obtained from a young (left panel) and a middle-aged woman (right panel). Note that each asterisk represents an LH peak as determined by *Cluster* analysis. In upper panels, LH pulses are shown by schematized deflections above data values. In middle panels, fitted deconvolution curves based on calculated secretion and clearance parameters are depicted. Deconvolution-resolved discrete LH secretory bursts are shown in lower panels. Reproduced with permission from Matt et al. Characterization of luteinizing hormone secretion in younger versus older premenopausal women. Am J Obstet Gynecol 1998;178:504–10.

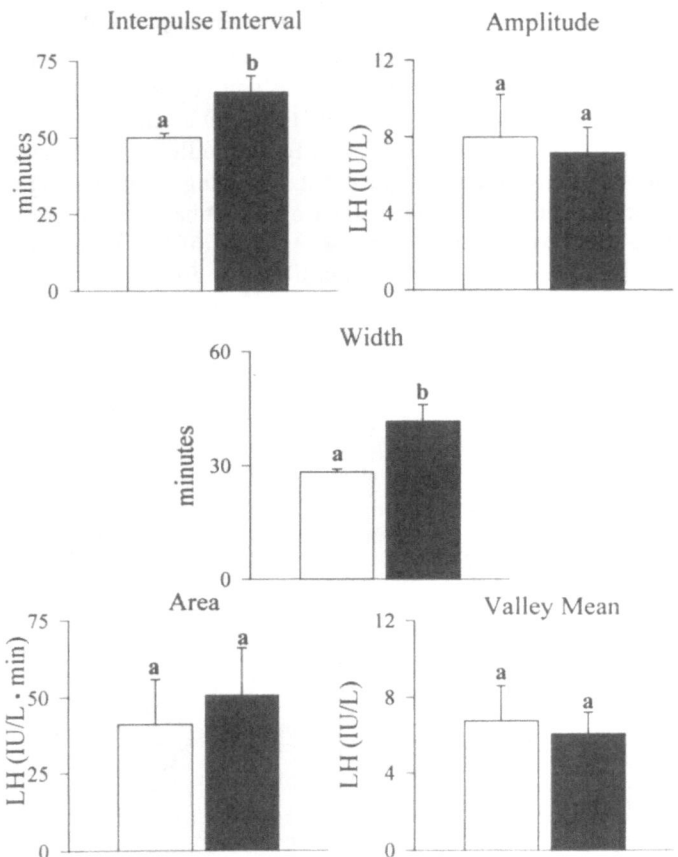

FIGURE 4.3. Comparison of the mean (± SE) LH interpulse interval (left upper panel), maximal pulse amplitude (right upper panel), pulse width (center panel), pulse area (left lower panel), and interpulse valley mean (right lower panel) as resolved by *Cluster* from concentration-time series obtained from 8 hours of sampling during the mid- to late-follicular phase in young (open bars) and middle-aged (solid bars) women. Bars with different superscripts differ significantly ($p < 0.05$). Note that middle-aged women showed a significantly greater interpulse interval and an increased pulse width compared with the young women. Despite significant differences in interpulse interval and pulse width, there were no differences in the other pulse characteristics between young and middle-aged women. Reproduced with permission from Matt et al. Characterization of luteinizing hormone secretion in younger versus older premenopausal women. Am J Obstet Gynecol 1998;178:504–10.

follicular or midluteal phases in middle-aged premenopausal as compared to young women.

In contrast, Reame et al. (14) found an increase in LH pulse frequency during the early follicular and late luteal phases in middle-aged premenopausal women as compared with young women with no differences observed during the midluteal phase. Wilshire et al. (15) reported a trend toward decreased LH pulses during the early follicular phase of premenopausal middle-aged women. It is difficult to reconcile the discrepancies in the findings in these studies, but further consideration should be given to the sampling intervals, menstrual phase of sampling, and pulse detection methods. It is noteworthy that each of these studies has utilized different pulse detection methods. A previous study in which several pulse detection algorithms were compared on the same LH data sets obtained from women during different phases of the menstrual cycle revealed significant differences in the estimates of pulse frequency, depending on the pulse detection method used (Fig. 4.4; 16).

It is clear from this comparative study that vastly different statistical conclusions can be drawn as a function of the pulse algorithm used rather than the actual underlying GnRH signal itself. Thus, prior to embarking into further studies of episodic LH release in premenopausal women, emphasis should be directed toward validating the pulse detection method considered. Validation of such methods requires independent knowledge of true events to which the pulse detection method can be tested. Use of computer simulations and independent in vivo methods will be required to test the validity of the pulse detection method. These approaches have been taken prior to using the *Cluster* algorithm in our study (10). Taken together, it is apparent that age-related changes in the GnRH pulse generator as inferred by LH concentration–time analyses are not easily identifiable. Notwithstanding, subtle changes in the pulse generator as a consequence of diminished hypothalamic homeostasis may be sufficient to contribute to the cascade of physiologic events that participate in the age-related loss of reproductive function observed in middle-aged premenopausal women.

Under well-controlled experimental studies, it has been established that alteration in GnRH pulse frequency is determinate on the relative release of pituitary LH and FSH (17). As such, the increased FSH secretion observed during the late luteal phase of normal menstrual cycling women is believed in part to be a result of a slowing in the GnRH pulse generator. The most conspicuous event heralding the menopausal transition is the monotropic rise in circulating FSH during the early follicular phase. We and others (13) have explored the possibility that this rise in FSH may be correlated to a slowing of the GnRH generator. These studies, however, have not found a significant correlation between early follicular phase FSH levels and LH pulsatility.

Conclusions

The possibility that age-related changes in the hypothalamic GnRH release mechanism may contribute to reproductive decline in all mammalian species including humans remains to be determined. Further detailed and intensive assessment of gonadotropin secretory events throughout the menstrual

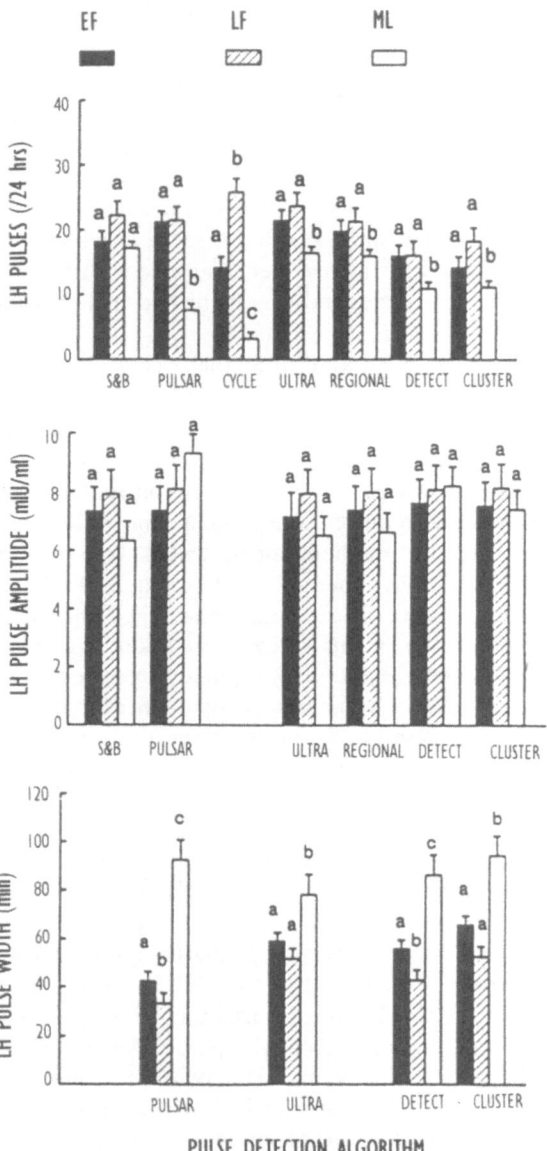

FIGURE 4.4. Comparison of the mean (± SE) number of LH pulses (upper panel), LH pulse amplitude (middle panel), and LH pulse width (lower panel) as determined from LH time-concentration series obtained from normal women during the early follicular (solid bars), late follicular (hatched bars), and midluteal (open bars) phases of the menstrual cycle and characterized by several pulse analysis algorithms. Bars with different superscripts are significantly ($p < 0.05$) different. Reproduced with permission from Evans et al. Contemporary aspects of discrete peak-detection algorithms. Endocrine Rev 1992;13:81–104. © The Endocrine Society.

cycle and during the entire menopausal transition, combined with studies to examine steroidal feedback effects at the level of the hypothalamic–pituitary axis, will be necessary to understand the potential role of the hypothalamus in the age-related changes in human reproduction.

References

1. Wise PM, Dueker E, Wuttke W. Age-related alterations in pulsatile luteinizing hormone release: effects of long-term ovariectomy, repeated pregnancies and naloxone. Biol Reprod 1988;39:1060–66.
2. Scarbrough K, Wise PM. Age-related changes in the pulsatile luteinizing hormone release precede the transition to estrous acyclicity and depend upon estrous cycle history. Endocrinology 1990;126:884–90.
3. Nass TE, LaPolt PS, Judd HL, Lu JKH. Alterations in ovarian steroid and gonadotrophin secretion preceding the cessation of regular oestrous cycles in aging female rats. J Endocrinol 1984;100:43–50.
4. Rubin BS, Lee CE, King JC. A reduced proportion of luteinizing hormone (LH)-releasing hormone neurons express Fos protein during the preovulatory or steroid-induced LH surge in middle-aged rats. Biol Reprod 1994;51:1264–72.
5. Lloyd JM, Hoffman GE, Wise PM. Decline in immediate early gene expression in gonadotropin-releasing hormone neurons during proestrus in regularly cycling, middle-aged rats. Endocrinology 1994;134:1800–5.
6. Chakravarti S, Collins WP, Forecast JD, Newton JR, Oram DH, Studd JWW. Hormonal profiles after menopause. Br Med J 1976;ii:784–86.
7. Judd JWW, Collins WP, Chakravarti S. Plasma hormone profiles after the menopause and bilateral oophorectomy. Postgrad Med J 1978;54(suppl):25–30.
8. Rossmanith WG, Scherbaum WA, Lauritzen C. Gonadotropin secretion during aging in postmenopausal women. Neuroendocrinology 1991;54:211–18.
9. Filicori M, Santoro N, Merriam GR, Crowley WF, Jr. Characterization of the physiological pattern of episodic gonadotropin secretion throughout the human menstrual cycle. J Clin Endocrinol Metab 1986;62:1136–44.
10. Urban RJ, Johnson ML, Veldhuis JD. In vivo biological validation and biophysical modeling of the sensitivity and positive accuracy of endocrine peak detection. I. The LH pulse signal. Endocrinology 1989;124:2541–47.
11. Veldhuis JD, Carlson ML, Johnson ML. The pituitary gland secretes bursts: appraising the nature of glandular secretory impulses by simultaneous multiple-parameter deconvolution of plasma hormone concentrations. Proc Natl Acad Sci USA 1987;84:7686–95.
12. Matt DW, Kauma SW, Pincus SM, Veldhuis JD, Evans WS. Characterization of luteinizing hormone secretion in younger versus older premenopausal women. Am J Obstet Gynecol 1998;178:504–10.
13. Klein NA, Battaglia DE, Clifton DK, Bremner WJ, Soules MR. The gonadotropin secretion pattern in normal women of advanced reproductive age in relation to the monotropic FSH rise. J Soc Gynecol Invest 1996;3:27–32.
14. Reame NE, Kelch RP, Beitins IZ, Yu MY, Zawacki CM, Padmanabhan V. Age effects on follicular-stimulating hormone and pulsatile luteinizing hormone secretion across the menstrual cycle of premenopausal women. J Clin Endocrinol Metab 1996;81:1512–18.

15. Wilshire GB, Loughlin JS, Brown JR, Adel TE, Santoro N. Diminished function of the somatotropic axis in older reproductive-aged women. J Clin Endocrin Metab 1995;80: 608–13.
16. Evans WS, Sollenberger MJ, Booth Jr RA, Rogol AD, Urban RJ, Carlsen EC, et al. Contemporary aspects of discrete peak-detection algorithms. II. The paradigm of luteinizing hormone pulse signal in women. Endocrine Rev 1992;13:81–104.
17. Wildt L, Hausler A, Marshall G, Hutchison JS, Plant TM, Belchetz PE, et al. Frequency and amplitude of gonadotropin-releasing hormone stimulation and gonadotropin secretion in the rhesus monkey. Endocrinology 1981;109:376–85.

5

Clinical and Scientific Significance of Perimenopausal Changes in Pituitary and Ovarian Hormone Secretion

Gerson Weiss

The perimenopause—the phase in a woman's life between the onset of cycle change or irregularity and the amenorrhea of menopause—has been shown to be characterized by dynamic changes in the hypothalamic–pituitary–ovarian axis. These changes clearly have clinical implications with regard to symptoms and effects on hormonally sensitive structures. One of the earliest observed changes, seen before cycle irregularity, is an elevation in circulating follicle-stimulating hormone (FSH) (1). FSH varies throughout the menstrual cycle. Clinicians traditionally measure FSH on day 3 of the cycle because early follicular gonadotropins generally vary less at that time than they do at later times of the menstrual cycle. Because the phases of the menstrual cycle may be of different lengths at different ages and in different women, there is greater standardization to day 3 FSH measurement. In practical terms, however, FSH measurements from day 2 to day 5 are clinically acceptable. Older reproductive-aged women have accelerated development of the dominant follicle (2). Samples later in the cycle will be in later phases, possibly after dominant follicle selection.

Early FSH levels generally rise before cycle irregularity occurs. There may be changes in FSH control dynamics even before there is an elevation in basal FSH. The clomiphene stimulation test has been used clinically as a predictor of fecundity (3). This test involves measuring serum FSH basally and again after 5 days of clomiphene citrate. Greater elevation of FSH after clomiphene citrate stimulation occurs even before there is a measurable basal increase in FSH; thus, there is a hyperactivity of FSH secretion earlier than there are observable changes in either ovarian steroids or pituitary gonadotropins. This indicates alterations in central control mechanisms. Low day 3 levels of inhibin B in serum correlates with poorer response to ovulation induction and a decreased pregnancy rate with in vitro fertilization (4). In contrast to FSH, luteinizing hormone (LH) secretion changes little in the perimenopausal period until there is a rise at the time of menopause (5). There are, however, clear changes in gonadotropin dynamics in the perimenopausal period. There is obviously a continued elevation in the FSH–LH ratio. There are changes in both frequency and amplitude of GnRH secretion, as witnessed by alterations in LH pulse frequency. In older women there is a greater LH pulse amplitude as well as an increase in pulse frequency (1,5). In the late perimenopause, the bioactive–immunoactive ratio of gonadotropins increases (6). Gonadotropins are under both positive and negative feedback control. Negative feedback is the long loop interplay of gonadotropins with estradiol. An opening of this loop by menopause or castration results in elevation in gonadotropins. This is similar to the long loop negative feedback control by TSH of thyroxine or by ACTH of cortisol. Unlike these other pituitary trophic hormones, gonadotropins are also under positive feedback control, the dynamics of which have been described by Knobil using the rhesus monkey as a model of human gonadotropin secretion (7). In brief, superimposed upon a negative feedback control, an elevation of estradiol above a threshold for a duration of at least 24 hours results in a gonadotropin surge similar to the midcycle preovulatory surge of gonadotropins that triggers ovulation. This mechanism has not been well studied in perimenopausal women. Some perimenopausal women with abnormal uterine bleeding, however, have a failure of this mechanism (8). These observations must be confirmed in normal perimenopausal women. This clinical data indicates clear differences in central nervous system gonadotropin control during this reproductive phase.

Regulation of FSH secretion is modulated both by the inhibitory feedback of inhibin from the ovary as well as by activin and follistatin within the central nervous system and pituitary itself. In the early perimenopause there is a fall in inhibin B (9). Inhibin A and activin A are also elevated in older women when compared with younger women (Santoro N, unpublished observation). The role and interaction of these substances in the changes that occur in the perimenopausal period remain to be elucidated.

Perhaps the most definitive demonstration of reproductive hormonal changes in the perimenopause is presented by Santoro et al. (10). She obtained daily first morning voided urine collections throughout the menstrual cycle in both 11 cycling women aged 43–52 years and in 11 younger, midreproductive cycling women. Urine was assayed for LH, FSH, estrone conjugates and pregnanediol glucuronide. Levels were normalized for creatinine. Compared with younger women, perimenopausal women had shorter follicular phases: 11 days versus 14 days. This resulted in shorter menstrual cycles. FSH excretion was higher in perimenopausal women than it was in younger women (range of means 4–32 vs. 3–7 IU/g Cr.). LH secretion was slightly greater in older women. The overall mean estrone conjugate excretion was higher in perimenopausal women than it was in younger women

(76.9 ng/mg Cr. vs. 40.7 ng/mg Cr.). There were estrogen elevations both in the follicular and luteal phases. In contrast, pregnanediol glucuronide excretion was lower in perimenopausal women compared with younger women (1.0–8.4 vs. 1.6–12.7 µg/mg Cr. in the luteal phase). As can be observed from the secretion patterns (Fig. 5.1), the perimenopause is clearly a hyperestrogenic state. These hyperestrogenic cycles coexist with decreased progesterone secretion in a pattern similar to what would be called luteal insufficiency in a younger woman because of decreased progesterone secretion and shorter luteal phases.

To summarize the changes in hormone secretion across the perimenopausal period, FSH continues to rise across the perimenopausal transition, but it is only in the latest perimenopausal phases that there is an LH rise. Cycle lengths vary in the early perimenopause. Cycle lengths usually shorten because of a shortened follicular phase. In the latest phases of the perimenopause, there are missed cycles and anovulatory cycles until the menopause is reached.

Hormone secretion patterns in ovulatory cycles vary across the menopausal transition. A phase is reached of still indeterminate length in which there is increased estradiol secretion across the menstrual cycle. These cycles are also marked by decreased progesterone secretion in the luteal phase. There is as yet little useful data as to the detail of longitudinal changes that

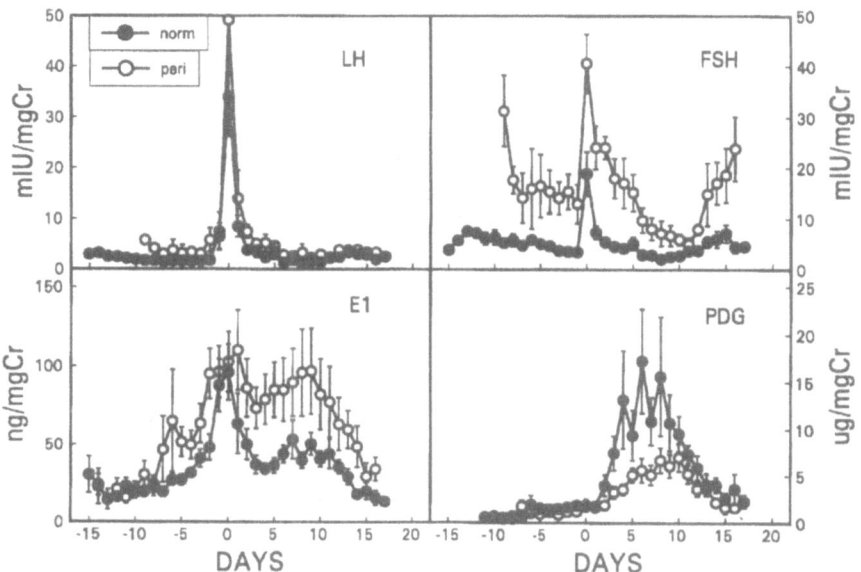

FIGURE 5.1. Mean ± SEM daily urinary gonadotropin and sex steroid excretion patterns in perimenopausal women and midreproductive-aged women. Data are standardized to day 0, the day of ovulation. Reproduced with permission from Santoro N et al. Characterization of reproductive hormonal dynamics in the perimenopause. J Clin Endocrinol Metab 1996;81:1495–501. © The Endocrine Society.

occur in ovarian peptides. It is certainly tempting to hypothesize the existence of a relationship between the hyperestrogenic state in the perimenopause and the increase in symptomatology that is experienced clinically at that time.

As the menopause is reached, hormone secretion is characterized by an opening of the long loop feedback of estrogen on gonadotropins and a decrease in inhibin secretion. Gonadotropins are at high serum concentrations. Progesterone is essentially no longer secreted and estradiol is at low levels in the circulation.

Fertility in the Perimenopausal Transition

To state the obvious, the biological significance of the menstrual cycle is reproduction of the species. It should follow from this that the onset of perimenopausal changes should be the stage at which there is a decline in fecundity based on hormonal and reproductive factors. Our working definition based on cycle irregularity is inadequate in defining this biologically important parameter. By the time a woman reaches the stage of cycle irregularity, our current definition of the perimenopause, fecundity is already minimal (11). By our current definitions, the normal and natural state of a perimenopausal woman is to be infertile.

Fecundity is highest before the age of 30. Starting at the age of 30, there is a decreased fecundity rate that progresses over time. At about age 38, there is a steep increase in the loss of fecundity. This rate of fecundity decrease accelerates further in the 42–44 year age group (11). After 45, it is fairly unusual for a pregnancy to occur naturally. The age-related decrease in fecundity is accompanied by an increase in the spontaneous miscarriage rate. By the early forties, roughly half of pregnancies are spontaneously miscarried (12). It has been estimated that after age 30, the probability of having a healthy baby decreases 3.5% per year (13). By combining a decreased ability to get pregnant with an increased miscarriage rate, it has been calculated that the chances of a woman age 35 having a healthy baby is half that of a woman 25 years old. It is estimated that a young couple having regular coitus has a greater than 20% chance of achieving pregnancy in a given menstrual cycle. This decreases to less than 5% at age 40.

The cause of the decreased fecundity with age is unknown. However, it is associated with progressive oocyte depletion. The fecundity decrease is likely to be related to decrease in function in the remaining ovarian follicles and oocytes. Egg donation programs that replace oocytes, but no other parts of the woman's reproductive system, result in high pregnancy success rates and low miscarriage rates. With increasing age in the forties, there is some evidence for decreased uterine receptivity, which is correctable by exogenous progestin therapy (14); however, the major defect responsible for the fecundity decline resides in the egg.

In general, FSH increases are associated with decreased fecundity. FSH, however, is a poor index of ovarian reserve because it has a low sensitivity, significant cycle to cycle variation, a very wide normal range, and fluctuations dependent on endogenous estradiol and ovarian peptide levels. By way of example, elevation in estradiol may lower FSH even in the presence of insensitive ovaries with poor fecundity potential. As a generalization, levels of FSH greater than 15 IU/L are associated with a poor success rate and poor ovulatory performance during in vitro fertilization cycles (15). Even though high levels of FSH are predictors of low fecundity, fecundity is decreased in older women even in the presence of low FSH.

Clinical Symptomatology of the Perimenopause

Although much is known of the pathology that occurs during the age of the perimenopause, there are surprisingly few clear pathophysiological correlations with the endocrinological changes. During this life phase there are changes in body habitus. Bone mineral density has been decreasing from the mid thirties due to unknown mechanisms even in the face of constant or even elevated estrogen levels (16). Vasomotor changes occur during the perimenopause. Intermittent episodes of hot flashes appear in some women several years before menstrual cessation. The mechanism of this change is poorly understood. It is quite possible that some women have these symptoms even during their hyperestrogenic episodes. It will take longitudinal study, such as the National Institute on Aging's Study of Women's Health Across the Nation (SWAN), to resolve these issues. During this phase, some women also experience an increase in breast symptoms and breast tenderness. Some women even complain of breast enlargement during the perimenopause. The explanation for this had been unknown. A reasonable hypothesis, however, is that the increase in breast symptoms and breast swelling and/or enlargement are related to increased ovarian estrogen secretion. Breast tissue is estrogen sensitive. Pubescent breast development is due to estrogen secretion. Even though the association awaits confirmation, it is highly likely based on our understanding of reproductive physiology that this is a valid association.

Uterine fibroids (leiomyoma uteri, myomas) are the most common benign tumors in women (17). Roughly 40% of women will have detectable fibroids during their reproductive years. Hysterectomy is the second most frequently performed major operation in American women with 590,000 procedures performed annually (18). Roughly one third of these are performed for uterine fibroids. Fibroids are hormonally responsive, and are stimulated by estrogens and androgens. Fibroids usually decrease in size and clinical significance after menopause or ovariectomy.

The frequency of fibroids varies with ethnicity. Fibroids are monoclonal with chromosome rearrangements observed in roughly half of studied tumors (19). In addition to steroids, a variety of additional growth factors have been implicated in their growth. Because fibroids are benign, their presence alone is an inadequate reason for surgical intervention; however, there are a variety of indications for fibroid removal. A large fibroid uterus will distort abdominal contour and impinge on adjacent structures, and quite frequently produce symptoms such as bowel and bladder pressure. Large fibroid uteri, which arise out of the pelvis, are therefore a frequent indication for surgery. Although the risk of malignant degeneration of a fibroid is low and less than the risk of the occurrence of uterine cancer, a rapid growth of fibroids has been cited as an indication for surgical intervention. A variety of pathologic occurrences can also affect myomas. Fibroids may undergo infarction. They may also undergo torsion or internal hemorrhage. All of these occurrences will be associated with severe pain warranting surgical intervention. Fibroids increase the size of the uterus and the surface area of the endometrium. The presence of fibroids under the endometrial lining may interfere with the myometrial constriction of the perpendicular blood vessels that supply the endometrium, which in turn may result in bleeding. One can also anticipate greater quantities of menstrual flow from a larger endometrial surface. Prolonged menses and increased menstrual flow may therefore be associated with fibroids. This may produce secondary anemia and may warrant medical or surgical intervention (20). The perimenopause is a time associated with the greatest frequency of hysterectomies, with many for symptomatic or enlarging fibroids. It is also clearly tempting to associate the growth of fibroids and their symptoms with the hyperestrogenemia of the perimenopause. It is, therefore, not a great stretch to anticipate that if more were understood about this phase of the cycle and its effect on tumor growth, then this knowledge could potentially lead to greater temporizing endocrine interventions to alter what may be transient changes in hormone secretion. In turn, this may decrease the incidence of hysterectomy in the United States.

Another cause of hysterectomy in the perimenopause is endometriosis. Endometriosis, the presence of endometrial tissue in aberrant sites, is a fairly common condition that might produce ovarian masses (endometriomata) or the occurrence of severe pelvic pain (21). The endometrium is also steroid sensitive. It is quite likely that control of ovarian steroid hypersecretion in the perimenopause would also alleviate the need for intervention for pelvic endometriosis.

Another common reproductive pathology in women during the perimenopause is abnormal uterine bleeding (dysfunctional uterine bleeding). This may either be ovulatory or anovulatory. Among causes of ovulatory abnormal bleeding are bleeding from endometrial polyps (20). Luteal phase abnormalities such as luteal insufficiency or a decreased luteal progesterone production may also be associated with dysfunctional bleeding. Anovulatory abnormal bleeding will most likely occur in the presence of estrogen. Estrogen secretion without predictable shedding due to changes produced by luteal phase progesterone can cause endometrial hyperplasia and, if prolonged, may be a precursor of endometrial adenocarcinoma. Women at greatest risk of endometrial adenocarcinoma are those exposed to high steroid levels and unopposed estrogen for long periods of time. This includes women with polycystic ovarian syndrome as well as obese women. An endocrine-induced endometrial hypertrophy results in a regular and sometimes heavy and prolonged surface endometrial shedding. These symptoms can require intervention (20).

The common thread of all of these conditions is disordered ovarian steroid secretion, estrogen hypersecretion, anovulation, and luteal insufficiency. This may be the natural state of the perimenopause. Thus, luteal insufficiency, hyperestrogenic secretion, and anovulation (common natural occurrences in the perimenopausal phase) may be causal to the abnormal bleeding seen in the perimenopause. Longitudinal study and hormonal correlates with these symptoms will be important to firm up these associations.

Summary

The perimenopausal period is marked by dramatic dynamic changes in the hypothalamic–pituitary–ovarian axis. Among these changes are alterations in inhibin, activin, and follistatin secretion, alterations in the neuroendocrine control of gonadotropin secretion, and gonadotropin dynamics. FSH and, to a lesser extent, LH levels are elevated. The resultant ovarian changes include luteal insufficiency characterized by lower levels of progesterone secreted for shorter periods of time compared with younger women. Estrogen secretion is uniformly elevated throughout the menstrual cycle during part of the perimenopausal transition.

The perimenopausal phase is also a time of many significant clinical alterations. Fecundity is a fraction of levels at younger ages. Successful fertility treatments usually have only minimal success. The decrease in fecundity correlates with age as well as elevation in FSH and estradiol secretion, which are also indexes of poor response to exogenous gonadotropin stimulation.

Hot flashes may occur even in the hyperestrogenic woman. Breast symptoms increase in many women and some note an increase in breast fullness. There is an increase in abnormal uterine bleeding and endometrial hyperplasia. Endometriosis symptoms progress in some women. Others experience a growth of uterine fibroids. Many of these symptoms are severe enough to result in hysterectomy or other surgical intervention.

It is likely that the hormonal changes of the perimenopause correlate with the clinical changes and may, in part, be causal of these changes. Thus, a better understanding of the physiology of this period will allow for greater understanding of the dynamics of the clinical changes and for more control of these symptoms. This, in turn, will alter clinical practices.

References

1. Reame NE, Kelch RP, Beitins IZ, Yu M-Y, Zawacki CM, Padmanabhan V. Age effects on follicle-stimulating hormone and pulsatile luteinizing hormone secretion across the menstrual cycle of premenopausal women. J Clin Endocrinol Metab 1996;81:1512–18.

2. Klein NA, Battaglia DE, Fujimoto VY, Davis GS, Bremner WJ, Soules MR. Reproductive aging: accelerated ovarian follicular development associated with a monotropic follicle-stimulating hormone rise in normal older women. J Clin Endocrinol Metab 1996;81:1038–45.

3. Loumaye E, Billion JM, Mine JM, Psalti I, Pensis M, Thomas K. Prediction of individual response to controlled ovarian hyperstimulation by means of a clomiphene citrate challenge test. Fertil Steril 1990;53:295–301.

4. Seifer DB, Lambert-Messerlian G, Hogan JW, Gardiner AC, Blazar AS, Berk CA. Day 3 serum inhibin-B is predictive of assisted reproductive technologies outcome. Fertil Steril 1997;67:110–14.

5. Matt DW, Kauma SW, Pincus SM, Veldhuis JD, Evans WS. Characteristics of luteinizing hormone secretion in younger versus older premenopausal women. Am J Obstet Gynecol 1998;178:504–10.

6. Schmidt PJ, Gindoff PR, Baron DA, Rubinow DR. Basal and stimulated gonadotropin levels in the perimenopause. Am J Obstet Gynecol 1996;175:643–50.

7. Knobil E. On the control of gonadotropin secretion in the rhesus monkey. Rec Prog Horm Res 1974;30:1–46.

8. Van Look PF, Lothian H, Hunter WM, Michie EA, Baud DT. Hypothalamic-pituitary-ovarian function in perimenopausal women. Clin Endocrinol 1977;7:13–31.

9. Hayes FJ, Hall JE, Boepple PA, Crowley WF, Jr. Differential control of gonadotropin secretion in the human: endocrine role of inhibin. J Clin Endocrinol Metab 1998;83:1835–41.

10. Santoro N, Brown JR, Adel T, Skurnick JH. Characterization of reproductive hormonal dynamics in the perimenopause. J Clin Endocrinol Metab 1996;81:1495–501.

11. Weiss G. Fertility in the older woman. Clin Consultation in Obstet Gynecol 1996;8:56–59.

12. Stein ZA. A women's age: childbearing and child rearing. Am J Epidemiol 1985;121:327–42.

13. VanNoord-Zaadstra BM, Looman CWN, Alsbach H. Delayed childbearing: effect of age on fecundity and outcome of pregnancy. Br Med J 1991;302:1361–65.

14. Meldrum DR. Female reproductive aging: ovarian and uterine factors. Fertil Steril 1993;59;1–5.

15. Toner JP, Philput CB, Jones GS, Muasher SJ. Basil follicle-stimulating hormone level is a better predictor of in vitro fertilization performance than age. Fertil Steril 1991;55:784–91.

16. Riggs BL, Melton LJ III. Age-related osteoporosis. N Engl J Med 1986;314:1678–86.

17. Weiss G. Management of uterine myomata. In: Keye WR, Chang RJ, Rebar RW, Soules MR, eds. Infertility: evaluation and treatment. Philadelphia: WB Saunders, 1994:412–24.

18. National Center for Health Statistics, Graves EJ. National hospital discharge survey: annual summary, 1990. Vital and health statistics. (Series) 13 No. 112. Wash-

ington DC; Government Printing Office, 1992 CDHAS publication no. (PHS) 92-1773.

19. Rien MS, Friedman AJ, Barbieri RL, Pavelka K, Fletcher JS, Morton CC. Cytogenetic abnormalities in uterine leiomyomata. Obstet Gynecol 1991;77:923–26.

20. Merrill JA, Creasman WT. Benign lesions of the uterine corpus. In: Scott JR, DiSaia PJ, Hammond CR, Spellacy WN, eds. Danforth's obstetrics and gynecology, sixth ed. Philadelphia: Lippincott, 1990:1023–54.

21. Talbert LM, Kauma SM. Endometriosis. In: Scott JR, DiSiai PJ, Hammond CB, Spellacy WN, eds. Danforth's obstetrics and gynecology, sixth ed. Philadelphia: Lippincott, 1990:845–52.

6

Follicular Depletion and the Menopausal Transition

Anne Newman Hirshfield and Jodi Anne Flaws

Every female mammal is born with a vast, but nonrenewable, reserve of dormant primordial follicles in her ovaries. The reserve is deposited during the embryonic period: the germ cells proliferate for a time, then leave the mitotic cell cycle and begin meiosis. Once mitosis has ceased, no new germ cells will ever be added to the original endowment. Throughout postnatal life, follicles reawaken and begin to grow. This results in a steady stream of growing follicles. At any given moment the ovaries contain large numbers of follicles in nearly every stage of growth. There are vast numbers of small growing follicles, and long periods of time are required to complete these stages of growth. There are fewer medium-sized and large follicles: These growth stages are traversed rapidly (1).

This image of a constant stream of growing follicles, accelerating in growth as they approach the end of their development, is essential for fertility. The enormous excess of growing follicles provides the basis for the mechanism that regulates, within narrow limits, the number of ova shed each estrous or menstrual cycle, and the length of time between cycles. The continuous stream of follicles through the developmental pipeline is maintained by the continuous entry of dormant follicles into the growing pool and their slow, steady passage through the first several generations of growth. Because the reserve of dormant primordial follicles is not renewable, the original endowment is gradually depleted with advancing age. By menopause the follicular reserve is completely exhausted. Long before all the follicles are gone, the human menstrual cycle becomes increasingly irregular. One explanation for this phenomenon lays the blame for the loss of fine control on the diminishing numbers of growing follicles in the developmental pipeline (2). This, in turn, could be explained by a paucity of primordial follicles remaining to prime the pipeline.

The decline of ovarian function with age has far-reaching consequences for women. The developing follicles are the primary source of the estrogens that are so important for maintaining bone density, protecting against cardiovascular diseases, preventing stress incontinence due to loss of pelvic diaphragmatic tone, and, perhaps, reducing neuronal cell death. Depletion of the follicular reserve means much more than cessation of menses: The negative impact of this phenomenon on other organ systems can lead to significant morbidity in later life (3–6).

Does the Size of the Primordial Follicle Endowment Affect Reproductive Function?

Despite their significance for ovarian physiology and aging, we know little, if anything, about the factors that govern the establishment and expenditure of the primordial follicle endowment. These factors may be critical determinants of the long-range fertility of the individual. The role of the primordial follicle endowment in the menopausal transition is a hotly debated issue with competing alternative hypotheses. One perspective is that impending exhaustion of the follicle endowment triggers the menopausal transition; the hypothalamic/pituitary changes that accompany the menopause are a consequence of diminished ovarian function (7,8). The alternative perspective is that age changes in the central nervous system (CNS) trigger the menopausal transition; the exhaustion of the follicle endowment is a consequence of altered CNS signals (9).

Observations Supporting the Hypothesis That the Ovary Drives the Menopausal Transition

Some investigators maintain that "there is little doubt that ovarian exhaustion of follicles is the pacemaker of reproductive senescence in women" (10). The primary mechanism by which the primordial follicle endowment is depleted is through the onset of growth. The rate at which quiescent primordial follicles enter the "growing pool" is therefore the critical variable determining the decay curve of the primordial follicle reserve. Many investigators believe that the number of follicles in the primordial endowment is in itself the major factor in determining the rate at which follicles begin to grow (11,12). In support of this contention are observations derived from animals that had been treated in various ways to diminish the primordial endowment. For example, exposure in utero to busulfan (a drug that destroys primordial germ cells of embryos) results in a greatly reduced number of primordial follicles at birth. When rats were exposed to various doses of busulfan in utero, there was an inverse correlation between the number of primordial follicles in the ovary at birth and the rate at which they began to grow. In rats given the highest dose of busulfan, all of the remaining follicles began to grow very early in life, exhausting the follicular reserve during the prepubertal period. When partially sterilized rats were allowed to mature, they began to have estrous cycles like normal animals (13). A Japanese study in rats showed that although busulfan depletes primordial follicles, the animals still produce estrogen. The busulfan-treated rats have early vaginal opening and their cycles start off abnormally, but they soon become similar to controls (14).

Similar effects are seen after postnatal exposure of animals or humans to toxicants (15–17) and radiation (18,19) that destroy oocytes. Studies on polyaromatic hydrocarbons (PAHs) indicate that in utero exposure of C57BL/6 or DBA/2 mice to cigarette smoke results in a reduced number of ovarian primordial follicles in female offspring (20). Direct exposure of female rodents to the three major PAHs in cigarette smoke—benzopyrene, 9:10-dimethyl-1:2-benzanthracene, and 3-methyl cholanthracene—similarly depletes primordial follicles in Sprague-Dawley rats and C57BL/6 mice within 14 days after a single intraperitoneal dose (21). This depletion of ovarian follicles results in premature ovarian failure, a condition similar to the early menopause observed in women actively or passively exposed to cigarette smoke (22).

Some of the industrial byproducts that have been shown to destroy primordial follicles and induce premature ovarian failure in laboratory animals are 4-vinylcyclohexene (VCH) and its epoxide metabolites (16). These chemicals are formed from the dimerization of 1,2 butadiene, which is a reaction that occurs during the production of synthetic rubber. In one experiment mice were given daily injections of VCH for 30 days. After 30 days, dosing was stopped and several reproductive parameters were measured. At 30 days, approximately 90% of the primordial follicles were depleted from the ovary. At this time, both FSH levels and cyclicity were similar to controls. By 240 days, FSH levels in VCH-treated animals increased to 130% of control. In addition, the number of regular estrous cycles in VCH-treated animals was reduced compared with controls. By 360 days, FSH levels in VCH-treated mice increased to 160% of controls. The treated animals also stopped cycling, whereas control animals continued to have regular cycles (23). These data suggest that VCH, which is a chemical that depletes primordial follicles, produces irreversible and premature reproductive senescence. This early senescence is not immediately observed because it takes several days for a loss of follicles to result in altered hormone levels or cyclicity (24).

Chemotherapeutic agents, such as cyclophosphamide (CPA), also induce premature ovarian failure in laboratory animals (25,26). Intraperitoneal injection of CPA reduces the numbers of primordial and antral follicles in both rats and mice within 24 hours of a single injection (25). This reduction in primordial follicles is thought to lead to premature ovarian failure and infertility observed in CPA-treated mice and women (27).

Although these studies demonstrate that chemicals cause premature ovarian failure in animal models, the association between chemical exposures and early menopause in women is largely unknown. The most consistent and convincing human data come from reports of cigarette smoking and age at menopause (22). Several studies have definitively linked both active and passive cigarette smoking with early menopause and subfecundity. In one study, the mean age at natural menopause was 49.7 years for smokers and 50.8 years for nonsmokers. In addition, the mean age at menopause of non-smokers whose spouses did not smoke was 51.9 years, whereas the mean age at menopause of nonsmokers whose spouses smoked was 49.8 years (28). In another report, the effect of smoking on age of menopause was examined in a large study of 7,828 women who were between 45 and 55 years old. In this study, smoking was associated with a significantly lower age at menopause (average decrease = 1.7 years) (29).

Even though the reduced age at menopause among smokers appears small, it is significant. The normal age at menopause is tightly regulated and generally occurs over a narrow age range (30). Further, even small shifts in age at menopause are thought to impact human health significantly (31).

The mechanism by which cigarette smoking induces early menopause is thought to involve depletion of the ovarian pool of follicles by the PAHs present in cigarette smoke (32,20). It also has been suggested that smoking decreases circulating levels of estradiol and increases follicle-stimulating hormone (FSH), leading to symptoms associated with menopause (33). In one study, FSH concentrations were increased 66% by active smoking and 39% by passive smoking as compared to women who were not actively or passively exposed to smoke. The ability of cigarette smoke to increase FSH concentrations and deplete ovarian follicles may lead to the subfecundity that is observed in smokers (34). In one study, the effect of smoking on fecundity was examined in 6,630 women who were randomly selected from census registers and electoral rolls in Denmark, Germany, Italy, Poland, and Spain. This study demonstrated that the risk for subfecundity was significantly increased in women who smoked more than 10 cigarettes/day compared with nonsmokers (35).

The implication of chemicals present in cigarette smoke in early menopause strongly suggests that other chemicals are likely to alter the age of menopause. Most population-based studies, however, have not assessed chemical exposures and age at menopause. Instead, many previous studies have focused on cancer, fertility, and pregnancy outcomes. The role of chemicals and age at menopause needs to be addressed so that it is possible to assess reliably whether some women are at an increased risk for early menopause and thus at an increased risk for morbidities and chronic diseases. Studies also are required to determine whether chemicals known to affect animals cause early menopause in humans. Further, such studies also are critical because the twentieth century has experienced an explosion of technological and industrial advancement. Even though this advancement has provided many benefits, it also may increase the risk of exposure to reproductive toxicants and place a greater number of women at risk for early menopause.

Observations Contradicting the Hypothesis That the Ovary Drives the Menopausal Transition

Some investigators have maintained that hypothalamic changes play a primary role in the etiology of menopause, and that the ovarian changes are simply a result of these CNS events. In support of this contention is the fact that, a few years before the menopause, loss of primordial follicles accelerates dramatically.

Approximately 25,000 follicles remain at the start of the last decade of menstrual cycling. These remaining follicles appear to undergo an accelerated rate of loss of over twice that of the first two decades of reproductive life (36). As a result, only several hundred or less follicles remain in the ovary at the time of menopause. The molecular and cellular mechanisms responsible for this apparent acceleration of follicular loss are virtually unknown. . . ."

If the rate of loss of primordial follicles remained constant, women would continue to cycle well into the eighth decade of life.

Associated with this acceleration in the rate of follicle loss is the appearance of premenopausal symptoms, which reflect changes in CNS function (hot flashes, disturbed sleep, etc.). These symptoms begin when there are still more than 25,000 follicles remaining in the ovary and cycling appears to be normal (9). In rats, too, there are changes in the CNS component of the reproductive axis long before the ovary has substantially reduced follicles. It could be postulated that destabilization of the ovarian–CNS feedback relationship, triggered by aging of the CNS, leads to conditions that cause degeneration of primordial follicles in situ (without growth) and depletion of the primordial endowment.

Several studies suggest that menopause results from the aging of the hypothalamus. In old animals, the hypothalamus is incapable of tightly regulating the secretion of gonadotropins from the anterior pituitary (37). Thus, there are transient increases in the levels of both luteinizing hormone (LH) and FSH (38). These transient increases in LH and/or FSH are thought to be toxic to the ovary and exacerbate normal depletion of the primordial follicle pool. In a previous study, we examined whether LH depletes the primordial follicle pool using transgenic mice that chronically express LH (bLHβCTP mice). This transgenic model dissociates age and pool size from hormone levels and enables us to elevate the level of a single gonadotropin. Our data indicate that excessive LH rapidly depletes primordial follicles, but not primary or small antral follicles (39). This suggests that transient elevations of LH during the premenopausal period may play an influential role in determining the size of the primordial pool and thus the timing of the menopause.

Other studies indirectly support a role for LH in determining the onset of menopause. In women, there are transient increases in LH levels during the perimenopausal period, a time concomitant with accelerated follicle exhaustion (8). In hypophysectomized mice (i.e., mice with very low LH), the rate of oocyte loss is less that that observed in control animals (40). In monkeys, there is a decline in ovarian activity at 21–25 years of age that is preceded by an increase in plasma LH (41). In addition, some chemicals that cause early menopause are thought to destroy ovarian follicles by increasing endogenous LH levels (24).

If the size of the primordial follicle endowment played a significant role in determining the duration of the fertile life span, then removal of one ovary (reducing the endowment by 50%) would be expected to hasten significantly the loss of regular cyclicity. However, when mice were hemicastrated early in life and then housed with males, there was "nothing in [the histological observations] to indicate that ULO uses up oocytes faster, or [that the size of the] pool of primordial [follicles] is related to reproductive failure." In fact, at 450 days of age, there were more than 500 oocytes left in the ovaries (42).

Studies in mice demonstrated that strains differed with regard to the endowment of oocytes at birth, the rates at which the oocytes were lost with age, and the overall level of fertility. Levels of fertility, however, could not be related to the size of the original endowment of oocytes, or to the rate at which they decreased over time (43).

Studies on caloric restriction also indicate that the timing of menopause may be regulated through the hypothalamic/pituitary axis (44). In rodents, caloric restriction increases the secretion of both luteinizing hormone and follicle stimulating hormone. The increased hormone levels are then thought to delay the onset of reproductive senescence because they enable the animal to continue estrous cyclicity and ovulation (45). Although it is unclear whether caloric restriction also alters ovarian function, a few studies show that numbers of follicles are similar in control rats and calorically restricted rats.

The rodent data on caloric restriction, however, contradict some observations on food deprivation in humans. In rodents, lack of calories prolongs reproduction, but in humans, it reduces the reproductive life span, at least in some circumstances.

Epidemiologic studies indicate that nutritional status influences the age at menopause in humans (30). Studies in the United States and Bangladesh show that leaner women reach menopause earlier than do heavier women. Studies in Papua New Guinea suggest that chronic malnutrition is associated with an early menopause. Although these data suggest that nutrition is an important determinant of age at menopause, they are confounded by a number of variables and thus provide only preliminary information. They also do not indicate whether the absence of nutritional factors results in toxicity to the ovarian or hypothalamic/pituitary axis.

New Directions

Although the ovary versus hypothalamus dichotomy has engaged our attention for many decades, it may be more productive to reframe the focal questions. Menstrual cyclicity is a consequence of a tightly controlled, oscillating feedback system between the ovaries and the hypothalamus. Age-related changes in both ovary and hypothalamus probably produce a subtle destabilization of the oscillating pattern, leading to further play in the system with progressive amplification over time (Fig. 6.1). Given this scenario, it is pointless to try to identify an initial event that signals the onset of the menopausal transition. We will always be limited by the sensitivity of our assays in detecting the earliest changes in the ovary or hypothalamus.

With respect to the ovary, several important questions remain to be addressed. What environmental or physiological conditions have a deleterious effect on the primordial follicles in the ovaries? What are the reproductive and other consequences of reducing the number of primordial follicles in the ovaries? Are there compensatory mechanisms that come into

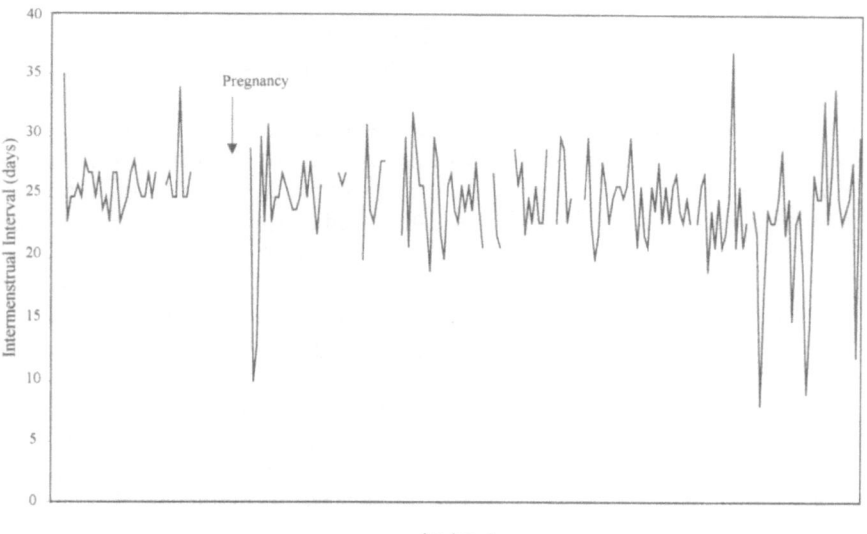

FIGURE 6.1. Menstrual history of one woman from age 30 to 45.

play to spare remaining follicles if the primordial follicle stockpile is reduced? What is the impact of the initial size of the primordial stockpile on reproductive function?

Transgenic technology will permit us to address these questions directly. For example, some transgenic and knockout mice that are born with a surplus of primordial follicles could be used to determine whether a surfeit of primordial follicles results in a longer reproductive lifespan (46). On the other hand, other transgenic and knockout animals that are born with deficiency of primordial follicles could be used to determine whether such deficiencies lead to early reproductive senescence (47).

To date, knockout mice have been produced that have a deficiency in primordial follicles compared to wild-type animals. Most of these animal models are lacking functional proteins that regulate programmed cell death. Mice that lack the CED-3/ICE homolog known as Ich-3 have approximately one third fewer primordial follicles compared to wild type animals (48). On postnatal day 4, ovaries from knockout mice contain 7,020 ± 1,043 primordial follicles, and controls contain 19,083 ± 2,239. The knockout animals also lack primary and small antral follicles, whereas control ovaries contain 613 ± 118 primary and 123 ± 60 small antral follicles. By postnatal day 42, there are no significant differences between the numbers of follicles in knockout and control animals. These data imply that even though Ich-3 knockout animals are born with relatively reduced primordial pool, follicular growth is altered so that the mice have the same number of follicles as do controls at adulthood. The implications of this for the timing of reproductive senescence are not known, and future studies should monitor the reproductive life span in these animals.

Mice that lack functional Bcl-2 also have fewer primordial follicles than do wild-type animals (47). On postnatal day 42, these knockout mice have about 40% fewer primordial follicles than controls. There is no difference in the number of primary and antral follicles between knockouts and controls at this time point. To date, there are no reports on whether these mice have an altered reproductive life span. Studies on the timing of acyclicity, alterations in hormone levels, and the onset of senescence in these animals will be useful for understanding the biological basis of the menopause.

Knockout animal models also have been produced that have a surfeit of primordial follicles compared with wild-type animals. These models also lack functional proteins that are involved in the regulation of programmed cell death. Mice that lack functional Caspase 2 (Ich-1/Nedd2) have approximately 150% more primordial follicles than do wild-type animals on postnatal day 4 (49). Mice lacking functional Bax protein have about threefold more primordial follicles than do wild-type mice at postnatal day 42 (46). We have observed that mice with growth hormone deficiency (Ames Dwarfs) have three times more primordial and primary follicles as controls at 2 years.

The consequences of a surfeit of primordial follicles are unknown. All of these animal models are fertile, but it is unclear whether their reproductive life span is increased compared with control animals.

Some transgenic mice have been designed to overexpress genes involved in programmed cell death. These mice may also be useful for studies on reproductive senescence. Mice that overexpress Bcl-2 in large antral follicles have a reduced level of ovarian apoptosis, greater ovulatory potential, and an increase in follicular growth compared to wild-type mice. The transgenic mice also have an increased incidence of benign ovarian teratomas compared with controls (50). These data suggest that Bcl-2 blocks cell death in antral follicles. Future studies could use these mice to determine whether alterations in the size of the antral follicle pool influence reproductive senescence.

We are currently in the process of characterizing a transgenic mouse line that we hope will have a greater-than-normal number of oocytes at birth. Vast numbers of germ cells are lost by apoptosis during the late embryonic period. If this germ-cell loss did not occur, human females would begin life with an endowment of more than 7 million oocytes rather than the 2 million that are actually found at birth. We are attempting to prevent this prenatal germ-cell loss in mice to determine the consequences, if any, in postnatal life with respect to reproductive function and the timing of reproductive senescence.

References

1. Hirshfield AN. Development of follicles in the mammalian ovary. Int Rev Cytol 1991;124:43–101.
2. Lacker HM, Beers WH, Meuli LE, Atkin E. A theory of follicle selection. Biol Reprod 1987;37:570–80.
3. Sowers MR, LaPietra MT. Menopause: its epidemiology and potential association with chronic diseases. Epidemiologic Rev 1996;17:287–302.
4. Gardsell P, Johnell O, Nilsson BE. The impact of menopausal age on future fragility fracture risk. J Bone Miner Res 1991;6:429–33.
5. Goemaere S, Ackerman C, Goethals K, Goethalsk K, DeKeyser F, Van der Straeten C. Onset of symptoms of rheumatoid arthritis in relation to age, sex, and menopausal transition. J Rheumatol 1990;17:1620–22.
6. Hui SL, Slemenda CW, Johnston CC, Appledorn CR. Effects of age and menopause on vertebral bone density. Bone Miner 1987;2:141–46.
7. Faddy MJ, Gosden RG, Gougeon A, Richardson SJ, Nelson JF. Accelerated disappearance of ovarian follicles in mid-life: implications for forecasting menopause. Hum Reprod 1992;7:1342–46.
8. Jaffe RB. The menopause and perimenopausal period. In: Yen SSC, Jaffe RB, eds. Reproductive endocrinology: physiology, pathophysiology, and clinical management. New York: WB Saunders & Co, 1986:406–23.
9. Wise PM. Aging of the female reproductive system: a neuroendocrine perspective. In: Muller EE, MacLeod RM, eds. Neuroendocrine perspectives, vol. 7. New York: Springer-Verlag, 1989:117–68.
10. Vom Saal FS, Finch CE. Reproductive senescence: phenomena and mechanisms in

mammals and selected vertebrates. In: Knobil E, Neill J, et al., eds. The physiology of reproduction. New York: Raven Press, 1988:2351–403.

11. Krarup T, Pedersen T, Faber M. Regulation of oocyte growth in the mouse ovary. Nature 1969;224:187–88.

12. Cran DG, Moor RM. The development of oocytes and ovarian follicles of mammals. Sci Prog 1980;66:371–83.

13. Hirshfield AN. The relationship between the supply of primordial follicles and the onset of follicular growth in rats. Biol Reprod 1994;50:421–28.

14. Kassuga F, Takahashi M. The endocrine function of rat gonads with reduced number of germ cells following busulphan treatment. Endocrinol Jpn 1986;33:105–15.

15. Cattanach BM The effect of triethylene-melamine on the fertility of female mice. Int J Rad Biol 1959;3:288–92.

16. Flaws JA, Doerr JK, Sipes IG, Hoyer PB. Destruction of pre-antral follicles in adult rats by 4-vinylcyclohexene diepoxide. Reprod Toxicol 1994;8:509–14.

17. Mattison DR. Morphology of oocyte and follicle destruction by polyaromatic hydrocarbons in mice. Tox Appl Pharm 1980;53:249–59.

18. Mandl AM. A quantitative study of the sensitivity of oocytes to X-irradiation. Proc Roy Soc B 1959;150:53–71.

19. Dobson RL, Felton JS. Female germ cell loss from radiation and chemical exposures. Am J Ind Med 1983;4:175–90.

20. Mackenzie KM, Angevine DM. Infertility in mice exposed in utero to benzo(a)pyrene. Biol Reprod 1981;24:183–91.

21. Mattison DR, West DM, Menard RH. Differences in benzo(a)pyrene metabolic profile in rat and mouse ovary. Biochem Pharm 1979;28:2101–4.

22. Mattison DR, Thorgeirsson SS. Smoking and industrial pollution, and their effects on menopause and ovarian cancer. Lancet 1978;1:187–88.

23. Hooser SB, Douds DA, Hoyer PB, Sipes IG. Long-term ovarian and hormonal alterations due to the ovotoxicant, 4-vinylcyclohexene. Reprod Toxicol 1994; 8:315–23.

24. Hoyer PB, Sipes IG. Assessment of follicle destruction in chemical-induced ovarian toxicity. Ann Rev Pharm Tox 1996;6:307–31.

25. Plowchak DR, Mattison DR. Reproductive toxicity of cyclophosphamide in the C57BL/6N mouse 1. Effects on ovarian structure and function. Reprod Toxicol 1992;6:411–21.

26. Shiromizu K, Thorgeirrsson SS, Mattison DR. Effect of cyclophosphamide on oocyte and follicle number in Sprague Dawley rats, C57BL/6N and DBA/2N mice. Ped Pharm 1984;4:213–21.

27. Warne GL, Fairly KF, Hobbs JB, Martin FIR. Cyclophosphamide-induced ovarian failure. N Engl J Med 1973;289:1159–62.

28. Everson RB, Sandler DP, Wilcox AJ, Schreinemachers D, Shore DL, Weinberg C. Effect of passive exposure to smoking on age at natural menopause. Br Med J (Clin Res Ed) 1986;293:792.

29. McKinlay SM, Bifano NL, McKinlay JB. Smoking and age at menopause in women. Ann Intern Med 1985;103:350–56.

30. Wood JW. Dynamics of human reproduction: biology, biometry, demography. New York: Aldine de Gruyter, 1994.

31. Goemaere S, Ackerman C, Goethals K, DeKayser F, Van der Straeten C. Onset of symptoms of rheumatoid arthritis in relation to age, sex, and menopausal transition. J Rheumatol 1990;17:1620–22.

32. Vahakangas K, Rajaniemi H, Pelkonen O. Ovarian toxicity of cigarette smoke exposure during pregnancy in mice. Tox Lett 1985;25:75–80.

33. Cramer DW, Barbieri RL, Xu H, Reichardt KV. Determinants of basal follicle-stimulating hormone levels in premenopausal women. J Clin Endocrinol Metab 1994;79:1105–9.

34. Cooper GS, Baird DD, Hulka BS, Weinberg C, Savitz DA, Hughes CL. Follicle-stimulating hormone concentrations in relation to active and passive smoking. Obstet Gynecol 1995;85:407–11.

35. Bolumar F, Boldsen J, European Study Group on Infertility and Subfecundity. Smoking reduces fecundity: a European multicenter study on infertility and subfecundity. Am J Epidemiol 1996;143:578–87.

36. Richardson SJ, Nelson JF. Follicular depletion during the menopausal transition. Ann NY Acad Sci 1990;592:13–20.

37. Wise PM, Scarbrogh K, Lloyd J, Cai A, Harney J, Chiu S, et al. Neuroendocine concomitants of reproductive aging. Exper Gerontol 1994;29:275–83.

38. Sherman BM, West JH, Korenman SG. The menopausal transition: analysis of LH, FSH, estradiol, and progesterone concentrations during menstrual cycles of older women. J Clin Endocrinol Metab 1976;42:629–36.

39. Flaws JA, Abbud R, Mann RJ, Nilson JH, Hirshfield AN. Chronically elevated luteinizing hormone depletes primordial follicles in the mouse ovary. Biol Reprod 1997;57:1233–37.

40. Meredith S, Kirkpatrick-Keller D, Butcher RL. The effects of food restriction and hypophysectomy on numbers of primordial follicles and concentrations of hormones in rats. Biol Reprod 1986;35:68–73.

41. Nozaki M, Mitsunaga F, Shimizu K. Reproductive senescence in female Japanese monkeys (Macaca fuscata): age- and season-related changes in hypothalamic-pituitary-ovarian functions and fecundity rates. Biol Reprod 1995;2:1250–57.

42. Jones EC, Krohn PL. The effect of unilateral ovariectomy on the reproductive lifespan of mice. J Endocrinol 1960;20:129–34.

43. Jones EC, Krohn PL. The relationships between age, numbers of oocytes and fertility in virgin and multiparous mice. Endocrinology 1960;21:470–95.

44. McShane TM, Wise PM. Life-long moderate caloric restriction prolongs reproductive life span in rats without interrupting estrous cyclicity; effects on the gonadotropin-releasing hormone/luteinizing hormone axis. Biol Reprod 1996;54:70–75.

45. Segall PE, Ooka H, Rose K, Timiras PS. Neural and endocrine development after chronic tryptophan deficiency in rats: brain monoamine and pituitary responses. Mech Ageing Devel 1978;7:1–17.

46. Knudson CM, Tung KSK, Flaws JA, Brown GAJ, Tilly JL, Korsmeyer SJ. Oocyte survival but spermatocyte cell death in bax deficient mice. Symposium on cell death and reproduuctive physiology. Serono Symposia USA 1996; Abstract 9, p. 33.

47. Ratts VS, Flaws JA, Kolp R, Sorenson C, Tilly, JL. Ablation of bcl-2 gene expression reduces the number of germ cells and primordial follicles established in the post-natal female mouse gonad. Endocrinology 1995;136:3665–68.

48. Flaws JA, Wang S, Miura M, Tilly JL, Yuan, J. Reduced ovarian follicle endowment and delayed activation of primordial follicle growth in mice lacking the Ced-

3/ICE homolog, Ich-3. Symposium on cell death and reproductive physiology. Serono Symposia USA 1996; Abstract 3, p. 30.

49. Bergeron L, Perez GI, Macdonald G, Shi L, Sun Y, Juriscova A, et al. Defects in regulation of apoptosis in caspase-2-deficient mice. Genes Devel 1998;12:1304–14.

50. Hsu SY, Lai J-M, Finegold M, Hseuh AJ. Targeted overexpresion of Bcl-2 in ovaries of transgenic mice leads to decreased follicle apoptosis, enhanced folliculogenesis, and increased germ cell tumorigenesis. Endocrinology 1996; 137:4837–43.

7

Ovarian Follicle Recruitment and Secretory Capacity in Women of Advanced Reproductive Age

Nancy A. Klein, David E. Battaglia,
and Michael R. Soules

Introduction

The final common pathway in reproductive aging in women is attrition of the initial endowment of ovarian follicles, predominantly through atresia. Mathematical models based on morphometric studies have been used to hypothesize that follicular depletion is bi-exponential, with an accelerated loss occurring after about age 38 (1). Among individual women, the age of menopause and the length of the reproductive life span (both ultimately dependent on follicle depletion) are quite variable, yet the controlling factors are unknown. There is an intriguing temporal relationship between the onset of the accelerated phase of follicle atresia, decline in fecundity, and the monotropic rise in follicle stimulation hormone (FSH). In addition, older reproductive-age women exhibit an earlier onset of the intercycle rise in FSH associated with earlier recruitment and ovulation of the dominant follicle. The relationship, if any, between these phenomena remains speculative. This chapter on studies of endocrine and ovarian changes in older reproductive-age women will attempt to describe what is known about dominant follicle recruitment and function in women of advanced reproductive age.

Reproductive Aging and Fertility

The fecundity of a couple is much more dependent upon the age of the female than of the male (2). Peak efficiency of the female reproductive system occurs at about age 24, which is when a woman experiences her maximum fertility potential, after which there is a general decline in fertility with a steep decline beginning after age 35. A classic study by Menken et al. reported effects of female age and marriage on fertility in seven historical population groups meeting the following criteria: (1) late marriage is relatively common; (2) contraception is rarely practiced; (3) marriage is not usually preceded by premarital conceptions; and (4) accurate birth records were maintained. The percentage of married women remaining childless rose steadily with age: 6% at age 20–24, 15% at age 30–34, 30% at age 35–39, and 64% at age 40–44 (2). Studies of donor insemination recipients, which control for age and fertility of the male partner and coital frequency, also reveal that cumulative pregnancy rates significantly decrease as women age (3–5). For example, data from the French registry examining conception rates in 2,193 donor insemination recipients with azoospermic husbands demonstrated significantly reduced cumulative conception rates after the age of 30 (3). The same trend is observed in the outcome of advanced infertility therapies such as in vitro fertilization (IVF). A study of women undergoing IVF ($n = 63,400$ oocyte recoveries) reported a sharp decline in the clinical pregnancy rate beyond age 38 (6). This decline in IVF success goes beyond the ovulatory response because only women with adequate follicular development who progressed to oocyte recovery were considered. Hull and colleagues examined the effects of age on embryo implantation in 561 IVF patients (7). Implantation rates declined as a function of female age: 18.2% at age 20–29, 16.1% at 30–34, 15.3% at 35–39, and 6.1% at 40–44 (7). Once an older woman becomes pregnant she also has a markedly increased risk of spontaneous abortion (8).

Clinical evidence suggests that the predominant effect of age on fertility is due to abnormalities present in the older oocyte. Oocytes from normal younger and older reproductive-age women examined at the second metaphase of meiosis exhibit distinct differences in microtubule and chromosome placement (9,10). In addition, higher rates of single chromatid abnormalities in oocytes (11) and aneuploidy in preimplantation embryos (12) are observed in older reproductive-age women. Finally, pregnancy and delivery rates after oocyte donation are much more strongly correlated with the age of the donor than they are the age of the recipient (13–16), largely (if not completely) reversing the age-related decline in fertility.

Physiology of Reproductive Aging

Ovarian Follicle Depletion

In women, the physiologic basis of the perimenopause and menopause can be largely attributed to changes within the ovaries. The female in utero at 5 months of fetal age has received her maximum endowment of primordial follicles (PFs), estimated to be about 2 million (17). The number of follicles in the cortex of the ovaries gradually declines from this point onward, until only a relatively few, poorly responsive follicles remain at the time of the menopause. Relative to atresia, ovulation accounts for the loss of only a few follicles over a woman's life: Only about 500 ovulations occur in the life span of a normal woman. The physiologic mechanism for this progressive loss of follicles of age is unknown, but it appears that programmed cell death (apoptosis) is integral to the process and is influenced by a number of promoting and inhibitory factors (18). Studies in mice raise the possibility that follicle atresia may be under pituitary control because hypophysectomy appears to prevent it (19). The normal age of menopause varies over a wide range (mean age 50±8 years), which implies that the initial endowment of PFs and/or the rate of follicle loss is highly variable between individuals. Current knowledge regarding PF depletion is based upon a combined data base consisting of three published studies (17,20,21). Altogether, these studies report morphometric estimates of PF number in 103 ovarian pairs from females age 6–55. Mathematical modeling has been used on this combined data base to estimate the number of PFs at birth (about 500,000) and to hypothesize that follicular depletion is bi-exponential with an accelerated loss occurring about age 38 (1,21,22).

Changes in Follicle Recruitment and Menstrual Cyclicity

The hypothalamic–pituitary–ovarian axis is a highly integrated feedback loop that results in the maintenance of regular, ovulatory menstrual cycles during the early stages of aging under the influence of a gradual increase in FSH secretion. In the premenopausal woman, the FSH rise is initially sufficient to compensate for decreased ovarian responsiveness and secretory capacity that result from declining follicle numbers. The ability to maintain regular, ovulatory cycles is usually preserved for several years after the onset of the accelerated decline in fertility.

As women age, the average length of the menstrual cycle progressively shortens. In a classic, prospective, longitudinal study, Treloar analyzed the menstrual records of more than 2,700 women, constituting 25,825 woman years. He demonstrated that there is a progressive, age-related shortening of the menstrual cycle from a median of 28 days at age 20 to 26 days at age 40 (23). Endocrine studies of steroid and gonadotropin secretion across the menstrual cycle demonstrate that the shortened cycle length is due to a progressive shortening of the follicular phase without a significant change in luteal phase length (24–26). Examples of this age-related shortening of the follicular phase are illustrated in Figure 7.1. The observed decrease in cycle length continues until the onset of oligomenorrhea (perimenopause—usually after the age of 43) when both average cycle length and the standard deviation begin to increase (23,27). Ultrasound and endocrine studies of normal older ovulatory women (age 40–45) confirm that this shortened follicular phase is associated with development and ovulation of an apparently

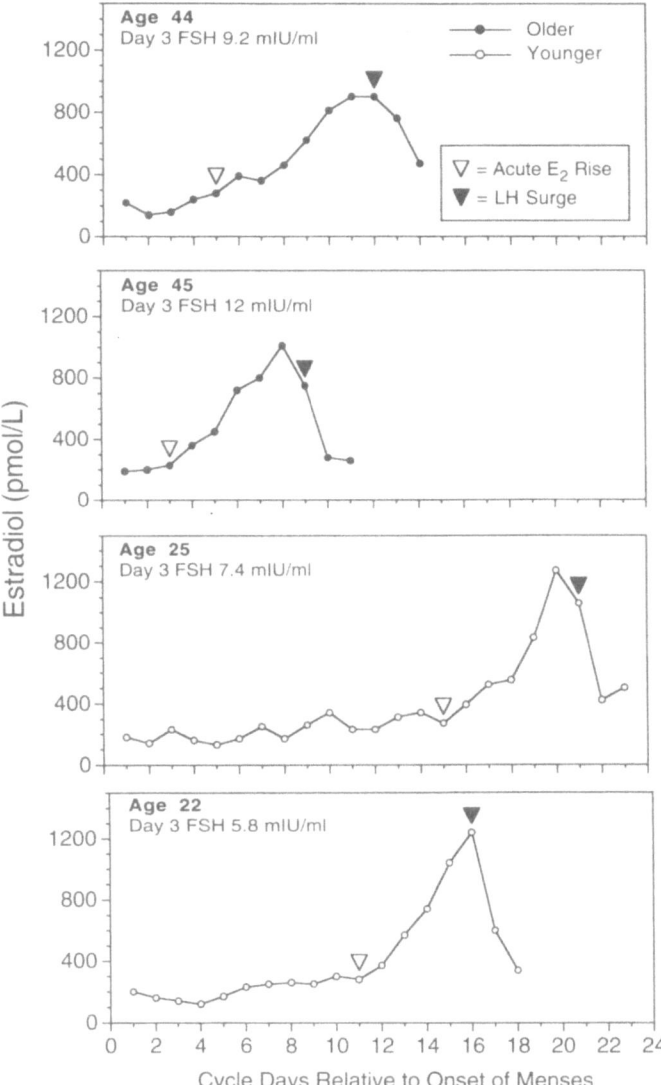

FIGURE 7.1. Follicular phase concentrations of estradiol in two younger and two older normal, ovulatory subjects, illustrating the decreased follicular phase length and earlier onset of the acute estradiol rise in older women. Reprinted with permission from Klein N et al. Reproductive aging. J Clin Endocrinol Metab 1996;81:1038–45. © The Endocrine Society.

normal follicle in the presence of elevated circulating levels of FSH (24,28). Metcalf has demonstrated that ovulatory frequency remains high (95%) in menstruating women age 40–55 until the onset of oligomenorrhea, when the percentage of ovulatory cycles drops significantly (29).

Gonadotropins

The most consistent endocrine finding in numerous investigations of reproductive aging has been a subtle rise in the concentration of FSH unaccompa-
nied by a rise in luteinizing hormone (LH) (the monotropic FSH rise)
(24,26,30–33). FSH is elevated throughout the entire menstrual cycle (most
pronounced in the early follicular phase) in older ovulatory women, initially
with no differences in circulating LH concentrations (Fig. 7.2) (24, 30). As
aging progresses, elevated circulating levels of LH are eventually also observed in ovulatory women, usually after age 45 (30,32,34). Although the
pattern of FSH secretion remains essentially unchanged across the menstrual
cycle, older ovulatory women demonstrate an earlier rise in FSH secretion
in the intercycle phase, with a significant and sustained rise being detectable prior to the onset of the menstrual period (24). The fact that FSH rises
earlier and to a greater degree in older women may lead to what appears to
be either earlier or accelerated dominant follicle recruitment. The early
monotropic rise in FSH is not related to a loss in bioactivity of the FSH
molecule, as there is no difference observed in the bioactive–immunoactive
FSH ratio between women in their early twenties compared with those in
their early forties (24). Frequent blood sampling in the early follicular and
midluteal phases have failed to detect a difference in either the LH pulse
frequency or amplitude, which argues against a functional change in the
GnRH pulse generator as the primary explanation for the monotropic FSH
rise in older reproductive-age women (35,36). One of these studies, however, did detect enhancement of LH pulse amplitude in the late luteal phase
(12±1 days after the LH surge), with a more subtle increase in LH pulse frequency in both the early follicular and late luteal phases in the older (age
40–50) compared with the youngest (age 19–35) age groups, despite no
differences in either estradiol or progesterone secretion (36). Whether there
is any age-related change in sensitivity of the hypothalamic/pituitary axis to
ovarian steroid feedback inhibition is unknown.

Ovarian Secretory Capacity

Ovulating women with regular menstrual cycles continue to have normal circulating concentrations of estradiol and progesterone after the age of 40 and the
onset of the monotropic FSH rise (Fig. 7.3) (24,30,36). In individual older ovulatory women near the end of this phase of regular ovulation, preovulatory estradiol levels tend to be even higher and rise earlier in the follicular phase than
those seen in younger women (28,34). Santoro and colleagues studied 6 cycling
women age 47 and older with daily determinations of urinary estrone conjugates
and pregnanediol glucuronide. Compared with women 19–38, these women had
elevated follicular and luteal phase estrone excretion (34). Earlier selection and
development of the dominant follicle lead to relative increases in early follicular
phase estradiol secretion (24,32). In fact, elevated early follicular phase (day 3)

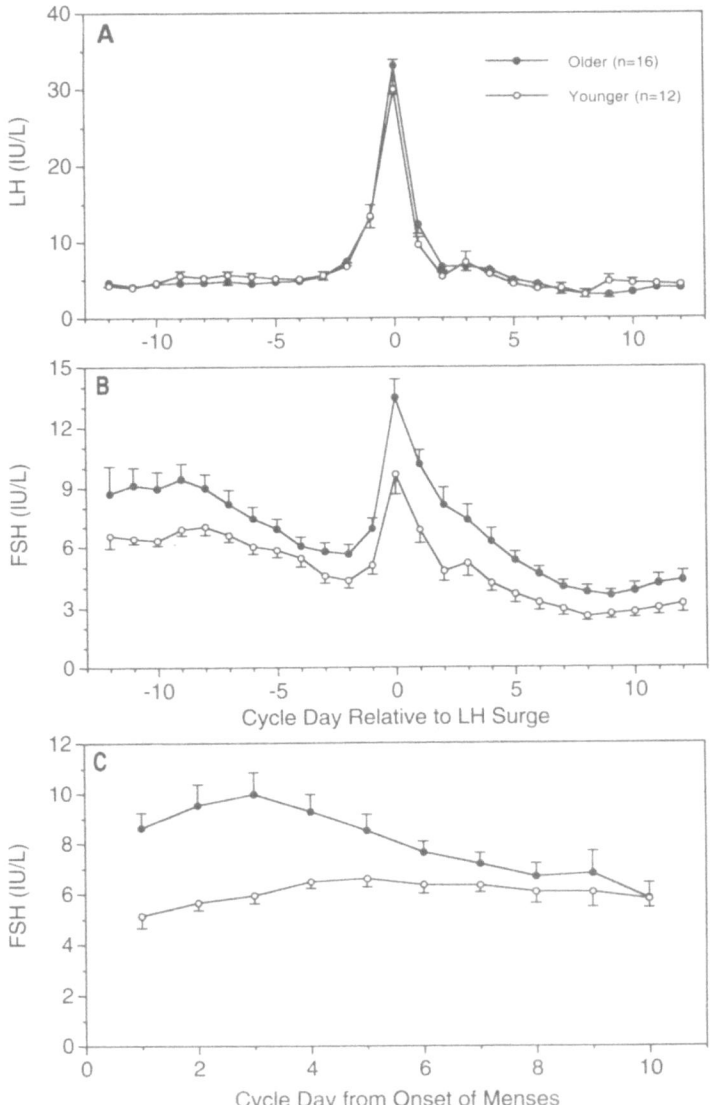

FIGURE 7.2. Figure 7.2A–B depict the mean ± SE LH (A) and FSH (B) concentrations across the menstrual cycle in older (40–45) and younger (20–25) normal women according to the day of the LH surge (Day 0). Data are shown for days −12 to +12 only. The pattern of secretion for both hormones is similar in the two age groups. Mean [LH] does not differ between the two groups, whereas [FSH] is elevated throughout the cycle in older subjects ($p < 0.01$). Figure 7.2C shows the follicular phase [FSH] according to the cycle day from onset of menses, illustrating that the most marked numerical difference in [FSH] occurs during the first 4 cycle days. Reprinted with permission from Klein N et al. Reproductive aging. J Clin Endocrinol Metab 1996;81:1038–45. © The Endocrine Society.

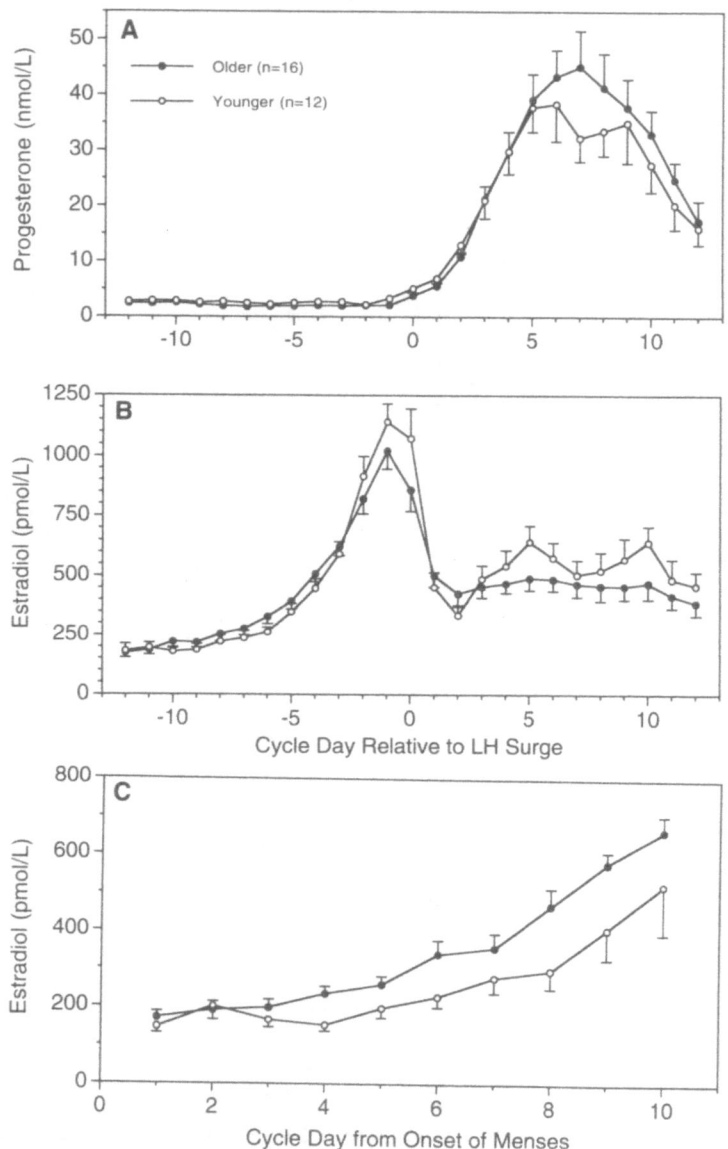

Figure 7.3. Figure 7.3A–B depict the mean ± SE concentrations of progesterone (A) and estradiol (B) relative to the LH surge (Day 0) in older (40–45) and younger (20–25) normal ovulatory women, revealing similar patterns of secretion with no significant differences throughout the cycle ($p > 0.05$). In contrast, Figure 7.3C illustrates higher follicular phase E secretion in older women when data are compared according to the cycle day from onset of menses ($p < 0.05$). Reprinted with permission from Klein N et al. Reproductive aging. J Clin Endocrinol Metab 1996;81:1038–45. © The Endocrine Society.

estradiol has been shown to predict a poor response to controlled ovarian hyperstimulation (37). The endocrine characteristics of the perimenopause have been elucidated through several small, descriptive, longitudinal studies (29,33,34,38,39). These studies reveal that as women approach the menopause, periods of hypoestrogenemia similar to that observed in postmenopausal women are interspersed with ovulatory and anovulatory episodes of estrogen secretion (34). Progesterone secretion is generally maintained in the normal range as long as ovulatory cycles are occurring at regular intervals (24,30). As reproductive age advances further, insufficient luteal phase progesterone secretion may be observed (32,34,38).

Inhibin is a peptide hormone secreted by the ovary that selectively inhibits FSH. Studies of normal perimenopausal and postmenopausal women have demonstrated that total immunoreactive inhibin falls in perimenopausal women, eventually becoming undetectable after menopause (40,41). Previous studies, however, which used nonspecific assays for total inhibin (including free alpha subunits and precursors) yielded conflicting results, with some (but not all) studies indicating a decline in ovarian inhibin secretion corresponding to the onset of the monotropic FSH rise. New, two-site specific assays for dimeric inhibin A and B have been developed, which demonstrates that these two hormones exhibit distinct and different patterns across normal ovulatory menstrual cycles (42,43). Inhibin A follows a pattern similar to that observed when total immunoreactive inhibin is measured: Levels are low in the early follicular phase, rise prior to ovulation, and are highest in the luteal phase (42). In contrast, inhibin B peaks in the early follicular phase and then falls throughout the remainder of the follicular phase (42). Inhibin B appears to be the predominant inhibin form in the follicular fluid (FF) of the developing early antral follicles, with concentrations decreasing in FF as the dominant follicle matures. In contrast, during dominant follicle development, the FF concentrations of inhibin A rise (42). Using specific assays for dimeric inhibin A and B, we have shown that early follicular phase secretion of inhibin B (but not inhibin A) is decreased in older women who demonstrate a monotropic FSH rise (Fig. 7.4) (44). One possibility is that this decrease in inhibin B concentration may be the result of fewer primordial and early antral follicles remaining in the ovaries of older women. Further studies are needed to elucidate the relative roles of inhibin A and B, as well as related regulatory peptides (e.g., activin, follistatin) throughout the menstrual cycle.

Dominant Follicle Function

It is unclear whether the changes in feedback signals (e.g., inhibin) from the ovary are due simply to a decrease in the number of follicles or whether there is a concomitant decline in the functional status of the follicular components due to aging. As previously noted, older ovulatory women recruit and develop a dominant follicle earlier in the cycle than younger women (24). Once selected, these follicles exhibit normal growth and collapse (24). We have aspirated the pre-ovulatory dominant follicle from normal

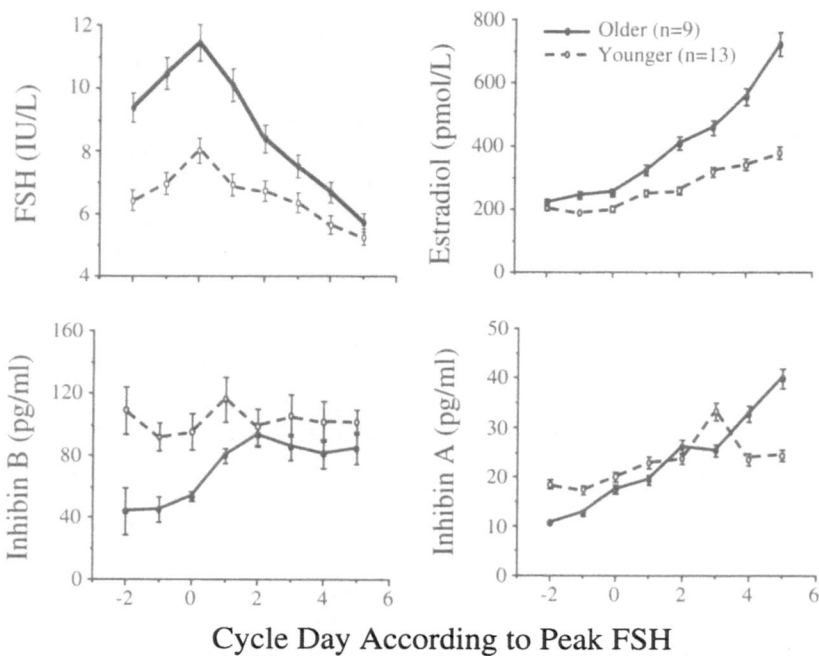

Cycle Day According to Peak FSH

FIGURE 7.4. Mean±SE concentrations of (clockwise from top left) FSH ($p < 0.01$), estradiol ($p < 0.01$), inhibin A ($p = 0.61$), and inhibin B ($p = 0.04$) according to the day of maximal FSH concentration (peak FSH=day 0) in normal older (40–45) and younger (20–25) women. Reprinted with permission from Klein N et al. Decreased inhibin B secretion is associated with the monotropic FSH rise in older, ovulatory women. J Clin Endocrinol Metab 1996;81:2742–45. © The Endocrine Society.

40–45-year-old women and observed that intrafollicular steroid concentrations are normal. There is a trend toward higher FF estradiol and lower androstenedione and testosterone concentrations (28), with no differences in intrafollicular concentrations of inhibin A or B (44). These findings are indicative of healthy FF environment in the dominant follicles of older women and may reflect a compensatory effect of FSH elevation. Normal follicular fluid hormone concentrations suggest that the dominant follicle is fully functional in terms of its secretory capacity. In the oocytes aspirated from these apparently normal preovulatory follicles, however, the alignment of chromosomes on the meiotic spindle at metaphase II is abnormal in the vast majority of the older women (9,10). Based on the information available from our studies to date, we would have to conclude that in older normal women: (1) the principal defect in the ovarian follicle resides in the oocyte; and (2) there is no evidence that, in early stages of reproductive aging, follicle development and granulosa cell function are compromised. In fact, increased pituitary FSH secretion and improved follicular fluid steroid profiles may represent compensatory mechanisms for the poor quality of the aging oocyte.

Other Ovarian Changes

Several studies have identified other evidence of generalized aging within the ovary. For example, the frequency of mitochondrial DNA mutations detected in the ovary after amplification with polymerase chain reaction increases with advancing age (45). The frequency of these mutations is further associated with menstrual status, occurring most frequently in postmenopausal women with amenorrhea of at least 1 year duration (45). The factors that lead to mitochondrial DNA mutations and corresponding effects on ovarian or oocyte function remain to be elucidated. Van Blerkom has reported an increased incidence of chromosomal abnormalities in oocytes obtained from follicles with reduced FF oxygen content, apparently as a function of blood flow as detected by ultrasound (46). The possibility that age-related changes in ovarian perfusion lead to hypoxic intrafollicular conditions which, in turn, are related to meiotic abnormalities is an attractive hypothesis. Whether the acceleration in follicle growth, estrogen secretion, and ovulation contribute to the abnormal development of the oocytes or whether they are intrinsically abnormal is unknown.

Summary

Reproductive aging in women is closely tied to the loss of ovarian follicles through atresia. The sentinel endocrinologic finding is the monotropic FSH rise, associated with a decline in ovarian inhibin B secretion. Fertility becomes significantly compromised long before overt clinical signs, such as cycle irregularity, occur. Compromised fertility is primarily related to oocyte dysfunction. As women with regular cycles near the end of the reproductive years, the following changes are usually manifested: (1) There is a selective rise in FSH, particularly in the early follicular phase; (2) there are decreased serum levels of inhibin B; (3) the selection and development of a dominant follicle occurs earlier; (4) there is earlier ovulation; (5) there is a short follicular phase and total cycle length; and (6) ovarian steroid secretion is normal. Figure 7.5 is a schematic representation of our model of the predominant age-related changes in the hypothalamus–pituitary–ovarian axis. The relationships between the monotropic FSH rise, accelerated follicular atresia, shortened follicular phase, and oocyte quality are yet to be defined. Further studies are needed to elucidate the factors contributing to the oocyte abnormalities in women of advanced reproductive age as well as the factors that determine the rate of follicle atresia and the length of the reproductive life span.

Monotropic FSH Rise—Inhibin Hypothesis

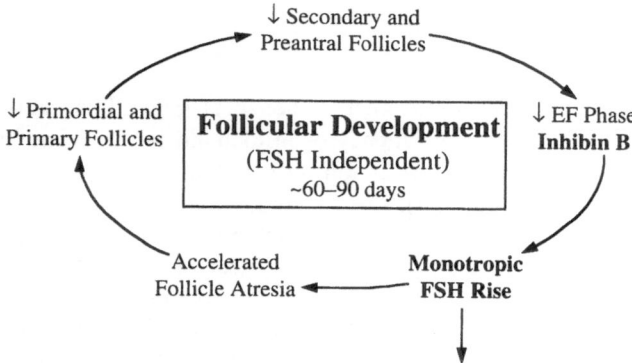

Monotropic FSH Rise—Inhibin Hypothesis

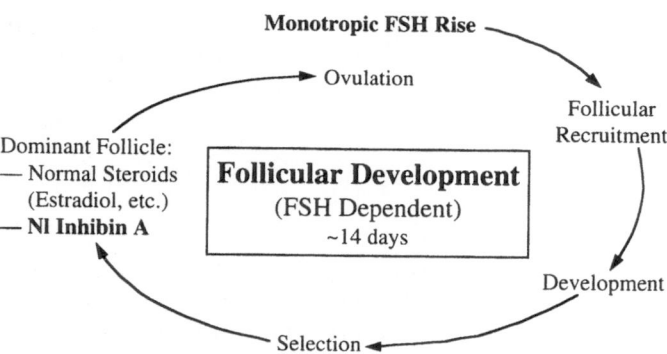

Figure 7.5. Schematic representation of our model of the predominant age-related changes in the hypothalamic–pituitary–ovarian axis: (Top) With advancing age, the number of primordial and primary follicles declines (through atresia), which results in decreased secondary and preantral follicles. This in turn results in decreased secretion of inhibin B, leading to a monotropic FSH rise. We speculate that elevated FSH may further accelerate the rate of follicle atresia. (Bottom) Under the influence of FSH elevation, older women demonstrate normal or accelerated follicle recruitment, development, and selection of a dominant follicle that is normal in terms of steroid and inhibin A secretion and ovulation.

References

1. Faddy M, Gosden R, Gougeon A, Richardson S, Nelson J. Accelerated disappearance of ovarian follicles in mid-life: implications for forecasting menopause. Hum Reprod 1992;7:1342–46.
2. Menken J, Trussell J, Larsen U. Age and infertility. Science 1986;233:1389–94.

3. Schwartz D, Mayaux M. Female fecundity as a function of age: results of artificial insemination in 2193 nulliparous women with azoospermic husbands. N Engl J Med 1982;306:404–6.

4. Stovall D, Toma S, Hammond M, Talbert L. The effect of age on female fecundity. Obstet Gynecol 1991;77:33.

5. van Noord-Zaadstra B, Looman C, Alsbach H, Habbema J, te Velde E, Karbaat J. Delayed childbearing: effect of age on fecundity and outcome of pregnancy. Br Med J 1991;302:1361.

6. FIVNAT (French In Vitro National). French national IVF registry: analysis of 1986 to 1990 data. Fertil Steril 1993;59:587–95.

7. Hull M, Fleming C, Hughes A, McDermott A. The age-related decline in female fecundity: a quantitative controlled study of implanting capacity and survival of individual embryos after in vitro fertilization. Fertil Steril 1996;65:783–90.

8. Smith K, Buyalos R. The profound impact of patient age on pregnancy outcome after early detection of fetal cardiac activity. Fertil Steril 1996;65:35–40.

9. Battaglia DE, Klein NA, Soules MR. Changes in centrosomal domains during meiotic maturation in the human oocyte. Molecular Hum Reprod 1997;2:845–51.

10. Battaglia D, Goodwin P, Klein N, Soules M. Influence of maternal age on meiotic spindle assembly in oocytes from naturally cycling women. Hum Reprod 1996;11:2217–22.

11. Angell R. Aneuploidy in older women. Hum Reprod 1994;9:1199–201.

12. Benadiva C, Kligman I, Munne S. Aneuploidy 16 in human embryos increases significantly with maternal age. Fertil Steril 1996;66:248–55.

13. Balmaceda J, Bernardini L, Ciuffardi I, et al. Oocyte donation in humans: a model to study the effect of age on embryo implantation rate. Hum Reprod 1994;9(11):2160–63.

14. Lydic M, Liu J, Rebar R, Thomas M, Cedars M. Success of donor oocyte in in vitro fertilization-embryo transfer in recipients with and without premature ovarian failure. Fertil Steril 1996;65(1):98–102.

15. Pantos K, Meimeti-Damianaki T, Vaxevanoglou T, Kapetanakis E. Oocyte donation in menopausal women aged over 40 years. Hum Reprod 1993;8(3):488–91.

16. Sauer M, Paulson R, Ary B, Lobo R. Three hundred cycles of oocyte donation at the University of Southern California: assessing the effect of age and infertility diagnosis on pregnancy and implantation rates. J Assis Reprod Gen 1994;11(2):92–95.

17. Block E. Quantitative morphological investigations of the follicular system in women: variations at different ages. Acta Anat 1952;14(suppl. 16):108.

18. Hsueh A, Billig H, Tsafriri A. Ovarian follicle atresia: a hormonally controlled apoptotic process. Endo Rev 1994;15:707–24.

19. Finch C. Neuroendocrine mechanisms and aging. Fed Proc 1979;38:178.

20. Gougeon A. Qualitative changes in medium and large antral follicles in the human ovary during the menstrual cycle. Ann Biol Anim Biochem Biophys 1979;19:1461–68.

21. Richardson S, Senikas V, Nelson J. Follicular depletion during the menopausal transition: evidence for accelerated loss and ultimate exhaustion. J Clin Endocrinol Metab 1987;65:1231–37.

22. Gougeon A, Ecochard R, Thalabard J. Age-related changes of the population of human ovarian follicles: increase in the disappearance rate of non-growing and early-growing follicles in aging women. Biol Reprod 1994;50:653–63.

23. Treloar A, Boynton R, Behn B, Brown B. Variation of the human menstrual cycle through reproductive life. Int J Fertil 1967;12:77–126.
24. Klein N, Battaglia D, Fujimoto V, Davis G, Bremner W, Soules M. Reproductive aging: accelerated ovarian follicular development associated with a monotropic follicle-stimulating hormone rise in normal older women. J Clin Endocrinol Metab 1996;81:1038–45.
25. Lenton E, Landgren B, Sexton L, Harper R. Normal variation in the length of follicular phase of the menstrual cycle: effect of chronological age. Br J Obstet Gynaecol 1984;91:6814.
26. Sherman B, Korenman S. Hormonal characteristics of the human menstrual cycle throughout reproductive life. J Clin Invest 1975;55:699–706.
27. Vollman R. The degree of variability of the length of the menstrual cycle in correlation with age of woman. Gynaecologia 1956;142:310–14.
28. Klein N, Battaglia D, Miller P, Branigan E, Guidice L, Soules M. Ovarian follicular development and the follicular fluid hormones and growth factors in normal women of advanced reproductive age. J Clin Endocrinol Metab 1996;81:1946–51.
29. Metcalf M. Incidence of ovulatory cycles in women approaching the menopause. J Biosoc Sci 1979;11:39–48.
30. Lee S, Lenton E, Sexton L, Cooke I. The effect of age on the cyclical patterns of plasma LH, FSH, oestradiol and progesterone in women with regular menstrual cycles. Hum Reprod 1988;3:851–55.
31. Metcalf M, Livesey J. Gonadotrophin excretion in fertile women: effect of age and the onset of the menopausal transition. J Endocrinol 1985;105:357–62.
32. Reyes F, Winter J, Faiman C. Pituitary-ovarian relationships preceding the menopause. I. A cross-sectional study of serum follicle-stimulating hormone, luteinizing hormone, prolactin, estradiol and progesterone levels. Am J Obstet Gynecol 1977;129:557–64.
33. Sherman B, West J, Korenman S. The menopausal transition: analysis of LH, FSH, estradiol, and progesterone concentrations during menstrual cycles of older women. J Clin Endocrinol Metab 1976;42:629–36.
34. Santoro N, Brown J, Adel T, Skurnick J. Characterization of reproductive hormonal dynamics in the perimenopause. J Clin Endocrinol Metab 1996;81:1495–501.
35. Klein N, Battaglia D, Clifton D, Bremner W, Soules M. The gonadotropin secretion pattern in normal women of advanced reproductive age in relation to the monotropic FSH rise. J Soc Gynecol Invest 1996;3:27–32.
36. Reame N, Kelch R, Beitins I, Yu M, Zawacki C, Padmanabhan V. Age effects on follicle-stimulating hormone and pulsatile luteinizing hormone secretion across the menstrual cycle of premenopausal women. J Clin Endocrinol Metab 1996;81:1512–18.
37. Licciardi F, Liu H, Rosenwaks Z. Day 3 estradiol serum concentrations as prognosticators of ovarian stimulation response and pregnancy outcome in patients undergoing in vitro fertilization. Fertil Steril 1995;64:991–94.
38. Metcalf M, Livesay J. Pregnanediol excretion in fertile women: age-related changes. J Endocrinol 1988;119:153–57.
39. Hee J, MacNaughton J, Bangah M, Burger H. Perimenopausal patterns of gonadotrophins, immunoreactive inhibin, oestradiol and progesterone. Maturitas 1993;18:9–20.

40. MacNaughton J, Banah M, McCloud P, Hee J, Burger H. Age related changes in follicle stimulating hormone, oestradiol and immunoreactive inhibin in women of reproductive age. Clin Endocrinol 1992;36:339.

41. Hee J, MacNaughton J, Bangah M, et al. Follicle-stimulating hormone induces dose-dependent stimulation of immunoreactive inhibin secretion during the follicular phase of the human menstrual cycle. J Clin Endocrinol Metab 1993;76:1340–43.

42. Groome N, Illingworth P, O'Brien M, Pai R, Mather J, McNeilly A. Measurement of dimeric inhibin B throughout the human menstrual cycle. J Clin Endocrinol Metab 1996;81:1401–5.

43. Groome N, Illingworth P, O'Brien M, et al. Detection of dimeric inhibin throughout the human menstrual cycle by two-site enzyme immunoassay. Clin Endocrinol 1994;40:717–23.

44. Klein N, Illingworth P, Groome M, McNeilly A, Battaglia D, Soules M. Decreased inhibin B secretion is associated with the monotropic FSH rise in older, ovulatory women: a study of serum and follicular fluid levels of dimeric inhibin A and B in spontaneous menstrual cycles. J Clin Endocrinol Metab 1996;81:2742–45.

45. Suganuma N, Kitagawa T, Nawa A, Tomoda Y. Human ovarian aging and mitochondrial DNA deletion. Horm Res 1993;39:16–21.

46. Van Blerkom J. The influence of intrinsic and extrinsic factors on the developmental potential and chromosomal normality of the human oocyte. J Soc Gynecol Invest 1996;3:3–11.

8

New Strategies in Studying Ovarian Aging

KUTLUK OKTAY, GUVENC KARLIKAYA,
AND AYKUT BAYRAKCEKEN

Introduction

Ovarian aging is a process that represents a quantitative decline in follicle reserve coupled with presumed qualitative changes in follicular cells (1). Follicular reserve is depleted at a steady rate until the age of 37. After that age, the rate of follicle depletion accelerates until the follicle supply is exhausted (2). In parallel to this acceleration, baseline FSH levels rise and the fecundity rates decline (3–6). Studies on ovarian aging, therefore, have to answer two fundamental questions: What causes the accelerated follicle depletion, and which mechanisms are responsible for the qualitative changes in the ovarian follicle? To address the problem of accelerated follicle depletion we must first determine what initiates primordial follicle growth. If the mechanisms underlying the initiation of primordial follicle are better understood, we can develop new strategies to retard the depletion of follicle reserve and postpone menopause.

Ovarian function requires an orchestration between stromal and follicular cells. Studying the follicle as a whole, therefore, is physiologically more relevant than studying individual follicular cells. Most studies on dynamics of follicular growth focused on cell culture systems or strictly relied on histological sections that did not provide in vivo or in vitro data on follicular development. The focus of our research has been to develop in vivo and in vitro experimental models to study the quantitative aspects of ovarian aging. Here, we will discuss new strategies developed in our laboratories to study the early phases of follicular growth. Among the strategies that will be discussed are proliferating cell nuclear antigen (PCNA) as sensitive marker of follicle growth initiation, isolation and culture of primordial and primary human follicles, organ cultures, human ovarian grafts in immunodeficient-hypogonadal (SCID-*hpg*) mice, and single-follicle PCR to determine the differential expression of genes in various stages of follicle growth.

PCNA: A Sensitive Marker of Follicle Cell Proliferation

In rodents, the time required for a primordial follicle to reach antral stage is less than 1 month (7), which compares with several months in humans (8,9). This faster growth enables us to perform some of the kinetic studies that could not otherwise be performed in humans. Short of a specific serum marker for early-stage follicles, labeling methods, such as systemic administration of tritiated thymidine, have been used. Because of the relatively slow rate of growth, Hirshfield described that over a week long infusions may be necessary to have maximum number of follicle cells labeled (10). Because the thymidine is taken up only during the S-phase of cell cycle, the sensitivity of thymidine-labeling technique is low in slowly growing cell types. In addition, the resolution of radioactive signals in diminutive granulosa cells of very small follicles can be poor. This radioactive method cannot be used in human studies. Because of the need for more sensitive and widely applicable cell proliferation markers, we have tested PCNA as a marker of follicle growth inititation (11). PCNA is a 35 kDa nuclear protein that forms a complex together with cyclin D, cyclin dependent kinase (CDK), and CDK inhibitor (CDKi) p21. PCNA is required for the cell commitment of proliferation (12). Its expression is detectable from the late G1, when the cell is committed to proliferate, and remains elevated through early M phase (13). Because PCNA can be detected even before the cell shows morphological evidence of growth, it is an early marker of cell proliferation. The technique is more sensitive because unlike thymidine incorporation, it labels cells prior to the S phase (12,13).

Using immunohistochemical methods we have found the PCNA is expressed at or prior to the earliest morphological sign of follicular growth both ir rodent and human follicles (8,11). It is more interesting to note that PCNA also marks the initiation of oocyte growth. This observation is interesting because the oocyte is arrested at the G1 phase of growth. PCNA is also a cofactor for DNA polymerase delta, an enzyme responsible for DNA repair, and this increase in PCNA expression may signal the ongoing repair process during oocyte growth. PCNA expression also can be upregulated by growth factors, and its detection in the oocyte may reflect paracrine effects of oocyte-specific growth factors such as GDF-9 (14).

PCNA marks growth prior to, and more precisely than, morphological changes, and can be combined with other methods that we will discuss shortly to increase the sensitivity of detecting follicle growth initiation.

In Vitro Growth Models

Subcutaneous Osmotic Minipumps in Laboratory Animals

To be able to test the effects of steadily elevated levels of FSH on follicle growth inititation rates, we subcutaneously inserted osmotic minipumps loaded to deliver a daily dose of 4 IU FSH to mice for 4 weeks (Alza Corp., Palo Alto, CA). Using this model, we have maintained FSH levels in menopausal range over 4 weeks. Although the FSH levels were comparable to levels in postmenopausal women, the rate of follicle depletion did not accelerate (15). This model can easily be adapted to test the long-term effects of other hormones and agents on the follicular growth. Larger pumps can be used in larger animals for continuous delivery up to 6 weeks.

Isolation and Culture of Human Ovarian Follicles

Follicle isolation provides the opportunity to study follicles individually in vitro. The isolation and culture of larger preantral and antral follicles is easier and it is better established (16–17), but the isolation of primordial and primary follicles is a challenging task. In the human, unlike the rodent, ovarian stroma is fibrous (18). A predigestion with collagenases is required to loosen up the stroma before the primordial and primary follicles can be mechanically isolated. We have described a method (Table 8.1) where human ovarian cortex is partially digested in collagenase type IA and the follicles are subsequently isolated manually under a dissection microscope (19) (Fig. 8.1). This technique is tedious, requires training, and is time-consuming. In addition, enzymes used for partial digestion tend to damage the basement membrane even though the majority of the follicles are deemed viable at the time of isolation (19). Once placed in culture, the follicles tend to extrude their oocytes and dismantle (20). It seems that the early stage follicles cannot survive in isolation from stromal influences.

The role of stroma, and specifically extracellular matrix–integrin interactions, in keeping the integrity and growth stimulation responsiveness of cells has been revealed within the past 5 years (21). The failure to respond to

TABLE 8.1. Isolation of human primordial follicles.

1. Obtain ovarian cortical tissue
2. Mince tissue into 2 × 2 × 1 mm blocks
3. Partially digest tissue in Leibovitz L-15 media containing 1 mg/mL of collagenase type IA (Sigma, St. Louis, MO) and 8 U/mL of DNAse (Sigma) for 1 hour, 45 minutes
4. Wash the tissue with serum supplemented media
5. Isolate follicles under 50–75 × using 26 gauge needles.

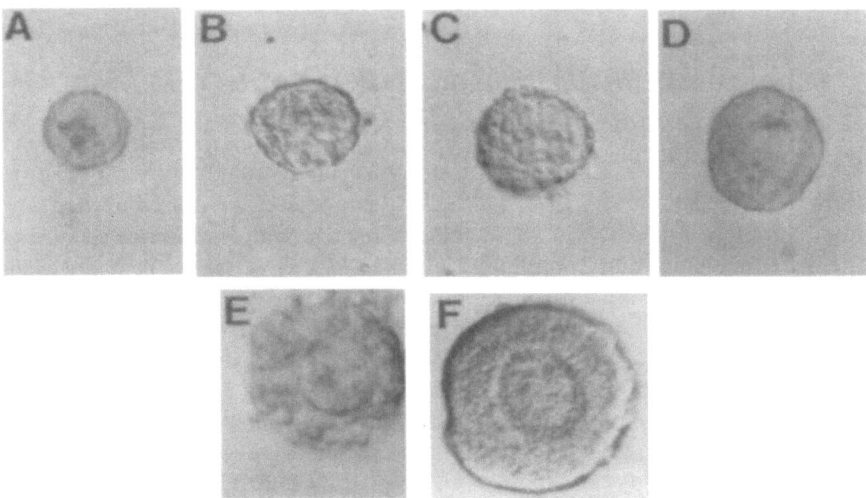

FIGURE 8.1. Isolated human ovarian follicles. (A) primordial follicle: a small oocyte surrounded by single layer of pregranulosa cells; (B) primary follicle with one cell enlarged (early primary); (C) primary follicle with single layer of enlarged granulosa cells; (D) two-layer follicle; (E) three-layer multilaminar (preantral) follicle; (F) large multilaminar (preantral) follicle. (× 300). Reproduced with permission of the © Endocrine Society from Oktay K, Briggs D, Gosden RG. Ontogeny of FSH receptor gene expression in isolated human ovarian follicles. J Clin Endocrinol Metab 1997;82:3748–51 (Ref. 27).

growth factors of isolated follicles may well be due to isolation from stroma. Following that logic, organ cultures or xenografts may be a solution for the extracorporal development of human follicles.

Organ Cultures

Primordial follicles of bovine and rodent ovaries initiate growth after relatively shorter periods of organ cultures (22,23). We have cultured 1 mm^3 ovarian cortical pieces for up to 7 weeks with few follicles surviving (unpublished). Follicle growth to primary and two-layer stages was seen, but we are still investigating the optimal conditions to achieve further growth in vitro. Although we have made progess with organ cultures, it is not yet possible to replicate the complex interactions between growth factors, hormones, and other influences in the living tissue. A next to ideal solution for studying in vivo follicle growth is using xenografts in immunodeficient laboratory animals.

In Vivo Models

Human Xenografts in Immunodeficient Mice (SCID): An In Vivo Model to Study Follicle Growth

We described a new approach to study in vivo human ovarian follicle growth using xenografts in *hpg*-SCID (hypogonadal–severe combined immunodeficiency) mice (8). These mice carry a double mutation, rendering them hypogonadal and completely anergic (24,25). Tissues from different species can be grafted into these mice without the fear of rejection (26). We grafted 1 mm^3 human ovarian cortical tissues from a young donor under the kidney capsules of 5-month-old *hpg*-SCID mice (8). The grafts were left in place for up to 17 weeks, and FSH was administered to one group for the last 6 weeks. We have found that the follicles reached antral stages and secreted estradiol in the FSH-treated group while halting growth at the two-layer stage in the control group (Fig. 8.2 and Table 8.2). We confirmed follicle growth by PCNA immunostaining of the grafts (Fig. 8.3). The SCID-xenograft model is potentially useful for studying various factors that are implicated in initiation of growth and follicular development.

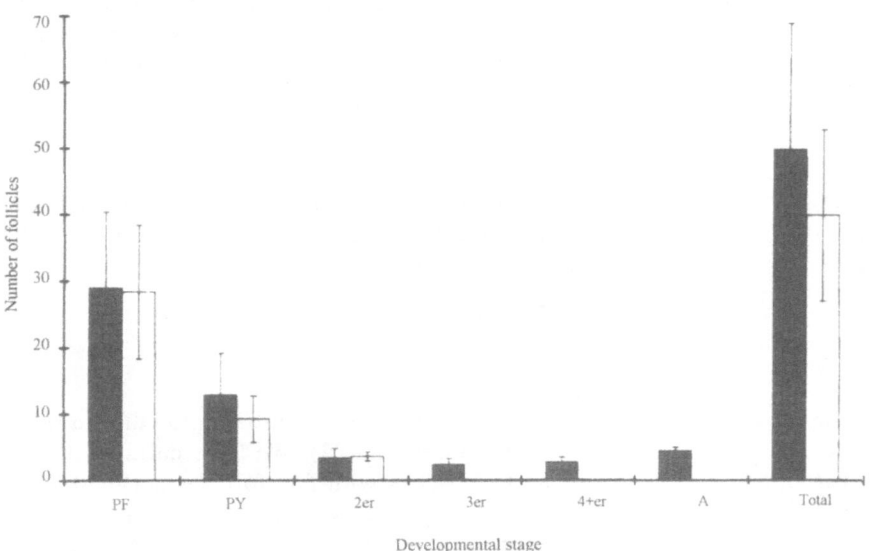

FIGURE 8.2. Relationship between the mean number of follicles ±SEM and the stage of development in ovarian xenografts: primordial (PF), primary (PY), two-layer (2er), three-layer (3er), ≥ three-layer (4+er), and antral (A) follicles. Grafts stimulated with FSH (black bars) compared with controls (blank bars). Reproduced from Ref. 8 with permission.

TABLE 8.2. Assessment of estrogen secretion by ovarian xenografts in SCID/*hpg* mice stimulated with FSH.

Host	Number of antral follicles (mm diameter)	Serum estradiol (pmol·L⁻¹)	Uterine weight* (mg)	Vaginal introitus
1	1 (5)	2070	212 (ballooned)	Patent
2	2 (3,4)	780	131	Patent
3	0	35	123	Patent
4	1 (2.5, hemorrhagic)	126	126	Patent

Reproduced with permission from Ref. 8.
*Median weight in unstimulated host: 28 mg.

Molecular Studies of Isolated Follicles

By using the isolation method we have previously described, we developed an approach to investigate gene transcription in individual follicles. Human follicles ranging from primordial to multilayer stages were isolated and the mRNA was extracted by magnetic means using Dynabeads (Dynal Inc., Lake Success, NY) (27). After reverse transcription, a nested PCR was performed using primers representing sequences in exons 7 and 9, and 8 and 9 of the hFSH receptor gene. Negative, positive, and actin controls were run in parallel. We showed that the FSH receptor gene is not expressed in the primor-

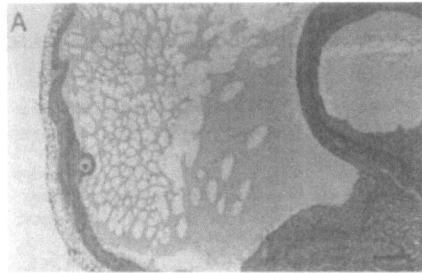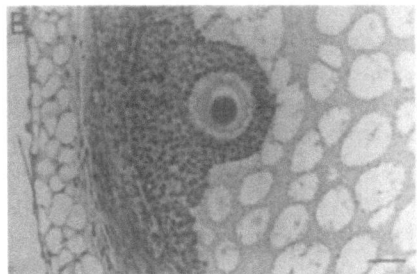

FIGURE 8.3. (A) Low magnification of a Graafian follicle (3 mm diameter) in human ovarian tissue xenografted 17 weeks previously and stimulated with FSH for the last 6 weeks (× 20). (B) Higher magnification of the same follicle showing a germinal vesicle oocyte surrounded by cumulus cells. Both the oocyte and the granulosa cells stain with PCNA (× 200) Hematoxylin. Reproduced from Ref. 8 with permission.

TABLE 8.3. FSH receptor mRNA expression in isolated follicles.

	Primordial[*+]	Primary[*]	Two-layer[*]	Multilaminar[*]	Total growing[+]
FSHr positive	0	4	3	4	11
FSHr negative	9	8	6	0	14

Reproduced with permission of the © Endocrine Society from Oktay K, Briggs D, Gosden RG. Ontogeny of FSH receptor gene expression in isolated human ovarian follicles. J Clin Endocrinol Metab 1997;82:3748–51 (27).
[*]$X^2 p = 0.01$. [+]Fisher's exact test, $p < 005$.

dial follicles. The expression was detected in only 33% of the one- and two-layer follicles. All follicles with three or more layers expressed the FSH gene (Table 8.3). The same method can be used to investigate the ontogeny of the expression of numerous genes and to determine the genes that are specific to individual stages of follicular growth. In that manner we can determine the factors regulating the expression of stage-specific genes expressed during follicle growth initiation.

Conclusions

We discussed our experience with studying the quantitative aspects of ovarian aging (i.e., primordial follicle depletion leading to menopause). Even though we are armed with better tools to investigate the factors that may regulate primordial follicle growth initiation, the process still remains enigmatic. This challenging field needs more investigators and creative ideas. The answer to what initiates follicle growth is more likely to lie in the paracrine factors that originate from stroma and follicular cells rather than endocrine influences. The process is likely to be complex and may involve interactions among various cell types. A multidisciplinary approach and close cooperation between clinicians and basic scientists is the key to a faster progress.

References

1. Burns WN, Oktay K, Tekmal RR, Nelson JF, Schenken RS. Diminished alpha-inhibin mRNA in in vitro fertiliztion-enbryo transfer poor responders reflects declining follicular reserve. Fertil Steril 1996;65:394–99.
2. Faddy MJ, Gosden RG, Gougen A, Richardson SJ, Nelson JF. Accelerated disappearance of ovarian follicles in mid-life: implications for forecasting menopause. Hum Reprod 1992;7(10):1342–46.
3. MacNaughton J, Banah M, McCloud P, Hee J, Burger H. Age-related changes in follicle stimulating hormone, luteinizing hormone, oestradiol and immunoreactive inhibin in women of reproductive age. Clin Endocrinol 1992;36:339–45.

4. Sherman BM, Korenman SG. Hormonal characteristics on the human menstrual cycle throughout the reproductive life. J Clin Invest 1975;55:699–706.
5. Sherman BM, West JH, Korenman SG. The menopausal transition: analysis of LH, FSH, estradiol, and progesterone concentrations during menstrual cycles of older women. J Clin Endocrinol Metab 1976;42:629–36.
6. Lenton EA, Sexton L, Lee S, Cooke ID. Progressive changes in LH and FSH and LH:FSH ratio in women throughout reproductive life. Maturitas 1988;10:35–43.
7. Hirshfield AN. Development of the follicles in the mammalian ovary. Int Rev Cytol 1991;124:43–101.
8. Oktay K, Newton H, Mullan J, Gosden R. Devlepoment of human primordial follicles to antral stages in SCID/*hpg* mice stimulated with follicle stimulating hormone. Hum Reprod 1998;13:1133–38.
9. Gougeon A. Dynamics of follicular growth in the human: a model from preliminary results. Hum Reprod 1986;1:81–87.
10. Hirshfield AN. Granulosa cell proliferation in very small follicles of cycling rats studied by long-term continuous tritiated thymidine infusion. Biol Reprod 1989;41:309–16.
11. Oktay K, Schenken RS, Nelson JF. Proliferating cell nuclear antigen (PCNA) marks the initiation of follicular growth in the rat. Biol Reprod 1995;53:295–301.
12. Xiong Y, Zhang H, Beach D. D type cyclins associate with multiple protein kinases and the DNA replication and repair factor PCNA. Cell 1992;71:505–14.
13. Xiong Y, Connolly T, Futcher B, Beach D. Human D-type cyclin. Cell 1991;65:691–99.
14. Dong J, Albertini DF, Nishimori K, Kumar TR, Lu N, Matzuk MM. GDF-9 is required during early ovarian folliculogenesis. Nature 1996;383:531–35.
15. Oktay K, Schenken RS, Nelson JF. The effects of long-term gonadotropin treatment on primordial follicle reserve in the mouse ovary. J Soc Gynecol Invest 1997;4(S):129A.
16. Gosden RG, Oktay K, Mullan J, Jenner L. *In vitro* maturation of human oocytes: basic aspects. Symposium on the ovary: dysfunction and treatment. Excerpta Medica Int Cong Ser 1996;1106:149–58.
17. Roy SK, Treacy BJ. Isolation and long-term culture of human preantral follicles. Fertil Steril 1993;59:783–90.
18. Oktay K, Newton H, Aubard Y, Gosden RG. Cryopreservation of human oocytes and ovarian tissue: an emerging technology? Fertil Steril 1998;69:1–7.
19. Oktay K, Nugent D, Newton H, Salha O, Chattergee P, Gosden R. Isolation and characterization of primordial follicles from fresh and cryopreserved human ovarian tissue. Fertil Steril 1997;67:481–86.
20. Oktay K. New horizons in assisted reproductive technologies. Assist Reprod Rev 1998;8:51–54.
21. Giancotti F. Integrin signaling: specificity and control of cell survival and cell cycle progress. Curr Opin Cell Biol 1997;9:691–700.
22. Wandji S-A, Srsen V, Voss AK, Eppig JJ, Fortune JE. Initiation *in vitro* of growth of bovine primordial follicles. Biol Reprod 1996;55:942–48.
23. Eppig JJ, O'Brien MJ. Development in vitro of mouse oocytes from primordial follicles. Biol Reprod 1996;54:197–207.
24. Bosma GC, Custer RP, Bosma MJ. A severe combined deficiency mutation in the mouse. Nature 1983;301:527–30.
25. Mason AJ, Hayflick JS, Zoeller RT, Young III WS, Phillips HS, Nicolics K, et al.

A deletion truncating the gonadotropin-releasing hormone gene is responsible for hypogonadism in the hpg mouse. Science 1986;234:1366–71.

26. Gosden RG, Boulton MI, Grant K, Webb R. Follicular development from ovarian xenografts in SCID mice. J Reprod Fertil 1994;101:619–23.

27. Oktay K, Briggs D, Gosden RG. Ontogeny of FSH receptor gene expression in isolated human ovarian follicles. J Clin Endocrinol Metab 1997;82:3748–51.

Part II

Effects of Reproductive Aging on Nonreproductive Tissues

9

Novel Mechanisms of Estrogen Action in the Developing Brain

C. Dominique Toran-Allerand

Introduction

Estrogen and the neurotrophins [e.g., nerve growth factor (NGF), brain-derived neurotrophic factor (BDNF), neurotrophin-3 (NT-3), and neurotrophin-4/5, (NT-4/5)] have been implicated in the development, survival, plasticity, and aging of neurons in mammalian forebrain regions that subserve cognitive functions. Neurons in overlapping forebrain regions of both sexes co-express estrogen and neurotrophin receptors, and they are the sites of estrogen and neurotrophin synthesis (1,2). An important question raised by these associations concerns the biological significance of receptor co-expression and the potential for interactions of their ligands.

Differentiative Effect of Estrogen

Estrogen has been shown to exhibit growth- or neurite-promoting properties for neurons of the developing forebrain. Toran-Allerand (3, for review) first demonstrated that estrogen elicits the selective enhancement of axon and dendrite (neurite) growth and differentiation in cultured slices of estrogen target forebrain regions of the developing rodent. Estrogen stimulation of neurite growth is developmentally regulated and not seen in the normal adult female brain. In the adult, however, as a result of the loss of trophic support, whether through estrogen deprivation, axotomy, or deafferentation, responsiveness to estrogen returns and estrogen can again be shown to influence the growth and differentiation of neurite-derived structures (e.g., axons, dendrites, dendritic spines, and synapses) (4). The discovery of a second estrogen receptor (ER) in the brain, ER-β (5), complicates matters considerably and raises the possibility of additional ER genes as well. Thus, it is not at all clear whether our understanding of the role and function of the "ER" in the brain applies to the "classical" receptor ER-α and/or ER-β, or to still other, as yet unidentified, subtypes.

This issue is of particular importance because ER-α and ER-β have been shown to exist both as homodimers and as heterodimers (6).

Traditional and Nontraditional Mechanisms of Estrogen Action

The traditional view of estrogen action postulates activation of the ER by tyrosine phosphorylation of both the unliganded ER and heat shock protein 90 (hsp90) with which it is complexed. Tyrosine phosphorylation elicits dissociation of the ER from hsp90, resulting in conformational changes within both molecules, dimerization of the activated ER, followed by nuclear translocation. The receptor dimers bind to specific estrogen response elements (EREs) within DNA, resulting in the regulation of gene expression (7). Such a mechanism, nonetheless, fails to explain adequately the complete and extensive range of estrogen's actions, including its ability to elicit transcriptional regulation of many genes that do not exhibit an apparent canonical ERE (8) and the very rapid effects of estrogen (9,10). The rapid time course, although inconsistent with transcriptional modulation via a classical nuclear receptor, is consistent with a receptor within the plasma membrane.

Increasing evidence suggests that there may be multiple types of ERs, including those associated with the plasma membrane (11–13), that can mediate extracellular signals rapidly in both a steroid-dependent and steroid-independent manner by means of growth factor signaling pathways (14,15). For example, in breast cancer and other cell lines, estrogen has been shown to elicit very rapid and transient responses (ranging from seconds to minutes) similar to those evoked by mitogenic peptide growth factors. This time course is inconsistent with transcriptional modulation, but it is consistent with the sharing of estrogen and growth factor signaling pathways for cell proliferation and ER phosphorylation (16–19). In MCF-7 cells, estradiol elicits maximal phosphorylation of src within 10 seconds (10), which is an effect that requires only 10–20% ER occupancy (20). This is a level of occupancy that corresponds to the percentage of ER reportedly estimated to be associated with the plasma membrane (11).

The Neurotrophins and Their Receptors

The neurotrophins are structurally and functionally related proteins with important growth and trophic actions on the development and survival of the nervous system (reviewed in 21). The biological activities of the neurotrophins are mediated by two structurally distinct classes of cell membrane receptors that are preferentially expressed in neural tissues (22). One class consists of members of the ligand-specific trk family of receptor tyrosine kinases (trkA/NGF, trkB/BDNF and NT-4/5; and trkC/NT-3), which initiates the signal transduction process through increased tyrosine autophosphorylation of the cognate receptor. The second receptor, p75, a 75 kDa transmembrane protein that binds all neurotrophins with low affinity, has a modulatory role on trk activity and function. The trks mediate responses to the neurotrophins, however, with or without the participation of p75.

Signal Transduction Pathways

Signals from polypeptide growth and trophic factors (e.g., the neurotrophins), are propagated by an essentially linear flow of sequential protein phosphorylation (by kinases) and dephosphorylation (by phosphatases) events. These intracellular protein cascades serve to funnel, amplify, and propagate signals generated at the cell surface through the intermediary of receptor tyrosine kinases into complex biological responses, including the regulation of target transcription factors and the induction of genes (23) (Fig. 9.1). In PC12 cells, treatment with NGF, the prototypical neurotrophin, elicits ty-

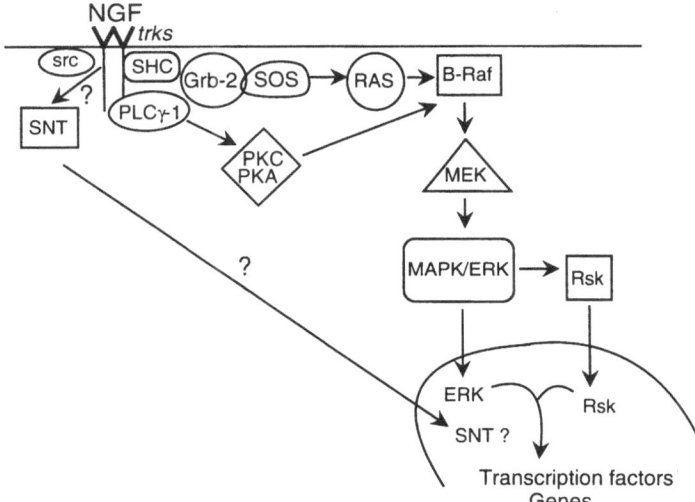

FIGURE 9.1. Some PC12 cell signaling pathways mediating differentiation. Treatment with NGF elicits tyrosine autophosphorylation and activation of its receptor trkA. Tyrosine autophosphorylation regulates interactions of the activated trkA with multiple intracellular proteins, including phospholipase Cγ-1, src, Shc, and Grb2. Shc and Grb2, with the GTP/GDP exchange protein SOS, connect activated trkA to the authentic initial signaling enzyme p21Ras. Ras then activates multiple signal transduction pathways, including the MAP kinase cascade. The MAP kinase cascade is triggered by Ras activation of Raf, followed by sequential activation of MEK and ERK (MAP kinase). Activated ERK then either translocates to the nucleus directly or first phosphorylates other kinases (e.g., Rsk), which then also translocate.

rosine autophosphorylation and activation of its receptor trkA. Tyrosine autophosphorylation regulates interactions of the activated trkA with multiple intracellular proteins that either have enzymatic functions (e.g., phospholipase C-γ1 and src) or are without intrinsic enzymatic activity (Shc and Grb2). Shc and Grb2 with the GTP/GDP exchange protein SOS connect activated trkA to the authentic initial signaling enzyme p21Ras. Ras, a small guanine nucleotide-binding protein, then activates multiple signal transduction pathways, including the MAP kinase cascade. The MAP kinase cascade is triggered by Ras activation of Raf, followed by sequential phosphorylation and activation of MEK and ERK (MAP kinase). Activated ERK then either translocates to the nucleus directly or first phosphorylates other kinases (e.g., Rsk), which then also translocate. Nuclear translocation of ERK and Rsk are the means by which signals that arise at the plasma membrane activate transcription factors and regulate genes (23,24). The MAP kinase cascade and SNT (25), an incompletely characterized, parallel, growth factor- and differentiation-specific pathway distal to trk, form two major pathways whose prolonged activation is necessary but not sufficient for PC12 differentiation. The signaling pathways mediating estrogen and neurotrophin-induced differentiative actions in the CNS are essentially unknown.

Cross-Coupling of Estrogen and Neurotrophin Signaling Pathways

The cascade of molecular events that follows estrogen activation of the neural ER is largely unknown. Estrogen and neurotrophin receptor co-expression suggested that their ligands may influence each other's actions by regulating receptor and/or ligand availability through reciprocal regulation at the level of gene transcription (3). Studies then documented differential and reciprocal regulation of estrogen and NGF receptor mRNA expression by their ligands in adult female rat sensory neurons in vivo (26) and the pheochromocytoma cell line PC12 (27). Colocalization of estrogen and neurotrophin receptors, however, may also lead to the sharing of similar, if not overlapping, sequences of intracellular biochemical events through convergence or cross-coupling of their signaling pathways. Cross-coupling of converging estrogen and neurotrophin signaling pathways may lead to nuclear endpoints that regulate the same broad array of genes involved in neurite growth and differentiation.

Findings document the potential for convergence of estrogen and neurotrophin signaling pathways by demonstrating for the first time in the brain that 17β-estradiol, like the neurotrophins, does in fact rapidly (within 5–15 minutes) induce the sustained phosphorylation and activation of ERK, which persisted above baseline for at least 2 hours (28) (Fig. 9.2). Because sustained activation and nuclear translocation of ERK in PC12 cells has been

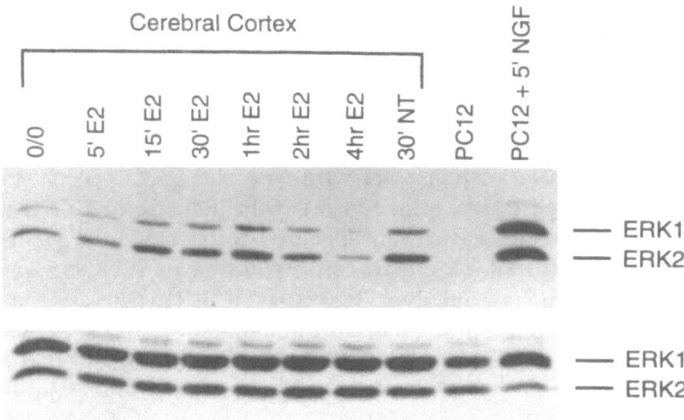

FIGURE 9.2. Estradiol-induced ERK phosphorylation. Lysates derived from cerebral cortical explants were probed with a phosphospecific ERK1/ERK2 antibody. Shown is a time course for: estradiol-induced ERK phosphorylation (upper panel) and the reprobed blot for ERK1 and ERK2 protein (lower panel). Explants were also treated with a 100 ng/ml neurotrophin cocktail (NGF, BDNF, NT-3, and NT-4/5), for a single 30 minute time point and served as the experimental positive control. Note the similarity in the intensity of the response to estrogen and the neurotrophins in the cortical explants. Untreated PC12 and NGF-treated PC12 cells served as negative and positive methodological controls, respectively.

associated with neuronal differentiation (21,29), the extended time course of ERK activation that follows estrogen exposure would appear to be consistent with its observed differentiative actions on neurites. In addition, 17α-estradiol, the transcriptionally inactive stereoisomer, also elicited the rapid and prolonged activation of ERK to a level perhaps even greater than that elicited by 17β-estradiol. Because 17α-estradiol does not apparently bind ER-α or ER-β (G.L. Greene, personal communication), these very preliminary findings question whether 17α-estradiol-induced ERK activation may be mediated by a third (non-α/non-β) ER.

ER-α OR ER-β?

In order to identify the ER subtype (e.g., ER-α or ER-β) that mediates estrogen activation of ERK, studies have been initiated on cerebral cortical explants derived from the ER-α-knockout (ERKO) mouse. Preliminary studies (M. Singh, G. Sétáló Jr, D. Frail, C.D. Toran-Allerand, unpublished observations) document that both estrogen and the neurotrophins elicited ERK activation in ERKO cortical explants and that the level of ERK activation was significantly greater than that seen with wild-type control cultures, in which both ER-α and ER-β mRNA expression is found. The pure ER antagonist ICI 182,780 inhibited estrogen-induced ERK phosphorylation only in the wild-type, but not in ERKO, cortical cultures.

These differential responses of wild-type and ERKO cortical explants to ICI 182,780 suggest that, although estrogen induction of ERK phosphorylation did not require ER-α and appeared to be mediated by ER-β (or by an as yet unidentified ER), estrogen regulation of ERK activation, nonetheless, must somehow involve ER-α as well. ERK activation may thus be either partially inhibited or otherwise modulated by the presence of ER-α, perhaps through the existence of ER-α/ER-β heterodimers that are present in wild-type but not ERKO tissue. The very ability of estradiol to elicit ERK phosphorylation in the ERKO cortical explants, however, also raises the possibility of involvement of a third ER, when taken in consideration with very preliminary studies in explants of wild-type developing cerebellum, a neurotrophin target region, where ER-β mRNA expression is very abundant (30), but ER-α mRNA and protein levels are low (C.D. Toran-Allerand, unpublished observations). Exposure of cerebellar explants to estrogen did not elicit ERK activation, although the neurotrophins were able to do so. Taken in conjunction with the reponses to 17α-estradiol, the cerebellar findings suggest that there may well be a third ER that is present in the cerebral cortex, but absent from the cerebellum. On the other hand, the apparent differential pattern or ratio of ER-α and ER-β expression in the cerebellum, as compared with wild-type and ERKO cerebral cortex, suggests that regional differences in the relative proportion of ER homodimers versus heterodimers may be important for estrogen-induced ERK responses.

Novel Estrogen Signaling Pathways

Because ERs colocalize with the neurotrophin ligands and receptors (2), exposure of the cortical explants to estradiol may have first elicited phosphorylation of the trk receptors, either directly or as a result of estrogen-induced neurotrophin release. Although estrogen reportedly elicits autophosphorylation of the EGF receptor (31), we found no evidence of estrogen-induced phosphorylation of trk, which might have explained rapid estrogen activation of ERK. In both wild-type and ERKO cortical cultures, however, estrogen- and neurotrophin-induced phosphorylation of ERK was inhibited to baseline levels in the presence of the MEK inhibitor PD 98059. These findings together suggest that convergence of estrogen and neurotrophin signaling pathways must occur at least at the level of MEK, the signaling protein immediately upstream of ERK, or perhaps further upstream, but certainly downstream of trk.

ER/Signaling Protein Complexes

Some additional clues concerning potential pathways for estrogen-induced ERK activation were initially suggested by observations in PC12 cells of a very large (>300 kDa), multimeric complex consisting of at least B-Raf and hsp90 (32). B-Raf and hsp90 were also found complexed in explants of the cerebral cortex (28). They were also physically associated with unliganded ER-α and src, a non-receptor tyrosine kinase. Src is of interest because estrogen reportedly elicits tyrosine phosphorylation of src within seconds in mammary tumour cells (10), leading to the suggestion that phosphorylated src may then serve as the kinase recruited by estradiol to elicit tyrosine phosphorylation its own receptor (20). Co-immunoprecipitation of ERK with hsp90 and with the ER has also been found in very preliminary studies (G. Sétáló Jr, M. Singh, C.D. Toran-Allerand, unpublished observations). Our novel findings of a direct association of the ER physically complexed with members of the MAP kinase cascade and src may delineate a pathway by which estrogen may rapidly activate ERK.

This putative multimeric complex may be similar to caveolar-associated complexes. Caveolae are plasma membrane microdomains that are involved in signal transduction and vesicular trafficking (33). All signaling components of the MAP kinase cascade are reportedly concentrated in the caveolae of quiescent cells (e.g., fibroblasts) (34). Addition of a growth factor such as PDGF, for example, to either intact cells or caveolae isolated from these cells stimulated tyrosine phosphorylation and activation of MAP kinases (34). Association of the ER with such caveolarlike complexes might serve as a plasma membrane receptor and mediate the rapid effects of estrogen.

The initial route taken by estrogen to elicit rapid activation of ERK may thus involve direct activation of the putative ER-containing multimeric complex (28) (Fig. 9.3). Neuronal exposure to estradiol induces tyrosine phosphorylation of src, the ER, and hsp 90, causing dissociation and conformational changes within this complex. This may in turn trigger sequential phosphorylation and activation of physically associated kinases (e.g., B-Raf). On the other hand, estradiol may induce a greater association of B-Raf and src into this complex. This recruitment could then rapidly activate further downstream members of the cascade through phosphorylation (e.g., MEK and ERK). Such multimeric complexes, consisting of the ER, members of the MAP kinase cascade, and src, could serve as nodal points, linking the estrogen and neurotrophin-signaling pathways and providing an alternative mechanism for some actions of estrogen in the developing brain.

Thus, regardless of whether the ligand is estrogen or a neurotrophin, tyrosine phosphorylation within the complex and dissociation of the ER from the physically associated kinases may result in both ER activation as well as in activation of ERK, leading in both instances to nuclear translocation. Although the dogma of estrogen action has been that the ligand-activated ER binds directly to EREs, our findings could also explain how, through the intermediary of nuclear translocation of ERK, estrogen and the neurotrophins

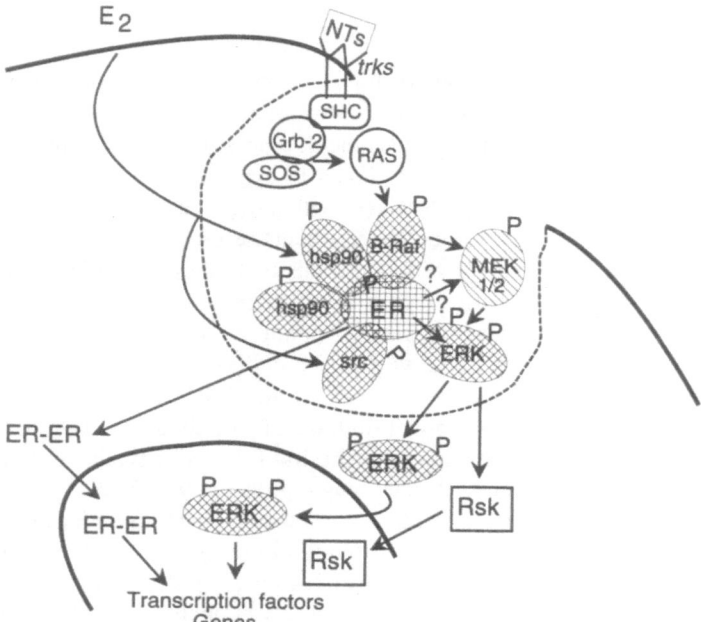

FIGURE 9.3. Direct activation of the putative multimeric complex by estrogen-induced tyrosine phosphorylation of hsp90, src, and the estrogen receptor (ER) may lead to conformational changes and rapid activation of the ER, MEK, and ERK, which results in the nuclear translocation of both the activated ER and ERK.

could each regulate the same broad array of ERE-containing genes such as GAP-43 (35,36) and tau-microtubule–associated protein (37,38) and non–ERE-containing genes such as β-tubulin (39,40) and MAP-2 (41,42) involved in neurite growth and differentiation.

The involvement of ERK in estrogen signaling, however, does not rule out estrogen activation of other signaling substrates as well (e.g., adenylate cyclase/cAMP or Ca^{+2} channels) (43). These substrates may also act, either in parallel or by converging onto the MAP-kinase pathway. Such routes may be secondarily important for estrogen-induced neuronal differentiation by recruiting and prolonging the initial rapid response through the variety of pathways with which these substrates interact.

Implications

One might well ask, What have developmental mechanisms to do with the postmenopausal state? In the first place, interactions of estrogen and the neurotrophins, which may be important only during neural development, may be recruited in the adult brain, following loss of trophic support, whether through brain injury, estrogen deprivation, or neurodegenerative disorders. Our earlier studies (1,3) have already provided a physiological rationale for the use of estrogen as a prophylactic treatment to prevent or retard the cognitive disorders associated with steroid deprivation and Alzheimer's disease. Characterizing estrogen signaling pathways in the brain could lead to the synthesis of estrogenlike drugs to duplicate estrogen's positive attributes for the brain, while eliminating many of its undesirable peripheral effects. Moreover, because many of the actions of testosterone in both the male and female brain depend upon its initial intraneuronal conversion through aromatization, by selectively activating only the neural pathways utilized by estrogen, such drugs, unlike estrogen proper, could benefit both sexes. By stimulating the synthesis of proteins required for neuronal differentiation, survival, and maintenance of function, estrogen and the neurotrophins, acting in concert as well as reciprocally, may have important and intertwined developmental roles and may also decrease the vulnerability of target neurons to the consequences of neurodegenerative disease processes (e.g., Alzheimer's disease), as well as increase the compensatory responses to degeneration of interacting neural systems.

Acknowledgments. Many thanks are due to Drs. Meharvan Singh and György Sétáló, Jr., for their contributions to the work described in this chapter; to Mr. Matthew Warren for his expert technical assistance; and to Dr. Lloyd A. Greene (Columbia University) for helpful discussions. The research was supported in part by grants from NIH, NIMH, NSF, the Alzheimer's Association/Burks B. Lapham grant and the Bader Foundation, and an ADAMHA Research Scientist Award to DT-A.

References

1. Toran-Allerand CD, Miranda RC, Bentham W, Sohrabji F, Brown TJ, Hochberg RB, et al. Estrogen receptors co-localize with low-affinity nerve growth factor receptors in cholinergic neurons of the basal forebrain. Proc Natl Acad Sci USA 1992;89:4668–72.
2. Miranda RC, Sohrabji F, Toran-Allerand CD. Estrogen target neurons co-localize the mRNAs for the neurotrophins and their receptors during development: a basis for the interactions of estrogen and the neurotrophins, Mol Cell Neurosci 1993;4:510–25.
3. Toran-Allerand CD. The estrogen/neurotrophin connection during neural development: is co-localization of estrogen receptors with the neurotrophins and their receptors biologically relevant? Dev Neurosci 1996;18:36–48.
4. Matsumoto A, Arai Y. Neuronal plasticity in the deafferented hypothalamic arcuate nucleus of adult female rats and its enhancement by treatment with estrogen. J Comp Neurol 1981:197–206.
5. Kuiper GGJM, Enmark E, Pelto-Huikko M, Nilsson S, Gustafsson J-A. Cloning of

a novel estrogen receptor expressed in rat prostate and ovary. Proc Natl Acad Sci USA 1996;93:5925–30.

6. Pettersson K, Grandien K, Kuiper GG, Gustafsson JA. Mouse estrogen receptor beta forms estrogen response element-binding heterodimers with estrogen receptor alpha. Mol Endocrinol 1997;11:1486–96.

7. Landers JP, Spelsberg TC. New concepts in steroid hormone action: transcription factors, proto-oncogenes and the cascade model for steroid regulation of gene expression. Crit Rev Eukaryotic Gene Expression 1992;2:19–63.

8. Sukovich DA, Mukherjee R, Benfield PA. A novel, cell-type-specific mechanism for estrogen receptor-mediated gene activation in the absence of an estrogen-responsive element. Mol Cell Biol 1994;14:7134–43.

9. Garcia-Segura LM, Olmos G, Tranque P, Naftolin F. Rapid effects of gonadal steroids upon hypothalamic neuronal membrane ultrastructure. J Steroid Biochem 1987;27:615–23.

10. Migliaccio A, Pagano M, Auricchio F. Immediate and transient stimulation of protein tyrosine phosphorylation by estradiol in MCF-7 cells. Oncogene 1993;8: 2183–219.

11. Watson CS, Pappas TC, Gametchy B. The other estrogen receptor in the plasma membrane: implications for the actions of environmental estrogens. Env Health Perspect 1995;103(suppl 7):41–50.

12. Pietras RJ, Szego CM. Specific binding sites for oestrogen at the outer surfaces of isolated endometrial cells. Nature 1977;265:69–72.

13. Karthikeyan N, Thampan RV. Plasma membrane is the primary site of localization of the nonactivated estrogen receptor in the goat uterus: hormone binding causes receptor internalization. Arch Biochem Biophys 1996;325:47–57.

14. Kato S, Endoh H, Masuhiro Y, Kitamoto T, Uchiyama S, Sasaki H, et al. Activation of the estrogen receptor through phosphorylation by mitogen-activated protein kinase. Science 1995;270:1491–94.

15. Read L, Greene G, Katzenellenbogen B. Regulation of estrogen receptor messenger ribonucleic acid and protein levels in human breast cancer cell lines by sex steroid hormones, their antagonists, and growth factors. Mol Endocrinol 1989;3: 295–304.

16. Ignar-Trowbridge DM, Nelson KG, Biwell MC, Curtis SW, Washburn TF, McLachlan JA, et al. Coupling of dual signaling pathways: epidermal growth factor action involves the estrogen receptor. Proc Natl Acad Sci USA 1992;89: 4658–62.

17. Bunone G, Briand P-A, Miksicek RJ, Picard D. Activation of the unliganded estrogen receptor by EGF involves the MAP kinase pathway and direct phosphorylation. EMBO J 1996;15:2174–83.

18. Patrone C, Ma ZQ, Pollio G, Agrati P, Parker MG, Maggi A. Cross-coupling between insulin and estrogen receptor in human neuroblastoma cells. Mol Endocrinol 1996;10:499–507.

19. Ignar-Trowbridge DM, Pimentel M, Teng CT, Korach KS, McLachlan JA. Cross talk between peptide growth factor and estrogen receptor signaling systems. Env Health Perspect 1995;103(suppl 7):35–38.

20. Migliaccio A, Di Domenico M, Castoria G, de Falco A, Bontempo P, Nola E, et al. Tyrosine kinase/p21^{ras}/MAP kinase pathway activation by estrogen-receptor complex in MCF-7 cells. EMBO J 1996;15:1292–300.

21. Barde Y-A. Trophic factors and neuronal survival. Neuron 1989;2:1525–34.

22. Chao MV. Neurotrophin receptors: a window into neuronal differentiation. Neuron 1992;9:583–93.
23. Marshall CJ. Specificity of receptor tyrosine kinase signaling: transient versus sustained extracellular signal-regulated kinase activation. Cell 1995;80:179–85.
24. Blenis J. Signal transduction via the MAP kinases: Proceed at your own RSK. Proc Natl Acad Sci USA 1993;90:5889–92.
25. Rabin SJ, Clehon V, Kaplan DR. SNT, a differentiation-specific target of neurotrophic factor-induced tyrosine kinase activity in neurons and PC12 cells. Mol Cell Biol 1993;13:2203–13.
26. Sohrabji F, Miranda RC, Toran-Allerand CD. Ovarian hormones differentially regulate estrogen and nerve growth factor mRNAs in adult sensory neurons. J Neurosci 1994;14:459–71.
27. Sohrabji F, Greene LA, Miranda RC, Toran-Allerand CD. Reciprocal regulation of estrogen and nerve growth factor receptors by their ligands in PC12 cells. J Neurobiol 1994;22:974–88.
28. Qui MS, Green SH. PC12 cell neuronal differentiation is associated with prolonged p21ras activity and consequent prolonged ERK activity. Neuron 1992;9:705–17.
29. Singh M, Sétáló G Jr, Warren M, Toran-Allerand CD. Estrogen-induced activation of MAP kinase (ERK) in cerebral cortical explants: cross-coupling of estrogen and neurotrophin signaling pathways. J Neurosci 1999;19:1179–88.
30. Shughrue PJ, Lane MV, Merchenthaler I. Comparative distribution of estrogen receptor-alpha and -beta mRNA in the rat central nervous system. Comp Neurol 1997;388:507–25.
31. Reddy KB, Mangold GL, Tandon AK, Yoneda T, Mundy GR, Zilberstein A, et al. Inhibition of breast cancer cell growth in vitro by a tyrosine kinase inhibitor. Cancer Res 1992;52:3636–41.
32. Jaiswal RK, Moodie SA, Wolfman A, Landreth GE. The mitogen activated protein kinase cascade is activated by B-Raf in response to nerve growth factor through interaction with p21[Ras]. Mol Cell Biol 1994;14:6944–53.
33. Couet J, Li S, Okamoto T, Ikezu T, Lisanti MP. Identification of peptide and protein ligands for the caveolin-scaffolding domain. Implications for the interaction of caveolin with caveolae-associated proteins. J Biol Chem 1997;272:6525–33.
34. Liu P, Ying Y, Anderson RG. Platelet-derived growth factor activates mitogen-activated protein kinase in isolated caveolae. Proc Natl Acad Sci USA 1997;94:13666–70.
35. Costello B, Meymandi A, Freeman J. Factors influencing GAP-43 gene expression in PC12 pheochromocytoma cells. J Neurosci 1990;10:1398–406.
36. Lustig R, Sudol M, Pfaff D, Federoff H. Estrogenic regulation and sex dimorphism of growth-associated protein 43 kDa (GAP-43) messenger RNA in the rat. Brain Res Mol Brain Res 1991;11:125–32.
37. Drubin D, Feinstein S, Shooter E, Kirschner M. Nerve growth factor-induced neurite outgrowth in PC12 cells involves the coordinate induction of microtubule assembly and assembly-promoting factors. J Cell Biol 1985;101:1799–807.
38. Guo JZ, Gorski J. Estrogen effects on histone messenger ribonucleic acid levels in the rat uterus. Mol Endocrinol 1988;2:693–700.
39. Black M, Aletta J, Greene L. Regulation of microtubule composition and stability during nerve growth factor-promoted neurite outgrowth. J Cell Biol 1986;103:545–57.

40. Ferreira A, Caceres A. Estrogen-enhanced neurite growth: evidence for a selective induction of tau and stable microtubules. J Neurosci 1991;11:293–400.
41. Fischer I, Richter-Landsberg C, Safaei R. Regulation of microtubule associated protein 2 (MAP2) expression by nerve growth factor in PC12 cells. Exp Cell Res 1991;194:195–201.
42. Lorenzo A, Diaz H, Carrer H, Caceres A. Amygdala neurons in vitro: neurite growth and effects of estradiol. J Neurosci Res 1992;33:418–35.
43. Aronica SM, Kraus WL, Katzenellenbogen BS. Estrogen action via the cAMP signaling pathway: stimulation of adenylate cyclase and cAMP-regulated gene transcription. Proc Natl Acad Sci USA 1994;91:8517–21.

10

Neuroprotective Effects of Estrogens

JAMES W. SIMPKINS, PATTIE S. GREEN, KELLY E. GRIDLEY, JIONG SHI, AND EILEEN K. MONCK

Postmenopausal estrogen replacement therapy (ERT) is associated with numerous overall health benefits, including reduced risk of osteoporosis and a decrease in mortality (1). Of particular interest to neurodegenerative disease, ERT correlates with a decreased incidence of Alzheimer's disease (AD) (2,3), reducing the onset of the disease by as much as 10 years in one study (3). Further, several small clinical studies support a role for estrogen therapy in improving cognitive function in AD patients (4–6). Epidemiological studies have also demonstrated a beneficial effect of ERT in reducing the mortality and morbidity associated with myocardial infarction and stroke in postmenopausal women (7–10).

Estrogen, specifically the naturally occurring 17β-estradiol (βE2), is a potent neuroprotective agent in multiple experimental models of neurotoxicity and neurodegeneration. In this chapter, we will first describe neuroprotective effects of estrogens in both cell culture and animal models, and then discuss possible cellular mechanisms of this protection.

Estrogen and Neuroprotection

In Vitro Models

Our laboratory first demonstrated that physiological doses of the potent estrogen, βE2, could exert direct cytoprotective effects on a neuronal cell line using a human neuroblastoma cell line, SK-N-SH, under the conditions of serum-deprivation (11). Treatment with βE2 did not increase ³H-thymidine uptake in these cells (11), verifying that this is a cytoprotective rather than a mitogenic effect of βE2. In addition, βE2 attenuates oxidative stress-induced toxicity, such as exposure of neurons to the Alzheimer plaque associated β-amyloid peptide (Aβ) (12–14) or exposure to H_2O_2 (12). βE2 treatment also attenuates cell death in rat primary hippocampal and cortical neurons due to excitotoxic insults, such as glutamate exposure (13,15,16) and anoxia/ reoxygenation (Zaulynov et al., unpublished observations).

Several lines of evidence suggest that these neuroprotective effects of βE2 are not mediated by a classical estrogen-receptor (ER) mediated mechanism. First, the structure–activity relationship for the neuroprotective effects of estrogens differs markedly from the structure–activity relationship for binding to the ER. We have demonstrated the 17α-estradiol (αE2) has similar neuroprotective efficacy and potency to the potent estrogen, βE2 (17), although αE2 binds only weakly to the ER and the αE2–ER complex binds only transiently to the estrogen responsive element (18,19). Both we (20) and Behl et al. (15) have demonstrated that estrogens with a hydroxyl group in the C3 position of the A ring (estratrienes) are neuroprotective (Fig. 10.1). If the phenolic nature of the A ring is removed by a 3-O-conjugation, the neuroprotective effects are abolished. The necessity of the phenolic A ring is demonstrated by the diphenolic estrogen mimic, diethylstilbesterol (DES). DES is neuroprotective, and retention of a single hydroxyl function on an aromatic ring is sufficient to retain neuroprotective activity (20). The di-O-methyl ether of DES does not demonstrate protective activity. Furthermore, steroids that lack a phenolic A ring, such as testosterone, progesterone, or cholesterol, do not demonstrate protective effects (20). This suggests that it is the possession of a phenolic A ring rather than binding to the ER that confers neuroprotective potential to estratrienes.

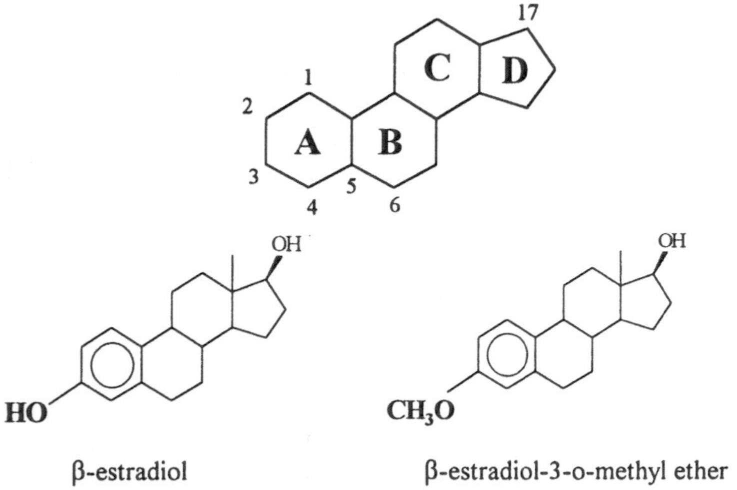

β-estradiol   β-estradiol-3-o-methyl ether

FIGURE 10.1. A structural representation of the cyclopentaphenanthrene ring (top), the neuroprotective estrogen, 17β-estradiol, and the nonprotective 17β-estradiol-3-methyl ether. The bold letters indicate ring designation, and the numbers are the carbon positions in the molecule. A hydroxyl group at the 3 position and an aromatic A ring are necessary for neuroprotective activity.

Second, ER antagonists fail to block the neuroprotective effects in SK-N-SH cells (17,21). Exposure to 2 nM 17β-estradiol (βE2) during 48 hours of serum deprivation increases live cell number by an average of two- to three-fold over vehicle controls and concurrent treatment with a 100-fold excess of tamoxifen, which is a mixed ER agonist/antagonist, does not significantly alter the degree of protection conferred by βE2 (17). Tamoxifen alone has no effect on cell viability in this assay system. Further, ICI 187,780, which is a pure ER antagonist that contains a phenolic A ring, is itself protective against Aβ toxicity (21). These results indicate that antagonism of the ER does not antagonize the protection conferred by estratrienes.

Finally, estrogens have been shown to protect a neuronal cell line that lacks an estrogen receptor. The HT-22 cell, which is a mouse hippocampal cell line, does not demonstrate specific ^3H-βE2 binding in crude nuclear extracts or whole cell preparations (22). Further, when HT-22 cells are transfected with an ERE- reporter plasmid construct, no increase in reporter plasmid expression was seen with estrogen exposure (12). Estrogens have been shown to protect these neuronal cells from the toxic effects of Aβ, glutamate, buthionine sulfoximine, and H_2O_2 (12,15,22). Further, we have shown that this protection can be achieved with physiologically relevant doses of estrogens (2 nM) (22).

Animal Models

In animal models, we and others have demonstrated that estrogens protect against events associated with ischemia (23–27). We have reported that the treatment with various forms of estrogens at a variety of time points exerts neuroprotective effects against the damages associated with middle cerebral artery (MCA) occlusion-induced focal ischemia in female rats. Twenty-four-hour pretreatment of ovariectomized rats with βE2-CDS, the brain-targeted chemical delivery system, or βE2 reduced post-MCA occlusion mortality by more than 50% compared with vehicle treated ovariectomized rats (23). In a separate study, ovariectomy similarly decreased 24-hour postocclusion survival from 87.5% in intact female rats to 76.5% (24). Pretreatment with the presumed inactive estrogen, αE2, consistently reduced mortality from 36 to 0%. The reduction in the ischemic lesion size may underlie this remarkable reduction in mortality. Twenty-four-hour pretreatment with βE2-CDS or βE2 in ovariectomized rats caused a reduction in ischemic lesion sizes from 25.6 ± 5.7% in ovariectomized rats to 9.1 ± 4.2% and 9.8 ± 4%, respectively.

Pretreatment with aE2 similarly reduced the ischemic lesion sizes by 55–81%. Of greater importance, treatment of ovariectomized rats with bE2 after onset of either temporary or permanent occlusion continued to reduce ischemic lesion sizes by about 50% (23; Shi et al., unpublished observations). βE2 treatment also reduced ischemic lesion size from 17 to 8% in intact male rats (25).

Potential Mechanisms of Action

Antioxidant Effects

Phenolic A ring estrogens may exert their neuroprotective actions through an antioxidant mechanism because lipophilic phenols are well known to be antioxidants (28). Mukai et al. have demonstrated that estrogens, specifically phenolic A ring estrogens, are potent antioxidants (29). 17β-estradiol also not only protects neurons from oxidative insults, such as hydrogen peroxide (12) and β-amyloid peptide (12–14) toxicity, but prevents the increase in lipid peroxidation that accompanies these toxicities (13,30). This is significant as increased lipid peroxidation is associated with a variety of neurodegenerative diseases, including ischemic/anoxic insults (31–32) and Alzheimer's disease (33–34).

Synergism with Glutathione

In attempts to reconcile our previous work, in which low concentrations of estradiol (2 nM) were protective against βAP 25-35-induced toxicity (14), with others where protection was obtained using estrogens at much higher concentrations (12,13), we identified that our work was done with reduced glutathione (GSH) present in the cell culture milieu, whereas others used culture media where GSH was absent. Further experimentation showed that when GSH was absent in the extracellular milieu, the EC_{50} for βE2 protection against Ab toxicity was 126 ± 89 nM for SK-N-SH cells (21) and 3.2 ± 408 μM for HT-22 cells (22). The presence of GSH (3.25μM) shifted the EC_{50} for βE2 neuroprotection to 0.03 ± 0.031 nM in SK-N-SH cells (21) and 5 ± 2 nM in HT-22 cells (22). We also evaluated the effect of GSH on the protection conferred by aE2 and estratriene-3-ol, which are two classically weak estrogens that we have formerly demonstrated to be as potent as βE2 in neuroprotection assays. GSH increased the neuroprotective potency of these estrogens by approximately 400-fold (22).

The low concentrations of GSH used in the aforementioned studies did not have any effect on cell viability in the absence of estrogen; however, GSH also has neuroprotective properties and protected SK-N-SH cells from Aβ toxicity with an ED_{50} of 82.6 ± 60 μM in the absence of estrogens. A physiologically relevant dose of βE2 (2 nM) significantly potentiated the neuroprotective potency of GSH to an ED_{50} 0.04 ± 0.02 μM (21). This implies a synergistic interaction as the neuroprotective potency of both molecules are markedly shifted by the presence of the other. We performed similar experiments in rat primary cortical neurons and obtained similar results (21). Although differences exist in the magnitude of the GSH-induced shift in the neuroprotective potency of E2 in these cell types, intracellular concentrations of GSH may play a role because we have determined that primary rat cortical neurons have higher intracellular GSH concentrations (172 ± 12 μM) than do SK-N-SH cells (15 ± 2) (21). In addition, this effect appears to be independent of the type of cytotoxic insult used because results obtained using serum deprivation and zinc toxicity (unpublished observations) were comparable to the data obtained from βAP 25-35 (21,22) and βAP 1-40–induced (22) toxicities.

The specificity of estrogens and glutathione for synergism in neuroprotection is supported by several lines of evidence. First, there are no apparent interactions noted between estrogen and the other thiols tested, lipoic acid or taurine, or any other antioxidants tested, ascorbic acid or α-tocopherol (21). Second, oxidized glutathione (GSSH) works in this model (21), and lends credence to the idea that the glutathione-estrogen interaction may involve the glutathione peroxidase/reductase system.

Estrogens are lipophilic and likely to partition to membrane constituents. They should associate their phenolic A rings with the charged hydrophilic head groups of the membrane phospholipids (Fig. 10.2). As a result, estratrienes are well placed to attenuate lipid peroxidation, and we predict the hydroxyl hydrogen of estradiol is donated to prevent the peroxidative cascade. This is further substantiated by the aforementioned structure-activity relationship with neuroprotection (15,17,20). Further, high potency of estratrienes may result from their ability to donate hydrogen ions from several positions on the A ring (35). An oxidized form of estrogen could result from this hydrogen ion donation that would be relatively stable, and glu-

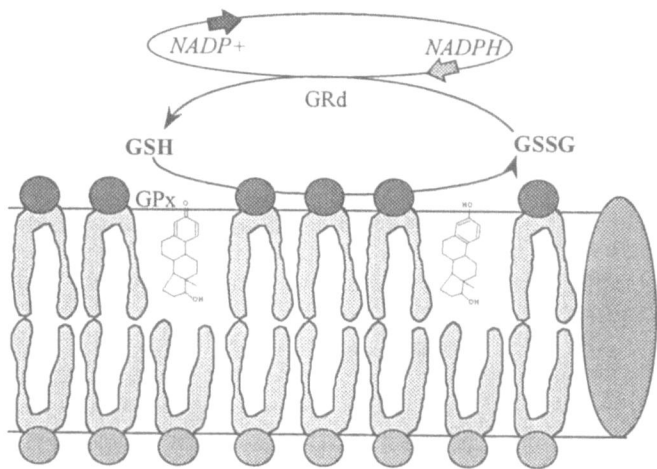

FIGURE 10.2. A schematic representation of the proposed relationship between glutathione and estradiol. The reduced and oxidized forms of estradiol and the reduced and oxidized forms of glutathione (GSH and GSSG, respectively) are depicted. Oxidative stress drives estradiol to the oxidized form, thereby protecting membrane lipids. The proposed role of glutathione is to reduce estradiol. GSSH is then reduced by glutathione reductase (GRd) using NADPH as a substrate.

tathione peroxidase would likely regenerate the reduced form of estrogen by using GSH as a substrate, thus explaining the synergy between the two molecules.

Effects on CREB

Estrogen's neuroprotective effects may also be mediated via interaction with cyclicAMP (cAMP) response element-binding protein (CREB), a constitutively expressed transcription factor that is activated by phosphorylation. Two independent studies have suggested that CREB may play an important role in neuroprotection. Walton et al. (36), using a hypoxic–ischemic injury model, demonstrated a decline in phosphorylated CREB (PO_4–CREB) immunoreactivity in CA1 pyrimidal cells, which do not survive mild hypoxia–ischemia injury. In contrast, the more resistant dentate granule cells and cortical cells showed an increase in PO_4–CREB immunoreactivity. We have previously reported that hypoglycemic seizure results in a selective reduction of CREB immunoreactivity (37). This decline in CREB immunoreactivity does not appear to be due to cell loss because the rats were sacrificed 90 minutes postseizure. It is interesting to note, however, that the regions of CREB decline correlated with regions that have previously been shown to have massive cell loss at 1 week after hypoglycemic seizure (38).

Estrogen replacement has been shown to reduce the seizure-induced decline in CREB immunoreactivity (38). We have examined the effect of serum deprivation on both CREB and PO_4–CREB levels in SK-N-SH cells (Green et al., unpublished observations). These cells showed a decline in both PO_4-immunoreactivity and CRE binding capacity at 4–6 hours of serum deprivation that was normalized by 12 hours. No loss of cell viability as determined by propidium iodide exclusion was detected at these time points. It is interesting that 10 nM βE2 prevented this decline in PO_4–CREB immunoreactivity and CRE binding capacity.

Zhou et al. (39) demonstrated that estrogen treatment of ovariectomized rats increases CREB phosphorylation within 15 minutes of injection in the preoptic area (POA) and bed nucleus of the stria terminalis. This rapid response suggests a non–ER-mediated effect of estrogen. We have demonstrated similarly that βE2 rapidly increases phosphorylation of CREB in SK-N-SH cells (unpublished observations). We observed a 50% increase in PO_4–CREB immunoreactivity within 1 hour using a physiological dose of βE2 (2 nM) (Fig. 10.3). We did not see a change in CREB immunoreactivity at this time point with βE2 doses of up to 10 μM. This rapid βE2-induced phosphorylation of CREB may be due to increased cAMP production because βE2 has been shown to activate adenylate cyclase activity in MCF7 breast tumor cells (40).

These data collectively suggest the possibility that estrogen regulates the phosphorylation state of CREB by interactions with signal transduction pathways other than the traditional estrogen receptor.

βE2 Dose (M)   0 10^{-9} 10^{-8} 10^{-7} 10^{-5}

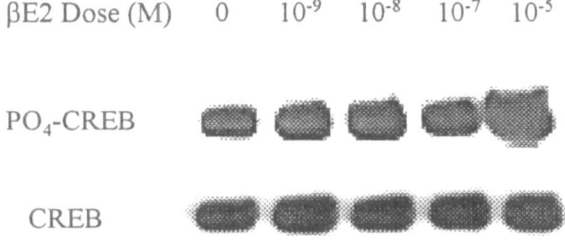

PO$_4$-CREB

CREB

FIGURE 10.3. Western blot analysis of CREB phosphorylation by 17β-estradiol. SK-N-SH cells were treated with the indicated dose of steroid for 1 hour. Nuclear protein was extracted and analyzed by Western blot using an antibody specific to PO$_4$–CREB (upper blot) and an antibody for CREB. 17β-estradiol increases immunoreactivity of PO$_4$–CREB while having no effect on overall CREB immunoreactivity.

Conclusion

Neuronal death due to normal aging, neurodegenerative diseases, and acute conditions (e.g., ischemia) is not currently treatable. There is mounting evidence that estrogens are potent cytoprotective compounds against a variety of neurotoxic insults. These actions of estrogens do not appear to require an ER and may involve antioxidant properties of the estratriene molecule as well as interaction with signal transduction pathways. Further, these neuroprotective actions may contribute to the beneficial effects of estrogens seen in both clinical trials and epidemiological studies of AD and/or stroke. As such, estrogen use should be further evaluated for treatment of various neurodegenerative conditions.

References

1. Paganini-Hill A. The risks and benefits of estrogen replacement therapy: leisure world. Int J Fertil 1995;40S:54–62.
2. Paganini-Hill A, Henderson VW. Estrogen deficiency and risk of Alzheimer's disease in women. Am J Epidemiol 1994;140:256–61.
3. Tang M, Jacobs D, Stern Y, Marder K, Schofield P, Gurland B, et al. Effect of oestrogen during menopause on risk and age at onset of Alzheimer's disease. Lancet 1996;348:429–32.
4. Fillit H, Weinreb H, Cholst I, McEwen B, Amador R, Zabriskie J. Observations in a preliminary open trial of estradiol therapy for senile dementia-Alzheimer's type. Psychoneuroendocrinology 1986;11:337–45.
5. Honjo H, Ogino Y, Naitoh K, Urabe M, Yasude J. In vivo effects by estrone sulfate on the central nervous system-senile dementia (Alzheimer's type). J Steroid Biochem 1989;34:521–25.
6. Ohkura T, Isse K, Akazawa K, Hamamoto M, Yaoi Y, Hagino N. Low-dose estrogen replacement therapy for Alzheimer disease in women. Menopause: J N Am Menopause Soc 1994;3:125–30.

7. Lafferty FW, Fiske, ME. Postmenopausal estrogen replacement: a long-term cohort study. Am J Med 1994;97:66–77.

8. Lobo RA. Cardiovascular implications of estrogen replacement therapy. Obstet Gynecol 1990;75:18s–25s.

9. Grodstein F, Stampfer MJ, Manson JE, Colditz GA, Willett WC, Rosner B, et al. Postmenopausal estrogen and progestin use and the risk of cardiovascular disease. N Engl J Med 1996;335:453–61.

10. Gaspard UJ. Evaluation of the cardiovascular impact of hormonal replacement therapy in menopausal women. J Gynecol Obstet Biol Reprod (Paris) 1996;25: 671–76.

11. Bishop J, Simpkins JW. Estradiol treatment increases viability of glioma and neuroblastoma cells in vitro. Mol Cell Neurosci 1994;5:303–8.

12. Behl C, Widmann M, Trapp T, Holspoer F. 17β-estradiol protects neurons from oxidative stress-induced cell death in vitro. Biochem Biophys Res Comm 1995;216: 473–82.

13. Goodman Y, Bruce AJ, Cheng B, Mattson MP. Estrogen attenuates and corticosterone exacerbates excitotoxicity, oxidative injury, and amyloid β-peptide toxicity in hippocampal neurons. J Neurochem 1996;66:1836–44.

14. Green PS, Gridley KE, Simpkins JW. Estradiol protects against β-amyloid (25-35)-induced toxicity in SK-N-SH cells. Neurosci Lett 1996;218:165–68.

15. Behl C, Skutella T, Lezoualch F, Post A, Widmann M, Newton CJ, et al. Neuroprotection against oxidative stress by estrogens: structure-activity relationship. Mol Pharmacol 1997;51:535–41.

16. Singer CA, Rogers KL, Strickland TM, Dorsa DM. Estrogen protects primary cortical neurons from glutamate toxicity. Neurosci Lett 1996;212:13–16.

17. Green PS, Bishop J, Simpkins JW. 17α-estradiol exerts neuroprotective effects in SK-N-SH cells. J Neurosci 1997;17:511–15.

18. Huggins C, Jensen EV, Cleveland AS. Chemical structure of steroids in relation to promotion of growth of the vagina and uterus of the hypophysectomized rat. J Exp Med 1954;10:225–43.

19. Korenman SG. Comparative binding affinities of estrogen and its relation to estrogenic potency. Steroids 1969;13:163–77.

20. Green PS, Gordon K, Simpkins JW. Phenolic A ring requirement for the neuroprotective effects of steroids. J Steroid Biochem Mol Biol 1997;63:229–35.

21. Gridley KE, Green PS, Simpkins JW. A novel, synergistic interaction between 17β-estradiol and glutathione in the protection of neurons against β-amyloid 25-35-induced toxicity in vitro. Mol Pharmacol 1998;54:874–80.

22. Green PS, Gridley KE, Simpkins JW. Nuclear estrogen receptor-independent neuroprotection by estratrienes: a novel interaction with glutathione. Neuroscience 1998;84:7–10.

23. Simpkins JW, Rajakumar G, Zhang YQ, Simpkins CE, Greenwald D, Yu CJ, et al. Estrogens reduce mortality and ischemic damage induced by middle cerebral artery occlusion in the female rat. J Neurosurg 1997;87:724–30.

24. Zhang Y-Q, Shi J, Rajakumar G, Day AL, Simpkins JW. Effects of gender and estradiol treatment on focal brain ischemia. Brain Res 1997;798:321–24.

25. Hawk T, Zhang Y-Q, Rajakumar G, Day AL, Simpkins JW. Testosterone increases and estradiol decreases middle cerebral artery occlusion lesion size in male rats. Brain Res 1998;796:2966–68.

26. Dubal DB, Kashon ML, Pettigrew LC, Ren JM, Finklestein SP, Rau SW, et al. Estradiol protects against ischemic injury. J Cereb Blood Flow Metab 1998;18:1253–58.
27. Toung TJK, Traystman RJ, Hurn PD. Estrogen-mediated neuroprotection after experimental stroke in males. Stroke 1998;29:1666–70.
28. Tappel AL. Lipid peroxidation damage to cell components. Fed Proc 1973;32: 1870–74.
29. Mukai K, Daifuku K, Yokoyama S, Nakano M. Stopped flow investigation of antioxidant activity of estrogens in solution. Biochem Biophys Acta 1990;1035: 348–52.
30. Gridley KE, Green PS, Simpkins JW. Low concentrations of estradiol reduce β-amyloid (25-35) induced toxicity, lipid peroxidation and glucose utilization in human SK-N-SH neuroblastoma cells. Brain Res 1997;778:158–65.
31. Flamm ES, Demopaulos HB, Seligman ML, Poser RG, Ransohoff J. Free radicals in cerebral ischemia. Stroke 1978;78:445–47.
32. Braugher JM, Hall ED. Central nervous system trauma and stroke. Biochemical considerations, oxygen radical formation and lipid peroxidation. Free Rad Biol Med 1989;6:389–401.
33. Smith CD, Carney JM, Starke-Reed PE, Oliver CN, Stadtman ER, Floyd RA, et al. Excess brain protein oxidation and enzyme dysfunction in normal aging and in Alzheimer's disease. Proc Nat Acad Sci USA 1991;88:10540–43.
34. Hensley K, Carney JM, Mattson MP, Aksenova M, Harris M, Wu JF, et al. A model for β-amyloid aggregation and neurotoxicity based on free radical generation by the peptide: relevance to Alzheimer's Disease. Proc Nat Acad Sci USA 1994;91: 3270–74.
35. Jellnick PH, Bradlow HL. Peroxidase-catalyzed displacement of tritium from regiospecifically labeled estradiol and 2-hydroxyestradiol. J Steroid Biochem 1990;35:705–10.
36. Walton M, Sirimanne E, Williams C, Gluckman P, Dragunow M. The role of the cyclic AMP-responsive element binding protein (CREB) in hypoxic-ischemic brain damage and repair. Molec Brain Res 1996;43:21–29.
37. Panickar KS, Purushotham K, King MA, Rajakumar G, Simpkins JW. Hypoglycemia-induced seizures reduce cyclic AMP response element binding protein levels in the rat hippocampus. Neuroscience 1998;83:1155–60.
38. Auer RN, Kalimo H, Olsson Y, Siesjo BK. The temporal evolution of hypoglycemic brain damage. Acta Neuropathol 1985;67:25–36.
39. Zhou Y, Watters JJ, Dorsa DM. Estrogen rapidly induces the phosphorylation of the cAMP response element binding protein in rat brain. Endocrinology 1996;137: 2163–66.
40. Aronica SM, Kraus WL, Katzenellenbogen BS. Estrogen action via the cAMP signaling pathway: stimulation of adenylate cyclase and cAMP-regulated gene transcription. Proc Natl Acad Sci USA 1994;91:8517–21.

11

Cognitive Changes and Aging: Is Estrogen a Factor?

BARBARA B. SHERWIN

Menopause is a universal reproductive event that occurs at a mean age of 50.8 years in industrialized countries and is diagnosed, retrospectively, by the absence of menstrual cycles for the preceding 12 months. Subsequent to the cessation of ovarian function at menopause, women are almost totally deprived of estrogen. During the 1990s, intensive research activity described the changes that occur in different organ systems as a function of the drastic decrease in estrogen production in the postmenopause and their possible prevention by exogenous estrogen administration. For example, it has been well demonstrated that estrogen replacement therapy prevents the decrease in bone density (1) and the atherogenic changes in lipoprotein lipid metabolism (2) that would otherwise occur in a considerable portion of women following the menopause. Since 1985, investigators have been asking whether there is any consequence for brain function when ovarian estrogen production ceases postmenopausally. Numerous mechanisms of estrogen that influence both the structure and function of brain areas known to be important for cognition (and reviewed in detail elsewhere in this book) provide biological plausibility for an estrogen–cognition relationship. Before discussing the relevant clinical studies that bear on this relationship, it is important first to address changes in cognitive functioning that occur with normal aging irrespective of the hormonal milieu.

Although there is some support for the idea that memory performance declines with increasing age (3,4), it appears that age-related deficits develop only in specific skills. Short-term memory, which deals with material that has just been presented and is still in the individual's conscious awareness, does not appear to be affected substantially by aging. For example, the Digit Span Test requires the individual to repeat a list of random digits immediately after presentation and shows only slight declines in performance with increasing age (5). In tasks that require working memory in which the participant must hold information in mind and carry out some decision-making activity based on that information, however, there are decreases in both speed and accuracy with normal aging (6,7).

In long-term memory (LTM) tasks, the participant must remember material that has left conscious awareness from seconds ago to years ago, and performance depends on both the efficiency of encoding (acquisition) and retrieval. Although all LTM functions decline with increasing age, the degree of impairment appears to depend on the availability of environmental cues. For example, differences in performance based on age are large in free recall, somewhat less in cued recall, and less again in recognition (8). Encoding and retrieval of new information, therefore, are most profoundly impaired in the aged when the new material is independent of any environmental context. It should be noted that none of these studies on memory and normal aging examined possible sex differences in memory impairments, and none considered the participants' hormonal milieu.

Observational Studies

Numerous observational studies have been carried out to assess whether postmenopausal estrogen users performed better on batteries of neuropsychological tests as compared with nonusers. In a cross-sectional study, estrogen users whose mean age was approximately 67 years had higher scores on a test of proper name recall than age-matched nonusers (9). Kampen and Sherwin (10) reported higher scores on a test of short- and long-term verbal memory in 65-year-old estrogen users compared with nonusers matched for age. These findings were subsequently confirmed by Kimura (11), although specific test results were not reported in that study.

Information from much larger longitudinal studies of estrogen and memory in healthy postmenopausal women has become available. Barrett-Connor and Kritz-Silverstein (12) administered a comprehensive battery of neuropsychological tests to 800 women between the ages of 65 and 95 who were the Rancho Bernardo cohort assembled in 1972–1974 to study heart disease risk factors. Almost half of this upper middle class cohort had used estrogen some time after the menopause, and one third were current users. Women who had used estrogen for at least 20 years had higher scores on the category fluency test, whereas those who were past users had significantly higher scores on the Mini-Mental State Examination, which assesses degree of dementia and in which a higher score indicates better cognitive functioning. No differences occurred, however, between the performances of past, current, or never estrogen users on other tests of verbal memory or on tests of visual memory.

Data on hormonal status and memory were examined in 288 postmenopausal women in the Baltimore Longitudinal Study of Aging (13). The mean score on the Benton Visual Retention Test (BVRT), which is a measure of short-term visual memory and visual perception, was significantly higher in the 116 estrogen users compared with the 172 nonusers. A subgroup of 18 women had been tested previously with the BVRT prior to initiation of estrogen replacement therapy (ERT). They were compared with an age-matched group of never users of estrogen for whom earlier BVRT scores were available. Women who began ERT during the interval between assessments maintained stable performance on the BVRT, whereas those women who never used ERT showed the predictable age-related increases in memory errors that typically occur with increasing age. This suggested that ERT may protect against memory decline in healthy postmenopausal women. The BVRT was the only neuropsychological test for which a complete data set was available in this study.

In a community-based epidemiological study of aging and dementia in northern Manhattan, New York, participants were followed longitudinally for an average of 2.5 years (14). Women who were current or past estrogen users had higher scores on the immediate and delayed recall of the Selective Reminding Task, on the similarities subtests of the Wechsler Adult Intelligence Scale, and on the Boston Naming Test compared with women who had never taken estrogen when age, education, and ethnicity were statistically controlled. Moreover, even though test scores had increased slightly over time in the estrogen users, they decreased slightly for never users, who were an average of 73 years of age.

In an epidemiological study, 2,542 nondemented elderly women were classified as estrogen never users, past users, or current users. Both the estrogen current users and past users had significantly higher scores on the Modified Mini-Mental State Examination compared with the never users after controlling for the effects of age, education, and APOE genotype (15).

Although, taken together, the results of these studies suggest that estrogen helps to maintain aspects of cognition in aging women, considerable inconsistency exists among them.

Experimental Studies

Seven prospective randomized studies and one nonrandomized study investigated whether exogenous estrogen affected aspects of cognition in naturally or surgically postmenopausal women (Table 11.1). Six of the eight studies found improvement in aspects of cognitive functioning in women who randomly received estrogen as compared with those given a placebo. Two of the studies, however, failed to include neuropsychological measures that specifically test memory (20,22). Of the six studies that did include tests of memory, one failed to find improvements in estrogen-treated subjects (17); however, this study was nonblind, nonrandomized, and nonplacebo-controlled (17), which makes these findings difficult to assess.

TABLE 11.1. Clinical trials of estrogen and cognition in nondemented women.

Author	Design	Drugs	Findings
Caldwell and Watson, 1952 (16)	Single-blind randomized	E_2^* benzoate 2 mg/wk IM*	⇑ on the verbal subtests of the Wechslee-Bellevue Intelligence scale
Rauramo et al., 1995 (17)	Nonblind, nonrandomized	E_2 2 mg/day orally	No improvement in Integration Memory Tests, Progressive Matrices, reaction time, and speed
Hackman and Galbraith, 1976 (18)	Double-blind randomized	Piperazine estrone sulphate 3 mg/day orally	⇑ Guild Memory scores in E-treated group
Campbell and Whitehead, 1977 (19)	Double-blind, randomized cross-over	CEE/1.25 mg/day orally	⇑ in memory scores on a visual analog scale in E-treated women
Fedor-Freybergh 1977 (20)	Double-blind randomized	E_2-17β valerate 2 mg/day orally	⇑ scores on Stroop Test, choice reaction time, visual search, 2 tests of attention in E-treated women
Sherwin 1988 (21)	Double-blind randomized cross-over	E_2 valerate 10 mg/month IM	⇑Digit span, paragraph recall, clerical speed, and accuracy in E-treated women
Ditkoff et al., 1991 (22)	Double-blind randomized	CEE 0.625 mg/day or 1.25 mg/day orally	No effect of estrogen on digit span or digit symbol
Phillips and Sherwin, 1992 (23)	Double-blind randomized	E_2 valerate 10 mg/month IM	⇑ Paragraph recall (immediate and delayed) in E-treated women

E_2 = estradiol; IM = intramuscular.

Three of these prospective, controlled studies that reported an estrogenic enhancement of memory used neuropsychological tests of memory with established reliability and validity. Hackman and Galbraith (18) administered the Guild Memory Tests, which included measures of immediate and delayed memory. Estrogen-treated postmenopausal women performed significantly better than did those given a placebo. In a prospective study of premenopausal women who underwent hysterectomy and bilateral oophorectomy and then received one of three hormone preparations or placebo, Sherwin (21) found higher scores in women treated with any of the hormonal preparations (estrogen, androgen, estrogen-androgen combination) compared to those who received a placebo on tests of immediate memory, attention, and abstract reasoning. Finally, Phillips and Sherwin (23) found improvements in the immediate paragraph recall subtest of the Wechsler Memory Scale (WMS) in hysterectomized and oophorectomized women who had randomly received estrogen compared with the group given placebo. It was also reported that scores on the immediate and delayed paired-associates subtest of the WMS decreased in the placebo-treated women but were maintained in those given estrogen following surgery. These effects occurred independent of mood or menopausal symptoms.

To summarize, there are a modest number of observational studies and fewer controlled, experimental studies of estrogen and memory currently available. Approximately 60% of the observational studies and 75% of the experimental studies reported a significant effect of estrogen treatment on memory in their postmenopausal subjects. These statistics, however, can be misleading because of the heterogeneity of the subject populations studied, the inconsistency between studies in the dose and route of administration of estrogen, as well as the variety of estrogen preparations used, the failure of some investigators to use validated tests of memory (19), and the failure of others to include tests that measured memory (17,20,22).

Because of these fundamental methodological problems inherent in the available literature, it is not possible to conclude with certainty whether or not estrogen helps to maintain memory in postmenopausal women. It is noteworthy, however, that the studies with the most rigorous methodologies did find that exogenous estrogen maintained and/or improved verbal memory in postmenopausal women.

Methodological Considerations

The potential effects of estrogen on brain function is a new and emerging area of investigation that has considerable clinical implications for the health and well-being of older women. The problems with the current literature suggest several avenues for conceptual and methodological refinements for future work in this area. In order to test whether estrogen affects cognitive functioning in women, it is first important to decide whether it is reasonable to think (1) that estrogen would enhance all cognitive functions (in which case, outcome would be defined as improved performance on every test in a comprehensive battery of neuropsychological tests) or (2), that estrogen enhances only one domain of cognitive function, such as memory (outcome would be measured by improved performance on tests of memory). Evidence from basic neuroscience suggests that the estrogenic effect on cognition may be specific rather than global. The basal forebrain and the hippocampus contain estrogen receptors, and these structures have been associated with learning and memory (24). Humans (25) and monkeys (26) with ischemic lesions that damaged the CA1 layer of the hippocampus show severe memory impairments compared with matched controls, which indicates that the CA1 area of the hippocampus is crucial for memory. Cognitive tasks most vulnerable to hippocampal damage involve explicit memory, or purposefulness accompanied by a feeling of familiarity about the past (27). Explicit memory, which is tested by neuropsychological tests of recall and recognition, shows the most impairment in hippocampal-lesioned animals and humans, and is the most impaired in normal aging (28). The fact that estrogen increases the production of acetylcholine in the frontal cortex and CA1 region of the hippocampus (29), increases the density of spines on the apical dendrites of the CA1 neurons (30), and colocalizes with nerve growth factor in the basal forebrain (31) all suggest that the primary impact of estrogen might be on the medial–temporal hippocampal system that is critically involved in memory.

Second, there is reason to believe that a specific kind of memory may be more affected by estrogen. There are known sex differences in cognitive functioning such that, on average, men excel in spatial and quantitative abilities and in gross motor strength, whereas women excel in verbal abilities, in perceptual speed and accuracy, and in fine motor skills (32). These differences are thought to occur as a result of differential exposure of the fetal brain to sex steroid hormones during prenatal life, and several studies of individuals exposed prenatally to abnormal levels of sex hormones found differences in their cognitive strengths consistent with this hypothesis (33,34). If this is true, then women should perform better on tests of verbal skills during phases of the menstrual cycle when estrogen is highest and would perform best on spatial tasks in the relative absence of estrogen. Indeed, several studies have confirmed these hypotheses (35,36). This would predict for postmenopausal women that verbal memory would be most vulnerable to estrogenic influences. Although, at this stage, it might be prudent to continue to test a more comprehensive range of cognitive functions in women receiving estrogen, the test battery must include reliable and valid tests of verbal and visual memory.

Third, it is well known that depression can cause impairments in memory and concentration (36), and there is a significant incidence of depression in elderly women. It is crucial, therefore, to administer a valid test of mood at every test session during which cognitive function is assessed and to control statistically for mood when analyzing cognitive data.

Another potential confound in studies of estrogen and cognition is the positive relationship between performance on cognitive tests and years of education and socioeconomic status because women who choose to take estrogen are often better educated and healthier than are nonusers (37,38). Higher socioeconomic status and better education are also thought to be associated with a lower risk of Alzheimer's Disease (39). The fact that estrogen use is not controlled in observational studies clearly poses a considerable source of bias. More definitive conclusions with respect to the effect of estrogen on cognition in postmenopausal women await the results of the prospective, randomized, double-blind, clinical trials that are currently under way.

Acknowledgments. The preparation of this chapter was supported by a grant from the Medical Research Council of Canada (No. MA-11623) awarded to B.B. Sherwin.

References

1. Weiss NS, Ure CL, Ballard JH, Williams AR, Daling JR. Decreased risk of fracture of the hip and lower forearm with postmenopausal use of estrogen. N Engl J Med 1980;303:1195–98.
2. Bush TL, Barrett-Connor E, Cowan LD, et al. Cardiovascular mortality and noncontraceptive use of estrogen in women: results from the Lipid Research Clinic Program Follow-up Study. Circulation 1987;75:1102–9.
3. Tulving E. Elements of episodic memory. Oxford: Oxford University Press, 1963.
4. Craik FIM. Age differences in human memory. In: Birren JE, Schaie KW, eds. Handbook of the psychology of aging. New York: Van Nostrand Reinhold, 1977: 384–420.
5. Parkinson SR. Performance deficits in short-term memory tasks. In: Cermal LS, ed. Human memory and amnesia. Hillsdale, NJ: Lawrence Erlbaum Associates, 1982:77–96.
6. Craik FIM, Morris RG, Gick ML. Adult age differences in working memory. In: Vallar G, Shallice T, eds. Neuropsychological impairments of short-term memory. Cambridge: Cambridge University Press, 1989:98–116.
7. Wingfield A, Stine EA, Lahar CJ, Aberdeen JS. Does the capacity of working memory change with age? Exp Aging Res 1988;14:103–7.
8. Craik FIM. Age differences in remembering. In: Squire LR, Butters N, eds. Neuropsychology of memory. New York: Guilford Press, 1984:3–12.
9. Robinson D, Friedman L, Marcus R, Tinkelberg J, Yesavage J. Estrogen replacement therapy and memory in older women. J Am Geriatr Soc 1994;42:919–22.
10. Kampen D, Sherwin BB. Estrogen use and verbal memory in healthy postmenopausal women. Obstet Gynecol 1994;83:979–83.

11. Kimura D. Estrogen replacement therapy may protect against intellectual decline in postmenopausal women. Horm Behav 1995;29:312–21.

12. Barrett-Connor E, Kritz-Silverstein D. Estrogen replacement therapy and cognitive function in older women. JAMA 1993;269:2637–41.

13. Resnick SM, Metter J, Zonderman AB. Estrogen replacement therapy and longitudinal decline in visual memory. A possible protective effect? Neurology 1997;49:1491–97.

14. Jacobs DM, Tang M-X, Stern Y, Sano M, Marder K, Bell KL, et al. Cognitive function in nondemented older women who took estrogen after the menopause. Neurology 1998;50:368–73.

15. Steffens DC, Worton MC, Tschanz JT, Wyse BW, Plassman B, Welsh-Bohmer KA, et al. Estrogen use enhances cognitive performance in non-demented, community-dwelling older women. 151st Annual Meeting of the American Psychiatric Association. Toronto, Canada, May 30, 1998, Abstr. # NR441, p. 187.

16. Caldwell BM, Watson RI. An evaluation of psychologic effects of sex hormone administration in aged women. Results of therapy after six months. J Gerontol 1952;7:228–44.

17. Rauramo L, Lagerspetz K, Engblom P, Punnonen R. The effect of castration and peroral estrogen therapy on some psychological functions. Front Horm Res 1975;3:94–104.

18. Hackman BW, Galbraith D. Six month study of oestrogen therapy with piperazine oestrone sulphate and its effects on memory. Curr Med Res Opin 1977;4:21–27.

19. Campbell S, Whitehead M. Oestrogen therapy and the menopausal syndrome. Clin Obstet Gynaecol 1977;4:31–47.

20. Fedor-Freybergh P. The influence of oestrogen on the well-being and mental performance in climacteric and postmenopausal women. Acta Obstet Gynaecol Scand 1977;64:5–69.

21. Sherwin BB. Estrogen and/or androgen replacement therapy and cognitive functioning in surgically menopausal women. Psychoneuroendocrinology 1988;13:345–57.

22. Ditkoff EC, Crary WG, Cristo M, Lobo RA. Estrogen improves psychological function in asymptomatic postmenopausal women. Obstet Gynecol 1991;78:991–95.

23. Phillips S, Sherwin BB. Effects of estrogen on memory function in surgically menopausal women. Psychoneuroendocrinology 1992;17:485–95.

24. Milner B. Amnesia following operation on the temporal lobe. In Whitty CWM, Zangwill OL, eds. Amnesia. London: Butterworth & Co., 1996:109–33.

25. Zola-Morgan S, Squire LR, Amaral DG. Human amnesia in the medial temporal region: enduring memory impairment following a bilateral lesion limited to field CA1 of the hippocampus. J Neurosci 1986;6:2950–67.

26. Alvarez-Royo P, Clower RP, Zola-Morgan S, Squire LR. Stereotaxic lesions of the hippocampus in monkeys: determination of surgical coordinates and analysis of lesions using magnetic resonance imaging. J Neurosci Methods 1991;38:223–32.

27. Squire LR. Memory and the hippocampus: a synthesis from findings with rats, monkeys, and humans. Psychol Rev 1992;99:195–231.

28. Craik FIM, Jennings JM. Human memory. In: Craik FIM, Salthouse TA, eds. The handbook of aging and cognition. Hillsdale, NJ: Lawrence Erlbaum Associates, 1994:51–110.

29. Luine VN, Khylchevskaya RI, McEwen BS. Effect of gonadal steroids on activi-

ties of monoamine, oxidase and choline acetylase in rat brain. Brain Res 1975;86: 293–306.

30. Gould E, Wooley C, Frankfurt M, McEwen BS. Gonadal steroids regulate dentritic spine density in hippocampal pyramidal cells in adulthood. J Neurosci 1990; 10:1286–91.

31. Toran-Allerand CD, Mirander RC, Bentham WDL, Sohrabbi F, Brown TJ, MacLusky NJ. Estrogen receptors co-localize with low affinity nerve growth receptors in cholinergic neurons of the basal forebrain. Proc Nat Acad Sci USA 1992;89:4668–72.

32. Jarvik LF. Human intelligence: sex differences. Acta Genet Med Gemellol (Rome) 1975;24:189–211.

33. Resnick S, Berenbaum S, Gottesman T, Bouchard T. Early hormonal influences on cognitive functioning in congenital adrenal hyperplasia. Dev Psychol 1986;22:191–98.

34. Nass R, Baker S. Androgen effects on cognition: congenital adrenal hyperplasia. Psychoneuroendocrinology 1991;16:189–201.

35. Komenick P, Lane DM, Dickey RP. Gonadal hormones and cognitive performance. Physiol Psychol 1978;12:1016–37.

36. Wickham M. The effects of the menstrual cycle on test performance. Br J Psychol 1958;49:34–41.

37. Barrett-Connor E. Postmenopausal estrogen and prevention bias. Ann Intern Med 1991;115:455–56.

38. Cauley JA, Cummings SR, Black DM, Mascioli SR, Seeley DG. Prevalence and determinants of estrogen replacement therapy in elderly women. Am J Obstet Gynecol 1990;163:1438–44.

39. Mortimer JA, Graves AB. Education and other socioeconomic determinants of dementia and Alzheimer's disease. Neurology 1995;45:1707–12.

12

Age-Related Changes and Hormonal Regulation of Mesenchymal Stromal Stem Cells from Human Vertebral Bone Marrow

Gianluca D'Ippolito, Paul C. Schiller, Camillo Ricordi, Bernard A. Roos, and Guy A. Howard

Introduction

Investigations of age-related osteopenia increasingly suggest that, in addition to enhanced resorption, defective bone formation also contributes to bone loss and osteoporosis. Altered formation could explain why skeletal mass decreases in certain young adults who exhibit neither hypogonadism nor excess resorption (1–3). Such defective/diminished bone formation activity could putatively arise from decreased osteoblast number and/or function. Although there has been much study of osteoblast function and matrix production and mineralization, efforts have begun to examine the regulation of osteoblast progenitors and their maturation. Hormones and growth factors known to play a role in regulation of osteoblast function might also be expected to participate in regulation of osteoblast progenitor proliferation and differentiation. Such a role could signify pathogenetic as well as therapeutic importance for such regulatory factors.

Osteoblasts originate from mesenchymal stem cells (MSCs) in bone marrow (4–8). When MSCs are plated in vitro, each of these previously quiescent stem cells transition into DNA synthesis, ultimately forming distinct colonies (colony forming unit-fibroblasts, CFU-F) (4,8). Depending on the hormones and growth factors in the culture medium, these bone marrow-derived fibroblastic colonies can differentiate into many cell types, including osteoblasts, chondroblasts, and adipocytes (4–10). The number of MSCs in bone marrow and how this number varies with donor age have not been clearly established, nor has a convenient method yet been validated to identify MSCs before they begin differentiation. Nonetheless, the methodology is evolving for such studies. Dexamethasone appears important for osteoprogenitor formation in vitro, and it is now well appreciated that with bone-derived CFU-F, dexamethasone-treated MSCs develop a high level of alkaline phosphatase (AP) activity (CFU-F/AP$^+$), which characterizes osteoblastic and other mineralizing cell types. Several animal studies on the influence of aging on number of CFU-F/AP$^+$ suggest that osteoprogenitor cell numbers decrease with aging (8,11–14). There is, however, discrepancy among similar studies of human bone marrow-derived CFU-F/AP$^+$, with some reports of decreasing numbers with age (15) and other reports of no change with donor age (16,17).

Although bone loss often is first noted in vertebrae in osteoporosis, the relative inaccessibility of vertebral bone has resulted in most studies of human bone stem cells being done with pelvic bone marrow aspirates and/or biopsy material. Despite some inconvenience, we have begun investigating vertebral bone stem cells with the goal of understanding what role vertebral stem cells and their derivatives might have in vertebral bone loss. As suggested by earlier animal studies, we have found that the number of vertebral MSCs with osteogenic potential (CFU-F/AP$^+$) decreases dramatically and, to our surprise, early during aging in humans.

Although decreased osteoprogenitors could ultimately explain age-related reduction in osteoblast number, prevention and management of this state will depend on elucidation of the role of various hormones and growth factors in causing and reversing these decreases. Although to date investigation of osteoblast origins and changes during aging have centered on the role of glucocorticoids as a determinant of osteogenic differentiation of MSCs and on the role of estrogen, androgen, and insulinlike growth factors in promoting their maturation, it is increasingly clear that other factors (e.g., PDGF and EGF) are involved in the expansion (but not differentiation) of stem cell populations (18). Moreover, our studies in cartilage indicate that hepatocyte growth factor (HGF) cooperatively with vitamin D (vit D) can promote both proliferation and differentiation of mineralized tissue progenitor cells (19).

We will describe in this chapter an age-related loss in human vertebral MSCs and a role for vit D and HGF cooperativity in promoting osteoprogenitor proliferation and maturation. These results provide an initial insight into age-related reduction in MSC proliferation and/or differentiation, and they suggest a novel approach to maintaining/restoring osteoprogenitor activity throughout life.

Materials and Methods

Isolation of Human Vertebral Bone Marrow Cells

Human bone marrow was isolated from postmortem thoracolumbar (T1–L5) vertebrae of donors (25 men and 16 women; 3–70 years old) immediately after death from trauma (20). Small bone chips were processed in HEPES-buffer X-Vivo-10 medium (Biowhittaker, Inc., Walkersville, MD) supplemented with human serum albumin, antibiotics, and heparin (20). Bone chips were separated from the cell suspension by filtering through two consecutive stainless-steel screens (450- and 180-μm pore size). The filtered cell suspension was centrifuged, then the cell pellet was resuspended in α-MEM with fetal bovine serum (FBS). Bone marrow cells that remained within bone trabeculae after the initial processing were released by two additional 30-minute cycles of gentle agitation of the bone fragments in α-MEM with FBS. All cell suspensions were then pooled, centrifuged, and resuspended in α-MEM with 10% FBS. The pooled suspension of bone marrow cells was filtered to remove any bone fragments that might have been trapped in the cell pellet and to separate any clumps of cells. Cultures were prepared with this final filtered cell suspension.

Cell Culture

Nucleated bone marrow cells were counted and plated at various concentrations into 10-cm dishes in α-MEM/10% FBS with antibiotics and ascorbic acid. Dexamethasone (Dex) was added on the second day of culture at a final concentration of 10 nM. After a week in culture, half of the growth medium was removed and replaced with fresh medium. Cultures were maintained for at least 13 days in a 100% humidified atmosphere of 95% air, 5% CO_2 at 37°C. For mineralization studies, cultures were maintained for 4–5 weeks in medium supplemented with 10 mM β-glycerophosphate.

AP Activity

At the indicated times, cells were fixed with cold 10% neutral-buffered formalin and then assayed for AP activity as previously described (21).

Colony Size and Number

After histochemical assay for AP activity on day 13, colonies with 50 or more cells (8) were scored visually as positive (i.e., blue stain, AP$^+$) using a dissecting microscope. AP$^+$ and AP$^-$ colonies were counted. Colony size (area) was determined by image analysis of 20 randomly chosen colonies per plate from a defined area of 15 cm^2 on each plate.

Matrix Ca^{+2} Content/Mineralization

At the times indicated mineralization was determined using alizarin red-S (AR-S, Sigma) as described by Stanford et al. (22). Extracellular matrix (ECM) mineral-bound stain was photographed under visible light microscopy.

Detection of AP$^+$ Colonies by Membrane Lifting

After 13 days in culture the medium was removed, and cells were washed with PBS. At this point a sterile nylon membrane was placed on top of the cells for not more than 5 seconds. The membrane was then removed and

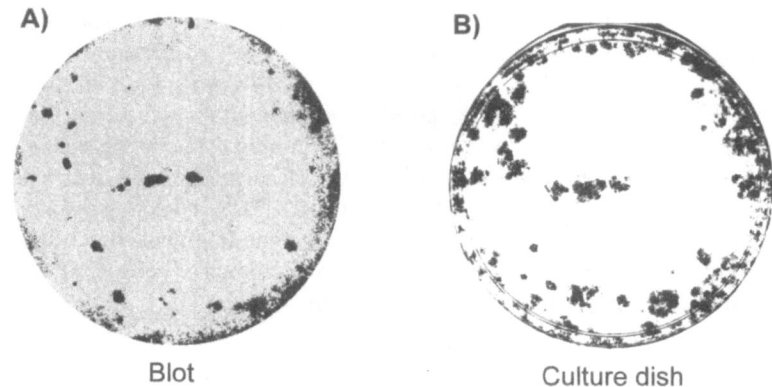

FIGURE 12.1. Detection of alkaline phosphatase activity in colonies formed from human bone marrow. (A) Alkaline phosphatase activity detected in a nylon membrane that had been placed on top of cells in culture, fixed, and then stained by a histochemical method (yellow in original). The blot shown in (A) aligns with the AP+ colonies on the plate (B) (blue in original).

placed cell-contact-side-up on top of blotting paper previously saturated with *p*-nitrophenyl phosphate, 4 mg/ml in glycine buffer, pH 10.4, a substrate for AP. The membrane and blotting paper were incubated for 30 minutes at 37°C. At the end of the incubation period, yellow spots indicated regions of AP activity on the membrane. We verified this indication of AP+ colonies by fixing the cells and then staining them for AP activity (Fig. 12.1A, B).

Cloning of AP+ Colonies

To test the AP+ colonies for osteogenic potential, individual positive colonies were isolated using cloning rings and replated in a single well of a 24-well cell culture plate. Cells were incubated for 21 days in α-MEM, 10% FBS, 100 mg/ml ascorbic acid, 10 mg/ml β-glycerophosphate, and 10 nM Dex. Medium was changed twice a week. Cultures were subsequently stained for mineralization and for AP activity.

Results

Characterization of CFU-F/AP+ in Vertebral Marrow Cell Cultures (23,24)

Isolated bone marrow cells were plated in culture as described in the Materials and Methods section. Two to 4 days in culture produced a heterogeneous population of mononuclear cells attached to the surface of the dishes, including fibroblastlike spindle-shaped cells, monocytes, macrophages, endothelial cells, and multinucleated osteoclasts. Serial observation suggested that only spindle-shaped cells proliferate: Distinct CFU-F colonies were noted by 6 days, and by 13 days roughly half the colonies were AP+. Colonies (CFU-Fs) that assayed positive for AP activity were designated CFU-F/AP+ and considered to have osteogenic potential (25). Dex increased the total number of colonies 30% regardless of donor age. Moreover, in the presence of Dex, more than 90% of the colonies were AP+ compared with about 50% in its absence. AP activity visually appeared to be consistently higher in Dex-treated cells. To verify the osteogenic potential (i.e., mineralizing capability) of the MSCs isolated from vertebral bodies, primary cultures of cells were grown for 4–5 weeks in α-MEM medium containing β-glycerophosphate, in the presence or absence of Dex from day 1. Mineralization occurred only in Dex-treated cultures.

To establish a source of cells for regulatory studies (see HGF and vit D studies), we developed a method of identifying osteogenic colonies for subculture. By blotting the live cultures under sterile conditions and then enzymatically staining the blot for AP activity, we were able to identify and selectively subculture AP+ colonies. An example of direct (blue in original) and blot (yellow in original) staining of AP+ colonies is shown in Figure 12.1. Subcultures of these initial CFU-F colonies revealed that AP+ colonies were stably transformed into an AP+ phenotype (mineralized matrix and AP+). In this way cells of a specific age donor that are AP+ can be expanded for additional studies (see later).

Effect of Plating Density

To determine the optimal cell plating number for colony formation, various numbers of nucleated bone marrow cells were plated with Dex. We found the highest number of total and CFU-F/AP+ colonies when 5×10^6 cells were plated per 10-cm dish. Bone marrow-derived cells were also cultured in the absence of Dex at 5×10^6 cells/10-cm dish for comparison with those cultures in the presence of Dex.

Age-Related Changes in CFU-F/AP+

The number of CFU-F/AP+ obtained after the 13-day culture period decreased with donor age regardless of the presence or absence of Dex. Figure 12.2 shows the range of results obtained when bone marrow cells were plated in DEX at 5×10^6 cells/dish; we noted a significant age-related decrease in the number of CFU-F/AP+ from donors up to age 40. No further decrease was noted for donors 41–70 years old. In the absence of Dex a similar significant decline was observed for the younger donors, although the level of enzyme activity and rate of decline (in CFU-F/AP+) for the younger donors was less steep (not shown). We observed no significant difference in colony number

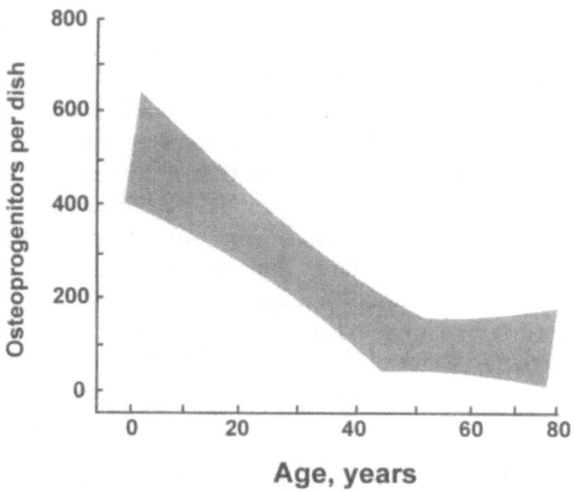

FIGURE 12.2. Regression analysis of the number of colonies obtained from younger (ages 3–36) and older (ages 41–70) donor groups. Cells were plated at $5 \times 10^6/10$-cm dish and allowed to proliferate in the presence of 10 nM Dex.

between men and women donors in either age group. CFU-F/AP⁺ obtained from all 41-year-old donors when 5×10^6 cells were plated in Dex averaged 38.6 colonies per 10^6 bone marrow cells plated. For younger donors (ages 3–36; $n = 19$), this value was higher (66.2 per 10^6 bone marrow cells); the older donors (ages 41–70; $n = 22$) generated only 14.7 colonies per 10^6 bone marrow cells.

CFU-F Colony Size Changes with Age

In the absence of Dex, colony size (taken as an index of proliferation rate) of cells from young and old donors was not significantly different. With Dex, colony size diminished in both old (41–70 years old) and young (3–36 years old) donors; however, the colonies from the older donors were even smaller than those of the younger donors (Table 12.1). We observed no significant

TABLE 12.1. Colony size (area) of CFU-F as a function of donor age.

	Control (%)	10 nM Dex (%)	Significance
Young	102	88	$p < 0.05$
Old	96	74	$p < 0.01$
Significance	NS	$p < 0.001$	
All donors	100[a]	85	$p < 0.001$

[a] All donors, control defined as 100%.

TABLE 12.2. Total number of MSCs (% of control) as a function of treatment.

Treatment[a]	Day 7	Day 11	Day 15
HGF	125	250	210
D_3	75	60	55
Dex	75	60	55
HGF + D_3	135	160	140
HGF + Dex	150	230	220
Dex + D_3	75	40	45

[a]Factor concentrations: 10 ng/ml HGF, 10 nM vit D, 100 nM Dex. No treatment was 100%.

difference in colony size between men and women donors in either age group (not shown).

Hormone and Growth Factor Regulation of Proliferation and Differentiation of MSCs

In our previous studies of cartilage stem cells, we ascertained that 10 nM 1,25-dihydroxyvitamin D_3 (vit D) and 10 ng/ml HGF (human recombinant material obtained from Dr. Ralph Schwall, Genentech, Inc., South San Francisco, CA) cooperatively stimulated AP (19). We tested these agents separately and combined in vertebra-derived MSC cultures (passages 3–9); Dex was also evaluated. As expected, vit D and Dex by themselves and in combination markedly inhibited proliferation (Table 12.2) and stimulated AP (Table 12.3), with vit D exhibiting the greater effect on differentiation. HGF by itself stimulated proliferation but had little effect on differentiation. HGF, together with Dex or vit D, could overcome the proliferation-inhibitory effects of the steroid hormones and at the same time promote differentiation as indicated by AP stimulation (Tables 12.2 and 12.3) and mineralization (not shown). In general, mineralization correlated with AP activity in Dex-treated cultures, and mineralization was greatly enhanced by Dex treatment. The ability of Dex to enhance mineralization in vit D-treated cultures, how-

TABLE 12.3. Alkaline phosphatase activity per total cell number (% of control).

Treatment[a]	Day 7	Day 11	Day 15
HGF	95	105	110
D_3	275	800	650
Dex	200	275	250
HGF + D_3	330	1000	800
HGF + Dex	120	300	320
Dex + D_3	420	1050	1050

[a]Treatments as in Table 12.2; no treatment was 100%.

ever, was probably not related to AP activity because Dex treatment increased AP activity only slightly, while dramatically increasing mineralization. Further, although vit D increased AP activity, it also increased osteocalcin (26); hence, the action of Dex to stimulate mineralization in cooperation with vit D may have occurred via suppression of vit D-induced osteocalcin, a mineralization inhibitor (27).

Discussion

In humans, bone mass increases, plateaus, and then decreases with aging (1,28). Several studies have reported a substantial premenopausal bone loss in lumbar spine (1–3). These data correlate well with our observation of a reduction in osteoprogenitor cells derived from the thoracolumbar spine of donors substantially younger than age 40. In fact, most skeletal mass is accumulated by age 18, and some skeletal sites begin to lose bone immediately after that age (including trabecular bone in the vertebrae) (29 and Refs. therein). Although resorption can increase vigorously in response to loss of steroid sex hormone or other influences, the view is increasingly emerging that the resorption-coupled response in terms of bone formation becomes progressively deficient with aging (30). Our studies suggest that relatively early in the aging process the number of MSCs residing in the bone marrow of human thoracic/lumbar vertebrae decreases. This MSC decrease could underlie some of the bone loss in humans, especially that occurring in the presence of high steroid sex hormone levels (28,31,32).

Our finding that CFU-F/AP$^+$ decreased with age agrees with findings from various animal studies (8,11–14) and with a report of human cells from a nonvertebral site (15). Such a decrease in CFU-F/AP$^+$ has not always been reported (16,17). At least one of these discrepancies, however, is more apparent than real because only cells from "older" (> 40 years) donors were used (16). We, too, found no changes (no further decline) in CFU-F/AP$^+$ after age 40 (Fig. 12.2).

Our studies suggest that MSCs from thoraciolumbar vertebral bodies are similar to those derived from ribs, femoral head, and iliac crest aspirates of animals and humans (8,25,26,33–36). These MSCs share a fibroblastic spindle shape and respond to Dex, shown by a higher number of CFU-F, a greater fraction of CFU-F/AP$^+$, transition to a more cuboidal morphology, and higher levels of AP activity, which are all consistent with the concept that glucocorticoids recruit uncommitted stem cells toward the osteogenic lineage (26,36–38). Colonies, especially those from older donors, appear to grow slower in Dex, which suggests that cell-cycle slowing and/or apoptosis has occurred in association with differentiation (25,34,36,38–41). Mineralization and its Dex enhancement occur in vertebra-derived human osteogenic MSCs in a manner comparable to that reported in human bone MSCs from other sites (24,27,32,34), but in contrast to a report of divergent effects of thyroid hormone on bone marrow cell cultures from vertebral and femoral sources (42). Expanded cultures of these MSCs show osteoprogenitor markers (Osf2/Cbfa, data not shown), and individual CFU-F/AP$^+$ colonies can be subcultured with virtually all the derived cells that show osteoblastic markers (AP and mineralization). These characteristics suggest that MSC differentiation toward osteoprogenitors is a relatively rapid and stable transformation (23,43–45).

Our results support the hypothesis that MSCs, and in particular MSCs with osteogenic potential, decrease during aging. Such a reduction could reflect a limited proliferation rate or potential and/or survival in the bone marrow as a result of cellular senescence. Whether this decrease reflects intrinsic properties of the MSCs and/or altered levels and responsiveness to growth factors and hormones remains to be investigated (1,46–49). The greater reduction in size of Dex-treated colonies from older donors (Table 12.1) suggests that MSCs in older persons proliferate slower during response to various differentiation signals. The reduced MSCs could also reflect age-related changes in the bone marrow microenvironment, resulting in cell-to-cell and cell-to-matrix interactions that may be unfavorable for MSC proliferation or which may favor MSC maturation toward a different lineage (e.g., adipogenic). The well-known inverse relationship between marrow adipocytes and osteoblasts with aging supports this concept (50–52).

Ducy et al. (43) have cloned a gene encoding a transcription factor, designated Osf2/Cbfa1, that has been identified as a key determinant for osteoblast differentiation and bone development (44,45). This factor is expressed in CFU-F and increases with evolution toward the osteogenic CFU-F/AP$^+$. One of the opportunities presented by our primary vertebral cultures of MSCs is the chance to ascertain what factors promote Osf2/Cbfa1 expression. After MSCs are exposed to Dex and commit toward the osteoblastic phenotype, we would expect Osf2/Cbfa1 expression to increase. We are now beginning to address the role of this factor as a mediator of hormone- and growth factor-induced osteoprogenitor growth and maturation. Consistent with a report that vit D is capable of inhibiting bone marrow stromal cell adipogenesis (53), our planned studies should reveal why vit D and specific growth factors such as HGF stimulate expansion of the osteoprogenitor population without risking de- or transdifferentiation to adipogenic or other phenotypes. Moreover, aging may alter the expression of Osf2/Cbfa1 under various basal and regulatory conditions.

We have been examining mineralized tissue stem cells in vitro in the hope of discovering novel approaches to treatment and prevention of mineralized tissue disorders (e.g., deficient bone and mineralization associated with osteoporosis and osteomalacia and excessive mineralization and osteophyte formation in osteoarthritis). We previously reported that vit D and HGF cooperatively stimulated mineralization-related parameters in cultures of growth-plate cartilage stem cells (19). We now find that vit D and HGF cooperatively enhance both growth and osteogenic maturation of the MSC-derived cultures. These effects are no doubt related to important cooperativity at the level of cell signaling and cell-cycle control. If cell-cycle slowing by vit D in the MSCs is under similar regulation as vit D cell-cycle regulation in other nonneoplastic systems, we would expect vit D-increased cyclin-dependent kinase inhibition via induction of p27 and/or p21 (39–41). HGF stimulates cell-cycle progression, most likely through *Ras* activation and MAP kinase cascade with cyclin D induction, the effect of which perhaps can overcome vit D's growth-inhibitory effects. Via other signaling pathways, however, both HGF and vit D must be exerting stimulation on AP and other osteogenic products. Thus, in contrast to EGF (5), which concurrently promotes growth and reduces osteogenic maturation, HGF appears to stimulate MSC growth and differentiation. Because of its high concentration in bone marrow, HGF could have an "endogenous osteogenic influence" on MSCs, as has been reported for other factors in other osteogenic cell systems (54). Age-related changes in HGF have not yet been studied.

In summary we have found that (1) cells derived from human vertebral bodies have characteristics in common with MSCs derived from other skeletal sites and can differentiate into osteoprogenitors (CFU-F/AP+) and be propagated as AP+ colonies in vitro; (2) vertebral MSCs and osteoprogenitor (CFU-F/AP+) colonies decrease with donor age; and (3) MSC-derived cells can be expanded and maintained in a differentiated state with hormone and growth factor combinations (vit D and HGF) that need not include Dex. This effect of HGF differs from that of the two growth factors, PDGF and EGF, considered critical to MSC proliferation because they promote growth accompanied by marked reduction in AP (18). Thus, our studies open the way to investigating the pathogenesis of age-related loss of MSCs and devising novel approaches to management of age-related decreases in bone formation.

Acknowledgments. We thank Mr. David Vazquez and Ms. Blanca Rodriguez for expert technical assistance in cell culture and cell assays, Mr. James McMannis and Ms. Topaz Kirlew from the Cell Transplantation Center (Diabetes Research Institute, University of Miami) for whole marrow isolation, Dr. Ralph Schwall of Genentech, Inc., for providing the recombinant human HGF, and Ms. Ginnie Roos for help in preparing the manuscript. We thank Dr. Robert Morgan for assistance with statistical analysis. Dr. Howard is the recipient of a VA Research Career Scientist award.

References

1. Riggs BL, Wahner HW, Melton LJ III, Richelson LS, Judd HL, Offord KP. Rates of bone loss in the axial and appendicular skeletons of women: evidence of substantial vertebral bone loss prior to menopause. J Clin Invest 1986;77:1487–91.
2. Theintz G, Buchs B, Rizzoli R, Slosman D, Clavien H, Sizonenko PC, et al. Longi-

tudinal monitoring of bone mass accumulation in healthy adolescents: evidence for a marked reduction after 16 years of age at the levels of lumbar spine and femoral neck in female subjects. J Clin Endocrinol Metab 1992;75:1060–65.

3. Matkovic V, Jelic T, Wardlaw GM, Ilich JZ, Goel PK, Wright JK, et al. Timing of peak bone mass in Caucasian females and its implication for the prevention of osteoporosis. Inference from a cross-sectional model. J Clin Invest 1994;93:799–808.

4. Friedenstein AJ, Chailakhjan RK, Lalykina KS. The development of fibroblast colonies in monolayer cultures of guinea-pig bone marrow and spleen cells. Cell Tissue Kinet 1970;3:393–403.

5. Owen M, Friedenstein AJ. Stromal stem cells: marrow-derived osteogenic precursors. In: Cell and molecular biology of vertebrate hard tissues. Chichester: Wiley (Ciba Foundation Symposium 136), 1988:42–60.

6. Wlodarsky KH. Properties and origin of osteoblasts. Clin Orthop Rel Res 1990;252:276–93.

7. Beresford JN. Osteogenic stem cells and the stromal system of bone and marrow. Clin Orthop Rel Res 1989;240:270–80.

8. Friedenstein AJ. Precursor cells of mechanocytes. Int Rev Cytol 1976;47:327–59.

9. Wakitani S, Saito T, Caplan AI. Myogenic cells derived from rat bone marrow mesenchymal stem cells exposed to 5-azacytidine. Muscle Nerve 1995;18: 1417–26.

10. Grigoriadis AE, Heersche JNM, Aubin JE. Differentiation of muscle, fat, cartilage, and bone from progenitor cells present in a bone-derived clonal cell population: effect of dexamethasone. J Cell Biol 1988;106:2139–51.

11. Tsuji T, Hughes FJ, McCulloch CAG, Melcher AH. Effects of donor age on osteogenic cells of rat bone marrow in vitro. Mech Aging Dev 1990;51:121–32.

12. Egrise D, Martin D, Vienne A, Neve P, Schoutens A. The number of fibroblastic colonies formed from bone marrow is decreased and the in vitro proliferation rate of trabecular bone cells increased in aged rats. Bone 1992;13:355–61.

13. Kahn AJ, Gibbons R, Perkins S, Gazit D. Age-related bone loss. A hypothesis and initial assessment in mice. Clin Orthop Rel Res 1995;313:69–75.

14. Bergman RJ, Gazit D, Kahn AJ, Gruber H, McDougall S, Hahn TJ. Age-related changes in osteogenic stem cells in mice. J Bone Miner Res 1996;11:568–77.

15. Majors AK, Boehm CA, Nitto H, Midura R, Muschler GF. Characterization of human bone marrow stromal cells with respect to osteoblastic differentiation. J Orthop Res 1997;15:546–57.

16. Glowacki J. Influence of age on human marrow. Calcif Tissue Int 1995;56(suppl 1):S50–51.

17. Rickard DJ, Kassem M, Hefferan TE, Sarkar G, Spelsberg TC, Riggs BL. Isolation and characterization of osteoblast precursor cells from human bone marrow. J Bone Miner Res 1996;11:312–24.

18. Martin I, Muraglia A, Campanile G, Cancedda R, Quarto R. Fibroblast growth factor-2 supports ex vivo expansion and maintenance of osteogenic precursors from human bone marrow. Endocrinology 1997;138:4456–62.

19. Grumbles RM, Howell DS, Wenger L, Altman RD, Howard GA, Roos BA. Hepatocyte growth factor and its actions in growth plate chondrocytes. Bone 1996;19:255–61.

20. Fontes PA, Ricordi C, Rao AS, Rybka WB, Dodson SF, Broznick B, et al. Human vertebral bodies as a source of bone marrow for cell augmentation in whole organ

allografts. In: Ricordi C, ed. Methods in cell transplantation. Austin, TX: RG Landes Co., 1995:619–28.

21. Liu C, Sanghvi R, Burnell JM, Howard GA. Simultaneous demonstration of bone alkaline and acid phosphatase activities in plastic-embedded sections and differential inhibition of the activities. Histochemistry 1987;86:559–65.

22. Stanford CM, Jacobson PA, Eanes ED, Lembke LA, Midura RJ. Rapidly forming apatitic mineral in an osteoblastic cell line (UMR 106-01 BSP). J Biol Chem 1995;270:9420–28.

23. D'Ippolito G, Schiller PC, Roos BA, Howard GA. Detecting the osteogenic potential of CFU-F in human bone marrow cultures. J Bone Miner Res 1997;12:S404.

24. D'Ippolito G, Schiller PC, Ricordi C, Roos BA, Howard GA. Age-related osteogenic potential of mesenchymal stromal stem cells from human vertebral bone marrow. J Bone Miner Res 1999;14:1115–22.

25. Owen ME, Cave J, Joyner CJ. Clonal analysis in vitro of osteogenic differentiation of marrow CFU-F. J Cell Sci 1987;87:731–38.

26. Cheng S-L, Yang JW, Rifas L, Zhang S-F, Avioli LV. Differentiation of human bone marrow osteogenic stromal cells in vitro: induction of the osteoblast phenotype by dexamethasone. Endocrinology 1994;134:277–86.

27. Cheng SL, Zhang SF, Avioli LV. Expression of bone matrix proteins during dexamethasone-induced mineralization of human bone marrow stromal cells. J Cell Biochem 1996;61:182–93.

28. Ritzel H, Amling M, Pösl M, Hahn M, Delling G. The thickness of human vertebral cortical bone and its changes in aging and osteoporosis: a histomorphometric analysis of the complete spinal column from thirty-seven autopsy specimens. J Bone Miner Res 1997;12:89–95.

29. Matkovic V. Editorial: skeletal development and bone turnover revisited. J Clin Endocrinol Metab 1996;81:2013–16.

30. Roholl PJM, Blauw E, Zurcher C, Dormans JAMA, Theuns HM. Evidence for a diminished maturation of preosteoblasts into osteoblasts during aging in rats: an ultrastructural analysis. J Bone Miner Res 1994;9:355–66.

31. Delling G, Amling M. Biomechanical stability of the skeleton—it is not only bone mass, but also bone structure that counts. Nephrol Dial Transplant 1995;10:601–6.

32. Parfitt AM, Mathews CHE, Villanueva AR, Kleerekoper M, Frame B, Rao DS. Relationships between surface, volume, and thickness of iliac trabecular bone in aging and in osteoporosis. Implications for the microanatomic and cellular mechanisms of bone loss. J Clin Invest 1983;72:1396–409.

33. Friedenstein AJ, Latzinik NV, Gorskaya YuF, Luria EA, Moskvina IL. Bone marrow stromal colony formation requires stimulation by haemopoietic cells. Bone Miner 1992;18:199–213.

34. Haynesworth SE, Goshima J, Goldberg VM, Caplan AI. Characterization of cells with osteogenic potential from human marrow. Bone 1992;13:81–88.

35. Jaiswal N, Haynesworth SE, Caplan AI, Bruder SP. Osteogenic differentiation of purified, culture-expanded human mesenchymal stem cells in vitro. J Cell Biochem 1997;64:295–312.

36. Beresford JN, Joyner CJ, Devlin C, Triffitt JT. The effects of dexamethasone and 1,25-dihydroxyvitamin D_3 on osteogenic differentiation of human marrow stromal cells in vitro. Arch Oral Biol 1994;39:941–47.

37. Bellows CG, Heershe JNM, Aubin JE. Determination of the capacity for proliferation and differentiation of osteoprogenitor cells in the presence and absence of dexamethasone. Dev Biol 1990;140:132–38.

38. Kamalia N, McCulloch CAG, Tenebaum HC, Limeback H. Dexamethasone recruitment of self-renewing osteoprogenitor cells in chick bone marrow stromal cell cultures. Blood 1992;79:320–26.

39. Campbell MR, Elstner E, Holden S, Uskokovic M, Koeffler HP. Inhibition of proliferation of prostate cancer cells by a 19-nor-hexafluoride vitamin D_3 analog involves induction of $p21^{waf1}$, $p27^{Kip1}$ and E-cadherin. J Mol Endocrinol 1997;19:15–27.

40. Kokontis JM, Hay N, Liao S. Progression of LNCaP prostate tumor cells during androgen deprivation: hormone-independent growth, repression of proliferation by androgen, and role for $p27^{Kip1}$ in androgen-induced cell cycle arrest. Mol Endocrinol 1998;12:941–53.

41. Robker RL, Richards JS. Hormone-induced proliferation and differentiation of granulosa cells: a coordinated balance of the cell cycle regulators cyclin D2 and $p27^{Kip1}$. Mol Endocrinol 1998;12:924–40.

42. Milne M, Kang MI, Quail JM, Baran DT. Thyroid hormone excess increases insulin-like growth factor I transcripts in bone marrow cell cultures: divergent effects on vertebral and femoral cell cultures. Endocrinology 1998;139:2527–34.

43. Ducy P, Zhang R, Geoffroy V, Ridall AL, Karsenty G. Osf2/Cbfa1: a transcriptional activator of osteoblast differentiation. Cell 1997;89:747–54.

44. Otto F, Thornell AP, Crompton T, Denzel A, Gilmour KC, Rosewell IR, et al. Cbfa1, a candidate gene for cleidocranial dysplasia syndrome, is essential for osteoblast differentiation and bone development. Cell 1997;89:765–71.

45. Komori T, Yagi H, Nomura S, Yamaguchi A, Sasaki K, Deguchi K, et al. Targeted disruption of Cbfa1 results in a complete lack of bone formation owing to maturational arrest of osteoblasts. Cell 1997;89:755–64.

46. Friedenstein AJ. Marrow stromal fibroblasts. Calcif Tissue Int 1995;56(suppl 1):S17.

47. Martin GM. Cellular aging—clonal senescence. A review (part I). Am J Pathol 1975;89:484–511.

48. Stanulis-Praeger BM. Cellular senescence revisited: a review. Mech Ageing Dev 1987;38:1–48.

49. Wheaton K, Atadja P, Riabowol K. Regulation of transcription factor activity during cellular aging. Biochem Cell Biol 1996;74:523–34.

50. Beresford JN, Bennett JH, Devlin C, Leboy PS, Owen ME. Evidence for an inverse relationship between the differentiation of adipocytic and osteogenic cells in rat marrow stromal cell cultures. J Cell Sci 1992;102:341–51.

51. Burkhardt R, Kettner G, Böhm W, Schmidmeier M, Schlag R, Frisch B, et al. Changes in trabecular bone, hematopoiesis and bone marrow vessels in aplastic anemia, primary osteoporosis, and old age: a comparative histomorphometric study. Bone 1987;8:157–64.

52. Meunier P, Aaron J, Edouard C, Vignon G. Osteoporosis and the replacement of cell populations of the marrow by adipose tissue. A quantitative study of 84 iliac bone biopsies. Clin Orthop Rel Res 1971;80:147–54.

53. Kelly KA, Gimble JM. 1,25-Dihydroxy vitamin D_3 inhibits adipocyte differentiation and gene expression in murine bone marrow stromal cell clones and primary cultures. Endocrinology 1998;139:2622–28.

54. Friedenstein AJ, Chailakhjan RK, Lalykina KS. The development of fibroblast colonies in monolayer cultures of guinea-pig bone marrow and spleen cells. Cell Tissue Kinet 1970;3:393–403.

13

Cellular and Molecular Mechanisms of Postmenopausal Osteoporosis

Stavros C. Manolagas

Introduction

Although the skeleton can survive for millions of years after death, living bones require periodic regeneration throughout life. During development and growth, bones are sculpted in order to achieve their shape and size by the removal of bone from one site and deposition at a different one; this process is called *modeling*. After maturity is reached, regeneration continues in the form of a periodic replacement of old bone with new at the same location. This process is called *remodeling* and is responsible for the complete renewal of the adult skeleton approximately every 10 years (1).

After menopause, the rate of bone remodeling is accelerated and bone, trabecular more than cortical, undergoes a phase of rapid loss that lasts between 5 and 10 years. This loss profoundly affects the spine, forearm, and femoral neck, leading to the syndrome of structurally weakened bones and fractures defined as postmenopausal osteoporosis (2). In this chapter, I will summarize our current understanding of the cellular and molecular mechanisms responsible for these changes. As menopause disrupts normal remodeling, I will start by highlighting the basic principles of this process.

The Basic Multicellular Unit: The Instrument of Bone Remodeling

Bone remodeling is carried out by a basic multicellular unit (BMU) and takes place mainly on the internal surfaces of bone (3). The BMU, approximately 1–2 mm long and 0.2–0.4 mm wide, consists of a team of osteoclasts in front, a team of osteoblasts following behind, a central vascular sinusoid, a nerve supply, and associated connective tissue. Osteoclasts adhere to bone and subsequently remove it by acidification and proteolytic digestion. After osteoclasts have left the resorption site, osteoblasts invade the area and begin the process of new bone formation by secreting osteoid, which is eventually mineralized into new bone. In cortical bone, the BMU tunnels through the tissue, but it moves across the trabecular surface forming a trench in cancellous bone (1). The average BMU has a life span of 6–8 months, and it replaces approximately 0.2 mm² of bone surface. As the BMU moves over the bone surface with an average speed of 25 µm/day, its cellular components maintain a well-orchestrated spatial and temporal relationship with each other. The life span of osteoclasts and osteoblasts, the executive cells of the BMU, is considerably shorter than the life span of the unit. The average life span of the osteoclast is 2 weeks, and the average life span of the active osteoblast is 3 months. As a result, the rate of supply of new cells and the time of apoptosis are the critical determinants of the initiation of new BMUs and/or extension or shortening of the lifetime of existing ones.

Birth and Death of Osteoblasts and Osteoclasts

Both osteoblast and osteoclast are derived from precursors that originate in the bone marrow. The precursors of osteoblast are multipotent mesenchymal stem cells that can also give rise to fibroblastic stromal cells, chondrocytes, as well as adipocytes and muscle cells (4). The precursors of osteoclast are hematopoietic cells of the monocyte/macrophage lineage (5). The process of osteoblast and osteoclast formation from their progenitors is controlled by growth factors and cytokines that are produced in the bone marrow microenvironment and by adhesion molecules that mediate cell–cell and cell–matrix interactions; these can be modulated by systemic hormones and mechanical signals (6).

After osteoclasts have eroded to a particular distance, either from the central axis in cortical bone or to a particular depth from the surface in cancellous bone, they die by apoptosis (7). Osteoblasts that have completed their bone forming task have one of three fates: they can become elongated "lining cells" that cover the newly formed bone surface; they can be entrapped in the mineralized matrix to become osteocytes—cells characterized by a striking stellate morphology, reminiscent of the dendritic network of the nervous system. Osteocytes represent the most common (~90%) cell type present in bone and are thought to be the ideal candidates for sensors of the local need for bone augmentation or reduction during functional adaptation of the skeleton, the detection of microdamage, and the transmission of signals that lead to bone repair by remodeling (8). The third and most common fate of the bone forming osteoblasts (50–70%) is death by apoptosis (9). The same locally produced cytokines, growth factors, and systemic hormones that control the birth rate of bone cells also regulate their death by apoptosis.

Coordination of Resorption and Formation

The life history of the BMU comprises separate stages of origination, progression, and termination, of which the second is by far the longest. As the cells that constitute the BMU advance during the progression phase, the same spatial relationships between the cells are maintained, and thus new cells are needed simultaneously to sustain the continued advance of the BMU. The simultaneous need for osteoclasts and osteoblasts during the progression phase of the BMU, and thereby simultaneous development of both osteoclast and osteoblast progenitors, seems inconsistent with the sequential appearance of osteoclasts followed by osteoblasts at the same remodeling site. This paradox can now be resolved from evidence indicating that mesenchymal cell differentiation toward the osteoblast phenotype and osteoclastogenesis are inseparably linked. Indeed, both are stimulated by the same factors (10), proceed simultaneously (11), and the former may not occur without the latter (12) because osteoclast development depends on support provided from stromal/osteoblastic cells. The mechanistic basis of this dependency has been established by the discovery of a membrane bound cytokinelike molecule, RANK ligand/osteoprotegerin ligand/TRANCE, which is expressed in mesenchymal cells and binds to a specific receptor (RANK) on hematopoietic osteoclast progenitors. Such binding is essential and, together with M-CSF, sufficient for osteoclastogenesis (13).

Osf2/Cbfa1, a transcription factor uniquely present in cells of the osteoblastic lineage, is an essential mediator of osteoblast differentiation from their pluripotent mesenchymal progenitors because it can directly activate several osteoblast specific genes (14). Lack of this factor prevents osteoblast development (15). It is of interest that lack of Osf2 leads also to a paucity of osteoclasts. Bone morphogenetic proteins (BMP), members of the TGFβ superfamily of growth factors, can induce Osf2 and had been heretofore implicated in the induction of osteoblast formation in the embryo and fracture repair (16). Studies from the author's laboratory have now elucidated that BMPs -2 and -4 are required for both osteoblast as well as osteoclast development in postnatal life. To be specific, they have shown that noggin, a BMP-antagonist, inhibits both osteoblast and osteoclast formation in bone marrow cell cultures from normal adult mice, which are effects that could be reversed by exogenous BMP-2 (17). Consistent with these observations, the genes for BMP-2 and -4, the BMP-2/4 receptor, as well as noggin, were expressed in the adult murine bone marrow and whole bone. Moreover, the promoter of the RANK ligand gene contains an OSF-2 binding site (18). Taken together, these findings indicate that a BMP→OSF-2→RANK ligand gene expression cascade in cells of the bone marrow stromal/osteoblastic lineage constitutes the molecular basis of the linkage between osteoblastogenesis and osteoclastogenesis. Moreover, these new observations imply that the balance between BMPs and noggin provides a tonic baseline control of osteoblastogenesis and osteoclastogenesis upon which other inputs (e.g. biomechanical, hormonal, etc.) operate.

IL-6, A Critical Mediator of the Effects of Estrogen Deficiency on Bone

The acceleration of the rate of bone remodeling and the bone loss that accompany menopause can now be explained as the results of the removal of an inhibitory control of estrogen on the production of cytokines and the responsiveness of bone marrow cell progenitors to such cytokines.

The best documented paradigm of a cytokine playing a critical pathogenetic role in the osteoporosis caused by loss of estrogen, as well as androgen, is interleukin-6 (IL-6). IL-6 is a member of a family of structurally related cytokines that use the gp130 signal transducer in their receptor complex. Besides IL-6, the family includes IL-11, oncostatin M (OSM), leukemia inhibitory factor (LIF), and cardiotropin 1. Binding of IL-6 to its specific cell surface receptor (gp80) causes recruitment and dimerization of gp130, which is then tyrosine phosphorylated by members of the JAK family of tyrosine kinases. This event results in tyrosine phosphorylation of several downstream signaling molecules, including members of the signal transducers and activators of transcription (STAT) family of transcription factors (19). Phosphorylated STATs in turn undergo homo- and heterodimerization, and translocate to the nucleus where they activate cytokine responsive gene transcription. The gp80 subunit of the IL-6 receptor also exists in a soluble form (sIL-6R), but unlike most soluble cytokine receptors, it functions as an agonist by binding to IL-6 and then interacting with membrane-associated gp130 to stimulate JAK/STAT signaling.

IL-6 is produced at high levels by cells of the stromal/osteoblastic lineage in response to stimulation by a variety of factors such as TGFß, PDGF, IL-1, TNF, and by PTH. Alone or in concert with other agents, IL-6 stimulates osteoclastogenesis and promotes bone resorption (10). OSM, IL-11, and LIF also stimulate osteoclast formation. The cells that mediate the actions of the IL-6 type cytokines on osteoclast formation are the stromal/osteoblastic cells, as stimulation of gp80 expression on these cells by pretreatment with dexamethasone allows them to support osteoclast formation in response to IL-6 alone (20); and gp130 activating cytokines increase the expression of RANKL in stromal osteoblastic cells. IL-6 does not seem to be required for osteoclastogenesis in vivo under normal physiologic conditions. In fact, osteoclast formation is unaffected in sex steroid replete mice treated with a neutralizing anti-IL-6 antibody, or in IL-6 deficient mice (21,22). This situation probably reflects the fact that the expression of the gp80 subunit of the IL-6 receptor in bone is a limiting factor for the effects of the cytokine.

In addition to their ability to stimulate osteoclast formation, IL-6 type cytokines stimulate the differentiation of osteoblasts. Receptors for these cytokines are expressed on a variety of stromal/osteoblastic cells (23), and ligand binding induces progression toward a more mature osteoblast phenotype (24). It is of interest that IL-6 type cytokines stimulate the differentiation of noncommitted embryonic fibroblasts (12–14 days of gestation) exclusively toward the osteoblast lineage (25). As in other cell types, the actions of IL-6 type cytokines on osteoblastic cells involve activation of both the JAK/STAT and the MAPK pathway (24) and are probably mediated by the cyclin-dependent kinase inhibitor p21 ^{WAF1,CIP1,SDI1} — a downstream effector of gp130/Stat3 activation (26). Consistent with the in vitro evidence, overexpression of IL-6 type cytokines increases bone formation in mice (10).

Estrogen as well as selective estrogen receptor modulators (SERMs), such as raloxifene and idoxifene, inhibit the production of IL-6 by cells of the stromal/osteoblastic lineage through receptor-mediated actions on the transcriptional activity of the IL-6 gene promoter (27–31). This effect does not require direct binding of the estrogen receptor to DNA. Instead, it is due to protein–protein interaction between the estrogen receptor and transcription factors such as NF-Kβ and C/EBP. This mechanism provides a model that best fits current understanding of the molecular pharmacology of estrogen and SERMs (32). Estrogen suppresses the expression of IL-6 and decreases in vitro the expression of both gp80 and gp130 in cells of the bone marrow stromal/osteoblastic lineage (33). Removal of these inhibitory controls following estrogen loss results in increased IL-6 production as well as an increase in the expression of gp80 and gp130 in the murine bone marrow (33) —findings that have been reproduced in humans. In fact, in postmenopausal women not receiving estrogen replacement therapy, IL-6 production in the marrow is at least 10-fold higher compared with women on estrogen replacement (34).

In direct support of the contention that IL-6 is responsible for the increased bone resorption that ensues following loss of sex steroids, injections of an IL-6 neutralizing antibody to gonadectomized female or male mice prevent the increase in osteoclastogenesis in the bone marrow and the increase in the number of osteoclasts in sections of trabecular bone (21,22). Furthermore, unlike wild-type controls, IL-6 knock out mice do not exhibit the cellular changes in the marrow and trabecular bone sections, and they are protected from the loss of trabecular bone following loss of sex steroids (22,29). In addition and consistent with the evidence that SERMs downregulate IL-6 production, the bone protective effects of raloxifene in ovariectomized rats correspond closely with the ability of this agent, as well as 17β-estradiol and ethinyl estradiol, to prevent the increase in serum IL-6 (A. Glasebrook, Eli Lilly Research Laboratories, personal communication).

The evidence that IL-6 plays a critical role in the bone loss caused by loss of gonadal function has been strengthened considerably by the demonstration of a similar role of IL-6 in several other conditions associated with increased bone resorption. Thus, in addition to estrogen-deficient women, increased local or systemic production of IL-6 and the IL-6 receptor has been documented in patients with multiple myeloma, Paget's disease, rheumatoid arthritis, Gorham-Stout or disappearing bone disease, hyperthyroidism, and primary and secondary hyperparathyroidism, as well as McCune Albright Syndrome (10).

Following the menopause both bone resorption and bone formation increase in women. This clinical observation can now be explained by the experimental demonstration that loss of ovarian function in mice increases the number of colony-forming units—fibroblasts (CFU-F)—and the colony forming units—osteoblasts (CFU-OB)—in the bone marrow, and that these changes are temporally associated with increased bone formation and parallel the increased osteoclastogenesis and bone resorption (11). The evidence that IL-6 type cytokines stimulate osteoblastogenesis and increase bone formation suggests that the overproduction of IL-6 and increased sensitivity to IL-6 and other members of the cytokine family that utilizes gp130 (IL-11, LIF, OSM, CNTF) may account for both the increased osteoclastogenesis as well as for the increased osteoblast formation that follows the loss of gonadal function. Consistent with this view, an increase in the number of a specific subset of CFU-F colonies in bone marrow cultures from ovariectomized mice can be replicated by adding IL-6 type cytokines to murine bone marrow cultures from estrogen sufficient mice (Jilka et al., unpublished observations).

The evidence that IL-6 type cytokines promote differentiation of osteoblastic progenitors and also stimulate osteoclastogenesis via their effects on the former cell type, taken together with the evidence that mesenchymal cell differentiation and osteoclastogenesis are tightly linked, strongly suggest that stimulation of mesenchymal cell differentiation toward the osteoblastic lineage following estrogen loss may be the first event that ensues following the hormonal change. Increased osteoclastogenesis and bone loss, therefore, could be downstream consequences of this change. Direct support for this contention is provided by work from the author's laboratory that indicates that estrogen suppresses the rate of replication of osteogenic stem cell progenitors in the murine bone marrow (35).

Other Cytokines and Growth Factors Implicated in Postmenopausal Bone Loss

TNF and IL-1 are potent stimulators of osteoclast formation and bone resorption. Unlike IL-6, which is produced at high levels by stromal/osteoblastic cells, IL-1 and TNF are produced at very low levels, if at all, by these cells. Instead, the main cellular source of IL-1 and TNF in bone is probably bone marrow monocytes and macrophages. Their ability to stimulate osteoclast formation, therefore, is probably mediated indirectly via their ability to increase the synthesis of other cytokines (e.g., IL-6) that are produced by stromal/osteoblastic cells and are directly responsible for the stimulation of osteoclastogenesis. In support of this, IL–1–induced osteoclast development can be blocked by an IL-6 superantagonist in human bone marrow cultures, or by the expression of a dominant negative form of STAT3 in murine cultures.

Estrogen suppresses the activity of the TNF gene promoter by 50% in a murine macrophage cell line following stimulation with the combination of IL and TNF by preventing binding of the transcription factor AP-1 (36). Estrogen also enhances the expression of the IL-1 decoy receptor gene, whereas it decreases the expression of the IL-1 signaling receptor in human osteoclasts (37); mice deficient in the functional IL-1 receptor do not lose bone after ovariectomy (38). In support of the contention that IL-1 and TNF play an important role in the bone loss caused by loss of estrogen, administration of IL-1RA and/or TNF-BP ameliorates the bone loss caused by ovariectomy in both rats and mice (39,40).

Unlike the case with IL-6, it remains unknown whether SERMs or androgens affect IL-1 or TNF production. Be that as it may, in contrast to IL-6, which stimulates osteoblast differentiation, IL-1 and TNF inhibit the expression of the osteoblast phenotype. In addition, IL-1 and TNF suppress the production of pro-osteoblastogenic cytokines (e.g., IGF and PDGF), and stimulate the expression IGFBP-4 that inhibits osteoblastogenesis (41–44). These properties therefore distinguish them from IL-6 and make them unlikely mediators of the increased osteoblastogenesis and thus the increased rate of bone remodeling that follows loss of sex steroids. Nonetheless, it is possible that IL-1 and TNF may arrest the differentiation of stromal/osteoblastic cells at a stage where these cells can stimulate osteoclastogenesis, thereby accounting for the evidence that IL-1RA and TNF-BP prevent loss of bone in estrogen deficiency.

Loss of estrogen may enhance the ability of IL-1 and TNF to stimulate the expression of M-CSF, a cytokine that is essential for osteoclastogenesis. Indeed, both TNF and IL-1 are capable of stimulating the expression of M-CSF by stromal/osteoblastic cells; ovariectomy increases the ability of IL-1 and TNF to stimulate M-CSF expression in ex vivo cultures of bone marrow cells. Addition of estrogen to these cultures, however, cannot suppress M-CSF (45). These findings are remarkably similar to the evidence that ovariectomy increases the number of the osteopontin-expressing cells in the bone marrow, whereas estrogen fails to influence the expression of this gene, which is required for osteoclast formation, at least in vitro (46). Thus, estrogen deficiency appears to cause an increase in the number of stromal/osteoblastic cells that are capable of producing M-CSF and osteopontin, possibly by changing the cell phenotype such that cytokine or receptor expression is altered, rather than by directly influencing the transcription of the M-CSF gene. This interpretation is consistent with the evidence that ovariectomy increases the number of cells of the osteoblast lineage.

TGFβ has been also implicated as mediator of the effects of estrogen loss on bone; however, whereas estrogen, as well as raloxifene and tamoxifen, stimulates TGFβ production by osteoblastic cells (47), there seems to be a paradoxical increase in the abundance of the TGFβ mRNA in bone following ovariectomy (48). TGFβ has inhibitory effects on osteoclast formation in vitro, and it stimulates early stages of osteoblast differentiation, but it does suppress the later stages of osteoblast formation both in vivo and in vitro (49). On the other hand, administration of TGFβ into the cancellous bone of rats, 2 weeks after ovariectomy, suppresses the increased number of osteoclasts, as well as the increased bone formation (50). It is possible that these in vivo effects are due to the proapoptotic effects of this cytokine on osteoclasts and osteoclast progenitors (51). In vivo overexpression of TGFβ in the osteoblasts of sex-steroid–sufficient mice unexpectedly caused high-turnover bone remodeling and osteopenia (52). On the other hand, mice with osteoblasts insensitive to TGFβ, which results from the expression of a dominant-negative TGFβ receptor, exhibit increased bone mass, but still lose bone following ovariectomy (53). In view of this evidence, the contention that TGFβ may be a pathogenetic factor in postmenopausal osteoporosis remains tenuous.

Effects of Estrogen on Bone Cell Apoptosis

In addition to the evidence for the modulating influence of estrogen on the birth of osteoclast and osteoblast, estrogen also affects the rate of apoptosis of these cells. Specifically, osteoclast apoptosis is stimulated by estrogen either directly or indirectly, via IL-6 or TGFβ (51), and perhaps via suppression of IL-6. Involvement of IL-6 in the prolongation of the life span of osteoclasts in estrogen deficiency is consistent with the potent antiapoptotic effects of this cytokine. In addition, it is consistent with the evidence that the estrogen receptor represses NFκB-activated gene expression and that NFκB induces genes that prevent apoptosis (54). Estrogen may have a directly opposite effect on the apoptosis of osteoblastic cells, as estrogen loss leads to increased osteocyte apoptosis in rats and humans (55,56). Increased life span of osteoclasts and decreased life span of osteoblasts following loss of estrogen could well explain the imbalance between resorption and formation and thereby the loss of bone.

Is Postmenopausal Bone Loss Due Solely to Estrogen Loss?

Observations from my group raise the possibility that postmenopausal bone loss may be due to estrogen deficiency as well as to changes in the activin–inhibin–follistatin hormonal system. Activins, like BMPs, are members of the TGFβ superfamily, and together with inhibin and follistatin they constitute a complex interactive regulatory loop in the anterior pituitary, where they regulate FSH production. Activin and inhibin are produced in the bone marrow, probably by mesenchymal cells of the stromal–osteoblastic lineage; and gonadal-derived inhibin may accumulate in the bone marrow. Activin has direct stimulatory effects on mesenchymal cell differentiation and osteoclastogenesis, which may be antagonized by gonadally derived inhibin and marrow-derived follistatin (57). Moreover, the activin/inhibin/follistatin counterregulatory system may well serve to modulate the tonic baseline control of osteoblastogenesis and osteoclastogenesis that is provided by the BMP2/4-noggin balance because activin can overcome the effects of noggin on osteoblastogenesis and osteoclastogenesis (58). It is therefore possible that the loss of gonadal inhibin in peri- and postmenopausal women increases local marrow activin tone, thereby stimulating osteoblastogenesis and osteoclastogenesis, increasing the rate of bone turnover, and contributing to the bone loss heretofore attributed solely to estrogen deficiency.

Summary

Several research breakthroughs and conceptual advances during the last few years have led to the appreciation of the fundamental fact that maintenance of an appropriate number of bone cells is essential for physiologic remodeling and skeletal homeostasis. Inappropriate increases or decreases in the production of bone cells, which reflect the frequency of cell division of the appropriate precursor cells, and changes in their life span, which reflect the timing of death by apoptosis, are responsible for the imbalance between bone resorption and formation that underlies not only postmenopausal osteoporosis, as well as senile (59) and glucocorticoid-induced osteoporosis (60).

References

1. Parfitt AM. Osteonal and hemi-osteonal remodeling: the spatial and temporal framework for signal traffic in adult human bone. J Cell Biochem 1994;55:273–86.
2. Lindsay R, Hart DM, Forrest C, Baird C. Prevention of spinal osteoporosis in oophorectomised women. Lancet 1980;2:1151–54.
3. Frost HM. Bone remodeling and its relationship to metabolic bone diseases. Springfield: Charles C. Thomas, 1973.
4. Aubin JE, Liu F. The osteoblast lineage. In: Bilezikian JP, Raisz LG, Rodan GA, eds. Principles of bone biology. San Diego: Academic Press, 1996;51–67.
5. Roodman GD. Advances in bone biology: the osteoclast. Endocrinol Rev 1996;17:308–32.
6. Manolagas SC, Jilka RL. Bone marrow, cytokines, and bone remodeling—Emerging insights into the pathophysiology of osteoporosis. N Engl J Med 1995;332:305–11.
7. Hughes DE, Boyce BF. Apoptosis in bone physiology and disease. J Clin Pathol Clin Mol Pathol 1997;50:132–37.

8. Aarden EM, Burger EH, Nijweide PJ. Function of osteocytes in bone. J Cell Biochem 1994;55:287–99.
9. Jilka RL, Weinstein RS, Bellido T, Parfitt AM, Manolagas SC. Osteoblast programmed cell death (apoptosis): modulation by growth factors and cytokines. J Bone Miner Res 1998;13:793–802.
10. Manolagas SC, Jilka RL, Bellido T, O'Brien CA, Parfitt AM. Interleukin-6-type cytokines and their receptors. In: Bilezikian JP, Raisz LG, Rodan GA, eds. Principles of bone biology. San Diego: Academic Press, 1996;701–13.
11. Jilka RL, Takahashi K, Munshi M, Williams DC, Roberson PK, Parfitt AM, et al. Loss of estrogen upregulates osteoblastogenesis in the murine bone marrow: evidence for autonomy from factors released during bone resorption. J Clin Invest 1998;101:1942–50.
12. Weinstein RS, Jilka RL, Parfitt AM, Manolagas S. The effect of androgen deficiency on murine bone remodeling and bone mineral density are mediated via cells of the osteoblastic lineage. Endocrinology 1997;138:4013–21.
13. Lacey DL, Timms E, Tan HL, Kelley MJ, Dunstan CR, Burgess T, et al. Osteoprotegerin ligand is a cytokine that regulates osteoclast differentiation. Cell 1998;93:165–76.
14. Ducy P, Zhang R, Goeffroy V, Ridall AL, Karsenty G. Osf2/Cbfa1: a transcriptional activator of osteoblast differentiation. Cell 1997;89:747–54.
15. Komori T, Yagi H, Nomura S, Yamaguchi A, Sasaki K, Deguchi K, et al. Targeted disruption of Cbfa1 results in a complete lack of bone formation owing to maturational arrest of osteoblasts. Cell 1997;89:755–64.
16. Rosen V, Cox K, Hattersley G. Bone morphogenetic proteins. In: Principles of bone biology. Bilezikian JP, Raisz LG, Rodan GA, eds. San Diego: Academic Press, 1996:606–71.
17. Abe E, Yamamoto M, Taguchi Y, Lecka-Czernik B, Economides AN, Stahl N, et al. Requirement of BMPs 2/4 for Postnatal Osteoblast as well as osteoclast formation: antagonism by Noggin. Bone 1998;23:S242.
18. O'Brien CA, Farrar NC, Manolagas SC. Identification of an OSF-2 binding site in the murine RANKL/OPGL gene promoter: a potential link between osteoblastogenesis and osteoclastogenesis. Bone 1998;23:S149.
19. Stahl N, Boulton TG, Farruggella T, Ip NY, Davis S, Witthuhn BA, et al. Association and activation of Jak-Tyk kinases by CNTF-LIF-OSM-IL-6 β receptor components. Science 1994;263:92–95.
20. Udagawa N, Takahashi N, Katagiri T, Tamura T, Wada S, Findlay DM, et al. Interleukin (IL)-6 induction of osteoclast differentiation depends on IL-6 receptors expressed on osteoblastic cells but not on osteoclast progenitors. J Exp Med 1995;182:1461–68.
21. Jilka RL, Hangoc G, Girasole G, Passeri G, Williams DC, Abrams JS, et al. Increased osteoclast development after estrogen loss: mediation by interleukin-6. Science 1992;257:88–91.
22. Poli V, Balena R, Fattori E, Markatos A, Yamamoto A, Tanaka H, et al. Interleukin-6 deficient mice are protected from bone loss caused by estrogen depletion. EMBO J 1994;13:1189–96.
23. Bellido T, Stahl N, Farruggella TJ, Borba V, Yancopoulos GD, Manolagas SC. Detection of receptors for interleukin-6, interleukin-11, leukemia inhibitory factor, oncostatin M, and ciliary neurotrophic factor in bone marrow stromal/osteoblastic cells. J Clin Invest 1996;97:431–37.

24. Bellido T, Borba VZC, Roberson P, Manolagas SC. Activation of the JAK/STAT signal transduction pathway by IL-6 type cytokines promotes osteoblast differentiation. Endocrinology 1997;138:3666–76.

25. Taguchi T, Yamate T, Mocharla H, Lin S-C, Vertino A, DeTogni P, et al. Interleukin-6 induces osteoblast differentiation in uncommitted embryonic fibroblasts (EF). J Bone Miner Res 1996;11:S101.

26. Bellido T, O'Brien CA, Roberson PK, Manolagas SC. Transcriptional activation of the p21 WAFI. CIPI. SDII gene by IL-6 type cytokines: a prerequisite for their prodifferentiating and anti-apoptotic effects on human osteoblastic cells. J Biol Chem 1998;273:21137–44.

27. Girasole G, Jilka RL, Passeri G, Boswell S, Boder G, Williams DC, et al. 17β-estradiol inhibits interleukin-6 production by bone marrow-derived stromal cells and osteoblasts in vitro. J Clin Invest 1992;89:883–91.

28. Pottratz ST, Bellido T, Mocharla H, Crabb D, Manolagas SC. 17β-estradiol inhibits expression of human interleukin-6 promoter-reporter constructs by a receptor-dependent mechanism. J Clin Invest 1994;93:944–50.

29. Bellido T, Jilka RL, Boyce BF, Girasole G, Broxmeyer H, Dalrymple SA, et al. Regulation of interleukin-6, osteoclastogenesis and bone mass by androgens: the role of the androgen receptor. J Clin Invest 1995;95:2886–95.

30. Galien R, Evans HF, Garcia T. Involvement of ccaat/enhancer-binding protein and nuclear factor-kb binding sites in interleukin-6 promoter inhibition by estrogens. Mol Endocrinol 1996;10:713–22.

31. Nuttall ME, Nadeau D, Prichett WP, Gowen M. Idoxifene, a tissue selective estrogen agonist/antagonist, has a mechanism of action in bone similar to estrogen and distinct from raloxifene. J Bone Miner Res 1997;12(S1):S170.

32. McDonnell DP, Norris JD. Analysis of the molecular pharmacology of estrogen receptor agonists and antagonists provides insights into the mechanism of action of estrogen on bone. Osteoporosis Int 1997;(suppl 1):S29–S34.

33. Lin SC, Yamate T, Taguchi Y, Borba V, Girasole G, O'Brien CA, et al. Regulation of the gp80 and gp130 subunits of the IL-6 receptor by sex steroids in the murine bone marrow. J Clin Invest 1997;100:1880–90.

34. Cheleuitte D, Mizuno S, Glowacki J. In vitro secretion of cytokines by human bone marrow: effects of age and estrogen status. J Clin Endocrinol Metab 1998;83:2043–51.

35. Di Gregorio GB, Manolagas SC, Jilka RL. 17β-estradiol inhibits osteogenic stem cell replication in the murine bone marrow: a potential mechanism for its antiremodeling effects. Bone 1998;23:S507.

36. Kimble RB, Srivastava S, Pacifici R. Estrogen inhibits macrophage TNF gene expression by modulating binding of transcription factors to the AP-1 binding site. J Bone Miner Res 1997;12(S1):S441.

37. Sunyer T, Lewis J, Osdoby P. Estrogen decreases the steady state levels of the IL-1 signaling receptor (Type I) while increasing those of the IL-1 decoy receptor (Type II) mRNAs in human osteoclast-like cells. J Bone Miner Res 1997;12(S1):S135.

38. Lorenzo J, Naprta A, Rao Y, Alander C, Glaccum M, Gronowicz G, et al. Mice deficient in the functional interleukin-1 receptor I (IL-1R1) do not lose bone mass after ovariectomy. J Bone Miner Res 1997;12(suppl 1):S126.

39. Ammann P, Rizzoli R, Bonjour JP, Bourrin SJ, Meyer M, Vassalli P, et al. Transgenic mice expressing soluble tumor necrosis factor-receptor are protected against bone loss caused by estrogen deficiency. J Clin Invest 1997;99:1699–703.

40. Kitazawa R, Kimble RB, Vannice JL, Kung VT, Pacifici R. Interleukin-1 receptor antagonist and tumor necrosis factor binding protein decrease osteoclast formation and bone resorption in ovariectomized mice. J Clin Invest 1994;94:2397–406.
41. Scharla SH, Strong DD, Mohan S, Chevalley T, Linkhart TA. Effect of tumor necrosis factor-alpha on the expression of insulin-like growth factor I and insulin-like growth factor binding protein 4 in mouse osteoblasts. Eur J Endocrinol 1994;131:293–301.
42. Xie J, Stroumza J, Graves DT. IL-1 down-regulates platelet-derived growth factor-α receptor gene expression at the transcriptional level in human osteoblastic cells. J Immunol 1994;153:378–85.
43. Tsukamoto T, Matsui T, Nakata H, Ito M, Natazuka T, Fukase M, et al. Interleukin-1 enhances the response of osteoblasts to platelet-derived growth factor through the α receptor-specific up-regulation. J Biol Chem 1991;266:10143–47.
44. Wetzler M, Talpaz M, Lowe DG, Baiocchi G, Gutterman JU, Kurzrock R. Constitutive expression of leukemia inhibitory factor RNA by human bone marrow stromal cells and modulation by IL-1, TNF-α, and TGF-β. Exp Hematol 1991; 19:347–51.
45. Kimble RB, Srivastava S, Ross FP, Matayoshi A, Pacifici R. Estrogen deficiency increases the ability of stromal cells to support murine osteoclastogenesis via an interleukin-1- and tumor necrosis factor-mediated stimulation of macrophage colony-stimulating factor production. J Biol Chem 1996;271:28890–97.
46. Yamate T, Mocharla H, Taguchi Y, Igietseme JU, Manolagas SC, Abe E. Osteopontin expression by osteoclast and osteoblast progenitors in the murine bone marrow: demonstration of its requirement for osteoclastogenesis and its increase after ovariectomy. Endocrinology 1997;138:3047–55.
47. Yang NN, Bryant HU, Hardikar S, Sato M, Galvin RJS, Glasebrook AL, et al. Estrogen and raloxifene stimulate transforming growth factor-β3 gene expression in rat bone: a potential mechanism for estrogen- or raloxifene-mediated bone maintenance. Endocrinology 1996;137:2075–84.
48. Cavolina JM, Evans GL, Harris SA, Zhang M, Westerlind K, Turner RT. The effects of orbital spaceflight on bone histomorphometry and messenger ribonucleic acid levels for bone matrix proteins and skeletal signaling peptides in ovariectomized growing rats. Endocrinology 1997;138:1567–76.
49. Bonewald L. Transforming growth factor-β. In: Principles of bone biology. Bilezikian JP, Raisz L, Rodan GA, eds. San Diego: Academic Press, 1996:647–59.
50. Beaudreuil J, Mbalaviele G, Cohen-Solal M, Morieux C, De Vernejoul MC, Orcel P. Short-term local injections of transforming growth factor-β₁ decrease ovariectomy-stimulated osteoclastic resorption in vivo in rats. J Bone Miner Res 1995; 10:971–77.
51. Hughes DE, Dai A, Tiffee JC, Li HH, Mundy GR, Boyce BF. Estrogen promotes apoptosis of murine osteoclasts mediated by TGF-β. Nat Med 1996;2:1132–36.
52. Erlebacher A, Derynck R. Increased expression of TGF-β2 in osteoblasts results in an osteoporosis-like phenotype. J Cell Biol 1996;132:195–210.
53. Filvaroff EH, Erlebacher A, Ye JQ, Gitelman SE, Lotz JC, Heilmann MR, et al. Increased trabecular bone in transgenic mice expressing a truncated, dominant-negative, type II TGF-β receptor in osteoblasts. J Bone Miner Res 1997;12(suppl 1): S118.
54. Van Antwerp DJ, Martin SJ, Kafri T, Green DR, Verma IM. Suppression of TNF-a-induced apoptosis by NF-kappaB. Science 1996;274:787–89.
55. Tomkinson A, Reeve J, Shaw RW, Noble BS. The death of osteocytes via apoptosis

accompanies estrogen withdrawal in human bone. J Clin Endocrinol Metab 1997;82:3128–35.

56. Tomkinson A, Gevers EF, Wit JM, Reeve J, Noble BS. The role of estrogen in the control of rat osteocyte apoptosis. J Bone Miner Res 1998;13:1243–50.

57. Coker JK, Ballew CB, Jilka RL, Manolagas SC, Gaddy-Kurten D. Inhibin and activin exert opposing effects on osteoblastogenesis and osteoclastogenesis. Abstract OR39-4, Eightieth Annual Meeting of the Endocrine Society, 1998.

58. Gaddy-Kurten D, Coker JK, Abe E, Stahl N, Manolagas SC. Activin substitutes for the BMP2/4 requirement for, and the noggin inhibition of, osteoblastogenesis and osteoclastogenesis in adult murine bone marrow cultures. Bone 1998;23:S166.

59. Jilka RL, Weinstein RS, Takahashi K, Parfitt AM, Manolagas SC. Linkage of decreased bone mass with impaired osteoblastogenesis in a murine model of accelerated senescence. J Clin Invest 1996;97:1732–40.

60. Weinstein RS, Jilka RL, Parfitt AM, Manolagas SC. Inhibition of osteoblastogenesis and promotion of apoptosis of osteoblasts and osteocytes by glucocorticoids: potential mechanisms of their deleterious effects on bone. J Clin Invest 1998;102: 274–82.

14

Changes in Cardiovascular Risk Factors During the Peri- and Postmenopausal Years

KAREN A. MATTHEWS, LEWIS H. KULLER,
AND KIM SUTTON-TYRRELL

Introduction

It is controversial whether or not menopause accelerates women's risk of cardiovascular disease. Some investigators have observed that the gender difference in coronary heart disease (CHD) mortality rates is largest around the time of the average age of menopause in Caucasians and to a lesser extent in African-Americans (1). Comparisons of pre- and postmenopausal women categorized into 5-year age groups show higher rates of cardiovascular disease among the postmenopausal women (2). These findings suggest that the loss of ovarian function increases women's cardiovascular disease risk.

On the other hand, Tunstall-Pedoe (3) argues that menopause is unlikely to accelerate women's risk of cardiovascular disease because of the absence of an abrupt increase in the rates of women's cardiovascular disease after the menopause. Because there is considerable variability in the age of last menses (i.e., from 40 to 55 years of age) (4), however, a general populationwide increase in cardiovascular disease rates is unlikely at a specific age. Atherosclerosis, which underlies CHD, develops over a long period of time. Even if the loss of ovarian function accelerates the development of atherosclerosis, CHD events or mortality would not occur immediately.

Another argument against a menopause risk hypothesis is that no differences in CHD rates were obtained between women who have gone through the natural menopause and premenopausal women when statistical adjustments are made for other cardiovascular risk factors, including age, cigarette smoking, cholesterol, blood pressure, weight, and education attainment (5). It is difficult to know how best to interpret these results because some of the cardiovascular risk factors also predict the age at menopause. For example, women who smoke cigarettes and are less educated are also likely to have an early menopause (6,7). Thus, the statistical adjustments for smoking and educational attainment may be controlling for part of the variance accounted for by changes in ovarian hormone status.

One approach to understanding the influence of the menopausal transition and changes in ovarian hormone exposure on the development of CHD is to describe changes in cardiovascular risk factors during the perimenopause in relation to the development of CHD. A severe limitation, however, is a consequence of women tending to have the onset of CHD 10–15 years after men (i.e., after 65 years of age). During the interval between menopause and CHD onset, many women are diagnosed with other important conditions (e.g., diabetes) or experience significant psychosocial events (e.g., bereavement) that affect the interpretation of the influence of menopause. One advance in cardiovascular epidemiology is the reliable measurement of subclinical disease. Subclinical disease can be measured early in the postmenopausal years and can be evaluated in relationship to pre-, peri-, and postmenopausal levels of risk factors. This chapter reviews the influence of the perimenopausal transition on cardiovascular risk factors and how risk factors in midlife predict subsequent subclinical disease. We begin by first describing methodological issues entailed in the epidemiological study of changes in cardiovascular risk during the menopause.

Methodological Issues in the Study of Menopause

Untangling the effects of menopause from aging of other systems has been difficult because of study design. Some studies have approached the study of menopause and midlife by using chronological age as a proxy for menopause, knowing that the median age of menopause is around 51. The variation in age at menopause, however, is substantial, and this approach is not sufficiently precise. A second, more precise approach is to use the date of the last menstrual period, but it is subject to memory distortions if information on bleeding patterns is not collected prospectively. A third approach uses prospective assignment of bleeding patterns and then compares the rate of change in women who change from pre- to postmenopause between examinations with age-adjusted changes in women who remain premenopausal. Even though this approach has been useful, the data it yields must also be interpreted with care. This is because changes in bleeding patterns are an approximate marker of ovarian aging and because clinical examinations are rarely conducted in relation to change in bleeding pattern.

Several other challenges are noteworthy. In attempting to understand ovarian aging, most of the large-scale epidemiological studies have recruited women who have been at least in their mid-forties. This means that a substantial number of younger women who had already ceased bleeding by their mid-forties were not included. Perhaps more important, by starting with a sample of women in their mid-forties, the many women who have had a surgical menopause by that time are not involved in the study (7). The major reasons for hysterectomy prior to the mid-forties are conditions related to high estrogen exposure (i.e., endometriosis, fibroids, and dysfunctional uterine bleeding) (8). If women who have had a surgical menopause are exposed to higher levels of ovarian hormones or are sicker in some systematic way related to risk of cardiovascular disease, the remaining women who experience a natural menopause may be less likely to show a worsening cardiovascular profile because they are healthier. Finally, an increasing number of women are electing to use hormone replacement therapy during the early postmenopausal period, which may mask some of the effects of ovarian aging. Ovarian aging does not stop with the last menses.

Longitudinal Studies of the Menopause

Table 14.1 shows the published longitudinal data as of August 1998 (9–19) on changes in risk factors during the perimenopausal transition, defined as the interval from menstruating on a regular basis to cessation of menses for a minimum of 6–12 months. Eliminated from the table are those studies conducted on small sample sizes (20–21) or those studies that did not have a premenopausal evaluation.

Prior to reviewing the data, we highlight a number of features of the existing studies. First, note that with the exception of the participants in one study, participants were European Caucasians. Risk factors for cardiovascular disease vary somewhat by ethnic group. For example, African-Americans have higher levels of blood pressure but lower levels of cholesterol than do Caucasians (22–23). Furthermore, central adiposity is not as highly correlated with insulin resistance, blood pressure, and weight among African-Americans as it is among Caucasians. Thus, the findings on the influence of menopause on cardiovascular risk among Caucasians should not be generalized to other ethnicities.

Second, with the exception of the Pittsburgh Healthy Women Study, the examinations are not scheduled in relationship to change in bleeding patterns. Annual assessments are satisfactory for examining changes in the menopausal status in relation to changes in risk factors, but most of the available data are based on repeat assessments across a long time interval in which many health and psychological changes can occur.

Third, none of the studies measured reproductive hormones at the time of the clinical examination, except for the Pittsburgh Healthy Women Study. In that study, too, the hormonal measurements were not conducted at study entry when the women were pre- or perimenopausal, but were conducted when women ceased cycling. Thus, the hypothesis underlying the menopausal risk hypothesis (i.e., ovarian hormone change) was not directly tested.

Fourth, a limited array of cardiovascular risk factors were evaluated (i.e., cholesterol, blood pressure, and weight). No information is available on clotting factors, and a full lipid profile was unavailable in most studies.

TABLE 14.1. Longitudinal studies of cardiovascular risk factor change during perimenopausal transition.

Investigators, year	Sample	Frequency of assessment/ length of follow-up	Analysis	Measures	Menopause results
Hjortland, McNamara, and Kannel, 1976(9)	N = 1,686 premenopausal, Caucasian, aged 40–51 from Framingham Heart Study	biennial/ 16 years	Comparisons of change between those becoming menopausal between two exams and those remaining premenopausal for at least 2 years	BP, total. cholesterol, weight, glucose, vital capacity, hemoglobin	Cholesterol ↑ for natural and surgical cases, no effect on any other risk factors.
Lindquist, 1982 (10) and Björkelund et al., 1996 (11)	1,302 aged 38–60 years mixed status from Gothenburg, Sweden	2 times/ 6 years	Matched by 2 year age groups, tests for significant change within each of 3 groups: those remaining pre-, changing to post-, and always postmenopausal in each age cohort excluding 60 year olds; WHR analysis based on 46 year olds	BP, weight, cholesterol, WHR	Cholesterol, triglycerides, and WHR ↑, no effect on any other risk factors
Anderson et al., 1987 (12)	1,090 aged 20–44, mixed status, Framingham Offspring Study	2 times/ 8 years	Multivariate analysis predicting risk factors adjusting for age, BMI, alcohol, smoking	LDL, HDL, VLDL	Total LDL, HDL ↑, no effect on VLDL
Matthews et al., 1989 (13) and Wing et al., 1991 (14)	541 premenopausal, healthy, aged 42–50, 90% Caucasian, Pittsburgh, PA	baseline, 3 months, 1, 2, 5, & 8 years postmeno-pause from 1983 to present	Comparison of change between those becoming menopausal and age-matched premenopausal; multivariate analysis in later papers	LDL, HDL, HDL2, triglycerides, BP, glucose, insulin, BMI, WHR	LDL ↑, HDL2 ↓, no effect on any other risk factors

Study	Population	Frequency/ years	Design	Measures	Results
van Beresteyn, van't Hof & deWaard, 1989 (15) and van Beresteijn et al., 1993 (16)	168 healthy peri- and post-menopausal, aged 49–56, stable weight, from Ede, the Netherlands	Annual/ 10 years	Slope of cardiovascular risk factors in years before/after last menstrual period in 42 perimenopausal; adjusted for age and BMI for cholesterol, adjusted for BMI for BP	Cholesterol, weight, BP	Cholesterol ↑, BP ↓, no effect on weight
Akahoshi et al., 1996 (17)	From 1,501 premenopausal, ≤ 50 years, 579 natural postmenopausal and 134 surgical postmenopausal	Biennial/ 16 years	Slope of cardiovascular risk factors in years before/after last menstrual period in women compared to slope of age-matched men	Cholesterol, SBP, BMI	Cholesterol ↑, no effect on SBP, BMI
Casiglia et al., 1996 (18)	525, mixed status from Mirano, Italy	2 times/ 16 years	Change in BP, rates of essential hypertension in those who changed menopausal status ($n=51$) compared with always postmenopausal, matched on age	BP, change in BP to cold pressor	No effect on age-adjusted BP, rate of hypertension, or BP response to cold pressor
Staessen et al., 1997 (19)	315 mixed status, aged 30–70	2 times/ 5 years	44 who changed from pre- to postmenopausal compared with men matched by age and rank of BMI and women who remained pre- or postmenopausal	BP, ambulatory BP	SBP, DBP ↑ in women who were and became menopausal, no change in BP in premenopausal women, DBP ↑ in men matched to menopausal women, incidence of hypertensives. Somewhat high in menopausal women

Turning to the results described in Table 14.1, we find that almost all the longitudinal studies of European Caucasians report increases in total or low density lipoprotein (LDL) cholesterol that are greater than the comparison groups of either premenopausal women or matched men. A more atherogenic lipid profile with ovarian aging is consistent with the results of clinical trials administering postmenopausal hormone replacement therapy. These studies show that estrogen alone or in combination with progestin alters lipid levels, especially high density lipoprotein (HDL) cholesterol (24).

The influence of the perimenopause on blood pressure levels is not clear: One study showed an increase, but another showed a decline, and two had no effect. No weight gain can be attributed to the perimenopause above and beyond chronological aging. One study reported that waist-to-hip ratio increased with the perimenopause due to a decline in hip measurement. Possible changes in fat distribution with the perimenopause are important to evaluate because of the atherogenic effect of central adiposity. Surgical menopause does appear to be associated with central adiposity (25), and estrogen replacement therapy in postmenopausal women reduces central adiposity (26).

Subclinical Cardiovascular Disease in Middle-Aged Women

Measures of subclinical cardiovascular disease are very useful for studies of the influence of the perimenopause because they can be used to evaluate the influence of change in established and suspected risk factors during the perimenopause, without waiting for later decades when women are likely to experience clinical events. Among the apparently healthy are women who have substantial subclinical disease and those who have little or none. Studies of the perimenopause can track both the changes in subclinical disease during the perimenopause as well as the factors that predict progression of no disease to subclinical disease. Measurements can be used to identify women who should have more aggressive treatments during the perimenopause or can constitute intermediate markers to evaluate the effectiveness of interventions instituted in the peri- or early postmenopausal periods. Finally, subclinical indicators can be used analytically to evaluate the pathophysiological development of cardiovascular disease.

Several measures are available to evaluate subclinical disease in epidemiological investigations of cardiovascular risk. The assessment of subclinical atherosclerosis in the carotid arteries can be accomplished noninvasively by means of B-mode ultrasound. This technique allows detection and measurement of carotid intimamedia thickness (IMT) and plaque in a highly reproducible fashion. Atherosclerosis in the carotid arteries is a marker of generalized atherosclerosis throughout the body, including the large coronary arteries (27). Individuals with atherosclerosis in the carotid arteries compared with those without have a significantly higher risk of death and cardiovascular events.

The extent of IMT and plaque in middle-aged women has been reported in several studies. Bonithon-Kopp et al. reported that among 517 women aged 45–54 years, 30.4% had IMT of ≥ 0.75 mm and 8.7% had plaque of $\geq$ 1.75 mm and localized protrusion in the lumen (28). The San Daniele Project studied 1,328 men and women 18–99 years of age. Among the women aged 50–59 years, 10% had IMT ≥ 1 mm and 8% had nonstenotic plaque (29). In the Atherosclerosis Risk in Communities (ARIC) Study, among women aged 45–54, the average IMT was 0.65 mm for premenopausal women and 0.66 mm for postmenopausal women (30). Among the women who had been postmenopausal for at least 5 years, the average IMT was 0.75 mm. Prevalence rates of carotid plaque were 18–23% in women aged 45–64 years.

We have compared the prevalence of IMT and plaque of premenopausal and postmenopausal women recruited into two separate studies at an average age of 47 years. The postmenopausal women were from the Healthy Women Study, an epidemiological study of the natural history of cardiovascular risk factors during the perimenopause (31). To enter the study, they had to be free from chronic disease requiring medications, be normotensive, not taking hormone replacement therapy, and menstruate in the last 3 months. Carotid ultrasound scans were performed when women were at least 5 years after the last menses. The premenopausal women were from The Women's Healthy Lifestyle Project (WHLP), a randomized study evaluating the effect of low fat, low cholesterol diet and weight loss on preventing the lipid increases during the perimenopause (32). To enter the study, women had to meet the same criteria as the Healthy Women Study. In addition, they had to be willing to be randomized and to have a body mass index between 20 and 34, fasting glucose ≤ 140 mg/dl, and LDL cholesterol between 80 and 160 mg/dl, as well as total cholesterol between 140 and 260 mg/dl. Carotid scans were performed on average a year after study entry.

Our results showed that the average IMT was 0.77 mm for the postmenopausal women, whereas the average IMT was 0.69 mm for the premenopausal women (33). Using the criteria of 0.75 mm as indicative of high IMT, 44.9% of postmenopausal women and 16.1% of premenopausal women had high IMT. The prevalence of any focal plaque was 54% of postmenopausal women and 25% of premenopausal women (see Fig. 14.1).

A second technique for measuring subclinical disease is called electron beam computed tomography (EBCT), which can be used to measure calcification in the coronary and aorta. Previous studies show that coronary calcification measured by EBCT predicts extent of coronary atherosclerosis and is greater among patients who had a myocardial infarction than comparison groups (34). In the Healthy Women Study, 134 women were evaluated by EBCT at their clinic examination five to eight years after the menopause. Thirty-nine percent of the women had some calcification in the coronary arteries, whereas 77% had calcification in the aorta (35). The prevalence of

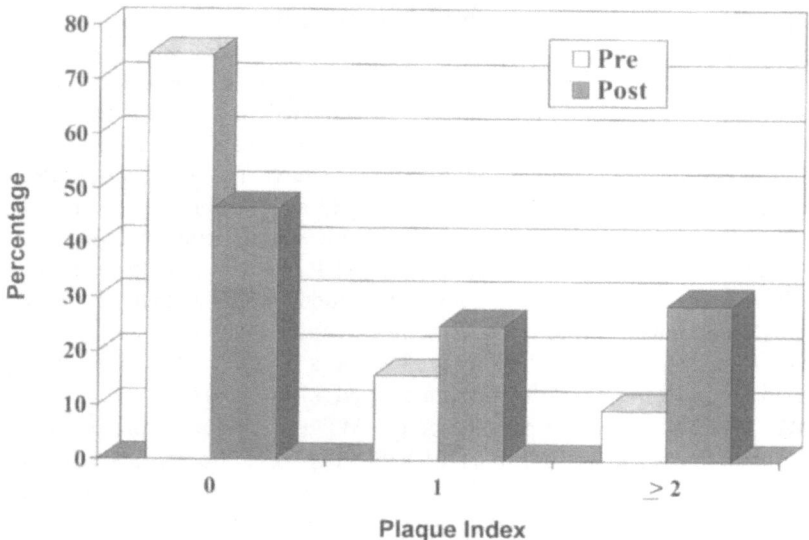

FIGURE 14.1. Prevalence of plaque (0 - none; 1 = 1 small plaque less than 30% of vessel diameter; 72 = at least 1 medium plaque between 30 and 50% of the vessel diameter or multiple small plaques) in premenopausal women from the women's Healthy Lifestyle Project and postmenopausal women from the Pittsburgh Healthy Women Study.

calcification in the coronary arteries was 30% in African-American women and 32% in Caucasian women aged 28–40 years in a substudy of CARDIA (36). The only study that has examined calcification (by radiography) according to menopausal status reported that women who had a bilateral oophorectomy had a fivefold greater risk of calcification in the abdominal aorta than premenopausal women; women who had natural menopause also had elevated risk (37).

Taken together, the cross-sectional data indicate that postmenopausal status is associated with the prevalence of plaque and high IMT in the carotid arteries and calcification in the aorta and coronary arteries. It will be important to establish the effects of ovarian aging on the development of carotid atherosclerosis and coronary calcification.

Cardiovascular Risk Factors During the Perimenopause and Subclinical Disease

We evaluated the relationship between indicators of subclinical disease and cardiovascular risk factors in participants from both the Healthy Women Study and WHLP (33). In WHLP, low HDL, high LDL-C, triglycerides, fasting insulin, body mass index, and systolic and diastolic blood pressure were correlated with high IMT in premenopausal women, whereas blood pressure was related to the plaque index.

In the Healthy Women Study, risk factors were measured when the women were premenopausal, in the first year after the cessation of menses, and at the time of the ultrasound scan, which occurred at least 5 years after the menopause (38). Extent of IMT in postmenopausal women was associated with lower HDL, higher glucose, systolic blood pressure, pulse pressure, and body mass index and being a smoker measured at each time point. Only premenopausal levels of triglycerides and postmenopausal levels of LDL-cholesterol levels were associated with IMT. Presence of plaque was also associated with elevated total cholesterol and smoking status measured at each time point and systolic blood pressure, pulse pressure, and triglycerides at the premenopausal evaluation only. It is of interest to note that more risk factors measured at study entry were significantly associated with plaque than were those measured at the time of the ultrasound scan.

Extent of calcification was also related to premenopausal risk factor levels among participants in the Healthy Women Study (35). Extent of coronary and aortic calcification was related to low HDL-C, and high LDL-C, systolic blood pressure, triglycerides, and apolipoprotein B. In addition, high body mass index was related to coronary calcification, whereas high glucose was related to aortic calcification.

Findings from the Muscatine Iowa Study also point to the ability of risk factors measured early to predict later calcification (39). In this study, extent of coronary artery calcification was measured in men and women between the ages of 29 and 37 years. They had been enrolled in a longitudinal study of risk factor assessment, and their risk factors were assessed at the ages of 15, 27, and 33. Weight, body mass index, skinfold thickness, systolic and diastolic blood pressure, triglyceride levels, and low HDL predicted subsequent calcification.

Taken together, these findings suggest that high levels of risk factors measured during the premenopausal years can identify women who are likely to have subclinical disease after the menopause. It is definitely not necessary to wait until the postmenopausal years to identify those women who are suitable candidates for pharmacologic or behavioral interventions. Interventions can be implemented in the premenopausal or perimenopausal years to prevent the development of atherosclerosis. The importance of early intervention is underscored by recent findings from a randomized clinical trial of continuous estrogen/progestin therapy among female coronary patients (40). This particular therapy was ineffective in preventing recurrent events.

A large cohort of women are approaching the usual age of the perimenopause (see Fig. 14.2; 41). Many American women are entering a time when the precursors of clinically important atherosclerotic disease are present, although the risk of a clinical event is low. Preventing the progression of subclinical to clinical disease is a high priority and can start in the perimenopausal years.

156   K.A. Matthews, L.H. Kuller, and K. Sutton-Tyrrell

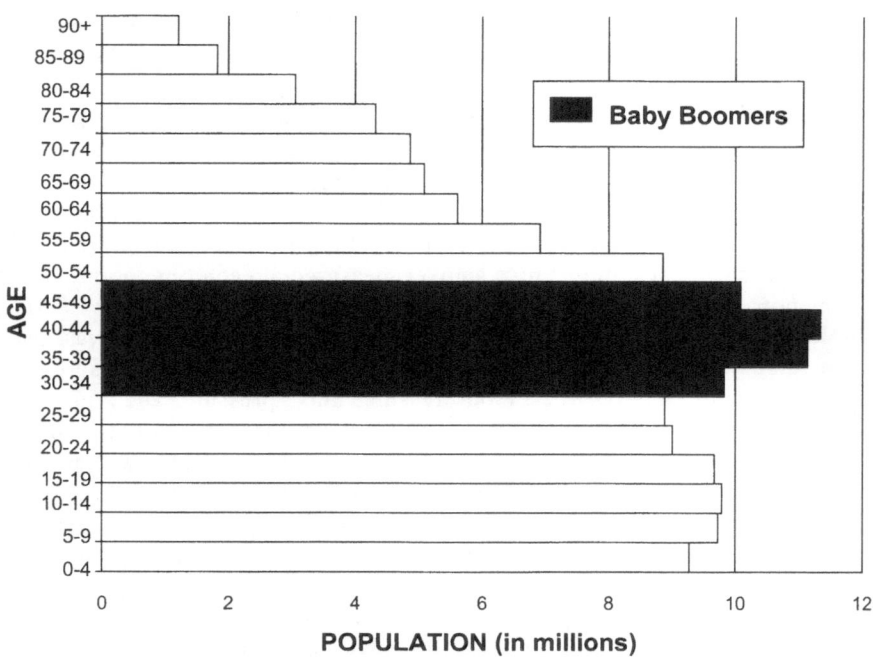

FIGURE 14.2. Age distribution of women in the United States in 2000 (Ref. 41).

References

1. Tracy RE. Sex differences in coronary disease: two opposing views. J Chron Dis 1966;19:1245–51.
2. Kannel WG, Hjortland MC, McNamara PM, Gordon T. Menopause and risk of cardiovascular disease: the Framingham study. Ann Intern Med 1976;85:447–52.
3. Tunstall-Pedoe H. Myth and paradox of coronary risk and the menopause. Lancet 1998;351:1425–27.
4. Matthews KA, Bromberger J, Egeland G. Behavioral antecedents and consequences of the menopause. In: Korenman SG, ed. The menopause. Norwell, MA: Serono Symposia USA, 1990:1–16.
5. Colditz GA, Wilett WC, Stampfer JJ, Rosner B, Speizer FE, Hennekens CH. Menopause and the risk for coronary heart disease in women. N Engl J Med 1987;316:1105–10.
6. McKinlay SM, Bifano NL, McKinlay JB. Smoking and age at menopause in women. Ann Intern Med 1985;103:350–56.
7. Wilcox LS, Koonin LM, Pokras R, Strauss LT, Xia Z, Peterson HB. Hysterectomy in the United States, 1988–1990. Obstet Gynecol 1994;83:549–55.
8. Carlson KJ, Nichols DH, Schiff I. Indications for hysterectomy. N Engl J Med 1993;328:856–60.
9. Hjortland MC, McNamara PM, Kannel WB. Some atherogenic concomitants of menopause: the Framingham study. Am J Epidemiol 1976;103:304–11.

10. Lindquist O. Intraindividual changes of blood pressure, serum lipids, and body weight in relation to menstrual status: results from a prospective population study of women in Goteborg, Sweden. Prev Med 1982;11:162–72.

11. Björkelund C, Lissner L, Andersoon S, Lapidus L, Bengtsson C. Reproductive history in relation to relative weight and fat distribution. Int J Obesity Rel Metab Dis 1996;20:213–19.

12. Anderson KM, Wilson PWG, Garrison FJ, Castelli WP. Longitudinal and secular trends in lipoprotein cholesterol measurements in a general population sample. The Framingham offspring study. Atherosclerosis 1987;68:59–66.

13. Matthews KA, Meilahn EN, Kuller LH, Kelsey SF, Caggiula AW, Wing RR. Menopause and risk factors for coronary heart disease. N Engl J Med 1989;321:641–46.

14. Wing RR, Matthews KA, Kuller LH, Meilahn EN. Weight gain at the time of menopause. Arch Intern Med 1991;151:97–102.

15. van Beresteyn ECH, van't Hof, deWaard H. Contributions of ovarian failure and aging to blood pressure in normotensive perimenopausal women: a mixed longitudinal study. Am J Epidemiol 1989;129:947–55.

16. van Beresteijn ECH, Korevaar JC, Huijbregts PCW, Shouten EG, Burema J, Kok FJ. Perimenopausal increase in serum cholesterol: a 10-year longitudinal study. Am J Epidemiol 1993;137:383–93.

17. Akahoshi M, Soda M, Nakashima E, Shimaoka K, Seto S, Yano K. Effects of menopause on trends of serum cholesterol, blood pressure, and body mass index. Circulation 1996;94:61–66.

18. Casiglia E, D'Este D, Ginocchio G, Colangeli G, Onesto C, Tramontin P, et al. Lack of influence of menopause on blood pressure and cardiovascular risk profile: a 16-year longitudinal study concerning a cohort of 568 women. J Hyperten 1996;14:729–36.

19. Staessen JA, Ginocchio G, Thijs L, Fagard R. Conventional and ambulatory blood pressure and menopause in a prospective population study. J Hum Hyperten 1997;11:507–14.

20. Fukami K, Koike K, Hirota K, Yoshikawa H, Miyake A. Perimenopausal changes in serum lipids and lipoproteins: a 7-year longitudinal study. Maturitas 1995;22:193–97.

21. Jensen J, Nilas L, Christiansen C. Influence of menopause on serum lipids and lipoproteins. Maturitas 1990;12:321–31.

22. Burt VL, Whelton P, Roccella EJ, Brown C, Cutler JA, Higgins M, et al. Prevalence of hypertension in the U.S. adult population: results from the Third National Health and Nutrition Examination Survey, 1988–91. Hypertension 1995;25:303–13.

23. National Center for Health Statistics. Health, United States, 1995. Hyattsville, MD: Public Health Services, 1996.

24. The Writing Group for the PEPI Trial. Effects of estrogen or estrogen/progestin regimens on heart disease risk factors in postmenopausal women. The Postmenopausal Estrogen/Progestin Interventions (PEPI) trial. JAMA 1995;273:199–208.

25. Sowers MF, Crutchfield M, Jannausch ML, Russell-Aulet M. Longitudinal changes in body composition in women approaching the midlife. Ann Hum Biol 1996;23:253–65.

26. Gambacciani M, Ciaponi M, Cappagli B, Piaggesi L, DeSimone L, Orlandi R, et al. Body weight, body fat distribution, and hormonal replacement therapy in early postmenopausal women. J Clin Endocrinol Metab 1997;82:414–17.

27. Kuller LH, Shemanski L, Psaty BM, Borhani NO, Gardin J, Haan MN, et al. Subclini-

cal disease as an independent risk factor for cardiovascular disease. Circulation 1995;92:720–26.

28. Bonithon-Kopp C, Scarabin PY, Taquet A, Touboul PJ, Malmejac A, Guize L. Risk factors for early carotid atherosclerosis in middle-aged French women. Arterioscler Thromb 1991;11:966–72.

29. Prati P, Vanuzzo D, Casaroli M, DiChiara A, DeBiasi F, Feruglio GA, et al. Prevalence and determinants of carotid atherosclerosis in a general population. Stroke 1992;23:1705–11.

30. Li R, Duncan BB, Metcalf PA, Crouse JR, Sharrett AR, Tyroler HA, et al. for the Atherosclerosis Risk in Communities (ARIC) Study Investigators. B-mode-detected carotid artery plaque in a general population. Stroke 1994;25:2377–83.

31. Matthews KA, Kelsey S, Meilahn E, Kuller LH, Wing RR. Educational attainment and behavioral and biologic risk factors for coronary heart disease in middle-aged women. Am J Epidemiol 1989;129:1132–44.

32. Simkin-Silverman L, Wing RR, Hansen DH, Klem ML, Pasagian-Macauley A, Meilahn EN, et al. Prevention of cardiovascular risk factor elevations in healthy premenopausal women. Prev Med 1995;24:509–17.

33. Sutton-Tyrrell K, Lassila HC, Meilahn E, Bunker C, Matthews KA, Kuller LH. Carotid atherosclerosis in premenopausal and postmenopausal women and its association with risk factors measured after menopause. Stroke 1998;29:1116–21.

34. Arad Y, Spadaro LA, Goodman K, Lledo-Perez A, Sherman S, Learner G, et al. The predictive value of electron beam CT of the coronary arteries: 19 month follow-up of 1173 asymptomatic subjects. Circulation 1996;93:1951–53.

35. Kuller LH, Matthews K, Edmundowicz D, Sutton-Tyrrell K. Subclinical atherosclerosis, coronary, aortic, carotid artery among women. Circulation 1998;98(suppl. 1):I-516–17. (Abstract)

36. Bild D, Folsom A, Lowe L, Sidney S, Kiefe C, Westfall A, et al. Coronary calcification in black and white young adults: the CARDIA study. Can J Cardiol 1997;13(suppl B):240B.

37. Witteman JCM, Grobbee DE, Kok FJ, Hofman A, Valkenburg HA. Increased risk of atherosclerosis in women after the menopause. Br Med J 1989;298:642–44.

38. Lassila HC, Sutton-Tyrrell K, Matthews KA, Wolfson SK, Kuller LH. Prevalence and determinants of carotid atherosclerosis in healthy postmenopausal women. Stroke 1997;28:513–17.

39. Mahoney LT, Burns TL, Stanford W, Thompson BH, Witt JD, Rost CA, et al. Coronary risk factors measured in childhood and young adult life are associated with coronary artery calcification in young adults: the muscatine study. J Am Coll Cardiol 1996;27:277–84.

40. Hulley S, Grady D, Bush T, Furberg C, Herrington D, Riggs B, et al. Randomized trial of estrogen plus progestin for secondary prevention of coronary heart disease in postmenopausal women. Heart and estrogen/progestin replacement study (HERS) research group. JAMA 1998;280:605–13.

41. U.S. Bureau of the Census. Current populations report, P251092. Population projections of the United States by age, sex, race, and Hispanic origin: 1992–2050. Washington, D.C.: Government Printing Office, 1992:48.

15

Estrogen and the Vascular Injury Response

C. Roger White and Suzanne Oparil

Introduction

Gender plays an important role in the modulation of cardiovascular risk and events. Cardiovascular disease is less prevalent in premenopausal women than it is in age-matched men, but there is an increase in coronary risk and events in women after menopause (1). Although a prominent antiatherogenic response to estrogen is a reduction in plasma low density lipoprotein cholesterol (LDL-c) and an elevation of high density lipoprotein cholesterol (HDL-c), other benefits are achieved by processes that are independent of changes in plasma lipid profiles (2–4). In this respect, estrogen therapy has proven to be effective in blunting the vascular injury response via both genomic and nongenomic mechanisms (5). Balloon injury of the rat carotid artery is an experimental paradigm that is commonly used to study mechanisms of atherogenesis. Estrogen minimizes the vascular injury response in this model through both direct antiproliferative effects and indirectly via the promotion of endothelial cell integrity and nitric oxide (NO) function (6–14). Additional protective effects of estrogen may be due to native antioxidant properties of the hormone (15). In this chapter, vasoprotective actions of estrogen exerted at the level of the arterial wall will be considered in the context of balloon injury models and hypercholesterolemia.

The Vascular Injury Response

Balloon injury models are characterized by endothelial cell denudation and the formation of a concentric fibromuscular lesion that encroaches on the arterial lumen (16–17). The association of macrophages and platelets with the vessel wall, the subsequent release of growth-promoting cytokines and chemoattractants, and the proliferation of vasculature smooth muscle cells (VSMC) in the subendothelial space are characteristic features of neointimal lesion formation (18). Cells in the developing neointima adopt a secretory phenotype, resulting in the excessive production of extracellular matrix. Because neointimal cells stain positively for smooth muscle α-actin, it has generally been assumed that they represent a phenotypic variant of VSMCs that migrate from the tunica media to the site of injury.

Evidence suggests, however, that injury to the vessel wall also leads to the activation of adventitial fibroblasts that become transformed to myofibroblasts (19–22). Immunohistochemical studies show that staining for bromodeoxyuridine (BrdU), an in vivo marker for cell proliferation, is increased in the adventitia during the acute phase after vascular injury and becomes localized to the media and neointima at later time points (21). These data suggest that myofibroblasts migrate/proliferate to the neointima in response to endoluminal injury. An increase in the expression of smooth muscle α-actin in proliferating myofibroblasts follows a similar time course (21). It has therefore been suggested that adventitial activation contributes to the neointimal proliferative response (Fig. 15.1).

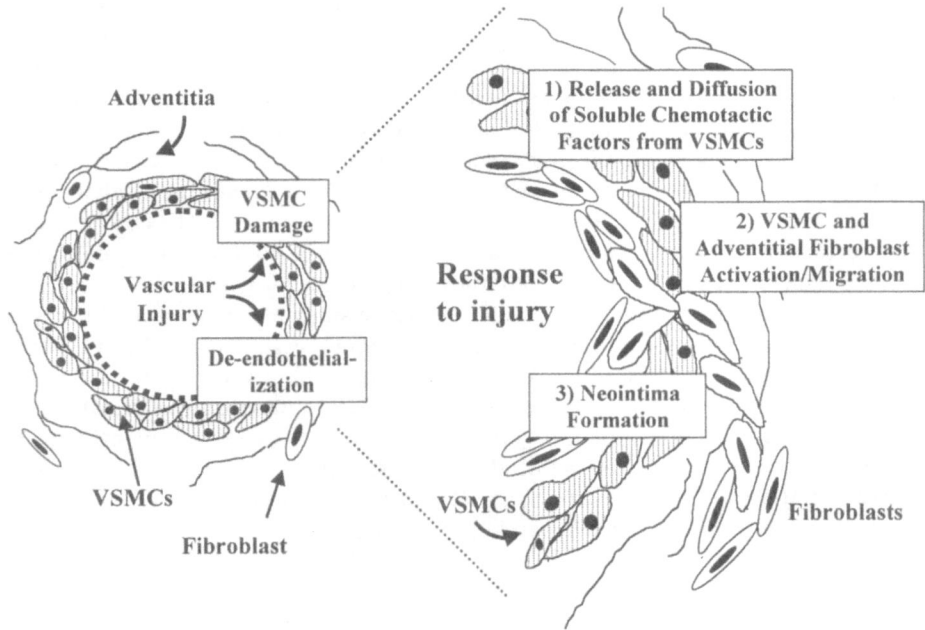

FIGURE 15.1. Neointima formation in balloon injured blood vessels. (Left panel) Balloon injury results in endothelial cell denudation (denoted by dashed line) and mechanical injury to the underlying VSMCs. (Right panel) Numerous factors contribute to the inflammatory response to balloon injury. These include release of chemotactic factors such as MCP-1 from VSMCs that attract circulating monocytes to the site of injury. VSMCs and fibroblasts also become activated and migrate to the site of injury, thus contributing to the developing neointima. Fibroblast migration may also be under the regulation of chemotactic factors released by the vessel wall.

These adaptations to balloon injury resemble the inflammatory lesions which develop in atherosclerotic blood vessels. The production of the oxidants superoxide anion (O_2^-) and hydrogen peroxide (H_2O_2) by adherent cells may be an additional stimulus for cell proliferation and migration in the injured vessel (23–24). As will be discussed later, NO may be a critical regulator of these cellular reactions. Under normal physiological conditions, NO limits vascular adhesion processes. With the loss of NO function in balloon injured vessels, however, inflammatory processes may proceed unchecked (25).

Antiproliferative Effects of Estrogen

Previous studies from our laboratory showed that estrogen supplementation reduces neointima formation and intima-to-media ratios in balloon-injured carotid arteries of ovariectomized and intact female rats at 2 weeks postinjury (6). This protective effect of estrogen was achieved at physiological plasma concentrations of the hormone. Estrogen treatment is without effect in intact males but it reduces the neointimal response in orchidectomized rats. These results suggest that even though estrogen prevents neointimal injury in females and gonadectomized males, the intact male is resistant to the vasoprotective actions of estrogen (26). Testosterone does not have direct stimulatory effects on neointima formation because its implantation in gonadectomized females does not enhance the neointimal response to injury (6).

Estrogen-mediated inhibition of neointimal growth correlates with a reduced expression of the protooncogene c-*myc*, which is a stimulus for VSMC hyperplasia (6). Cellular proliferation in atherosclerotic blood vessels, as assessed by BrdU labeling, is reduced in estrogen-treated female rabbits to a greater extent than in vehicle-treated controls (27). Similar results have been described in a rabbit model of balloon injury, where estrogen supplementation reduced ^{3}H thymidine incorporation and inhibited neointima formation in injured vessels (7). The inhibitory effect of estrogen on VSMC proliferation has also been confirmed under in vitro cell culture conditions (28).

The hyperproliferative response to vascular injury may be subject to regulation by progestins. Although treatment of gonadectomized balloon-injured rats of both sexes with the synthetic progestin medroxyprogesterone acetate (MPA) did not enhance neointimal proliferation, MPA administered in conjunction with estrogen blocked the antiproliferative effects of estrogen (29). This inhibitory action of MPA was unrelated to serum estrogen levels. Results of clinical studies suggest that MPA attenuates the beneficial effect of estrogen on plasma HDLc (30). Other data suggest an inhibitory effect of MPA on estrogen-stimulated NO production, as detected by the measurement of the NO metabolites nitrate and nitrite in plasma of postmenopausal women (31). Progesterone also blunts the vasoprotective effects of estrogen in experimental models of atherosclerosis. In surgically postmenopausal *cynomolgus* monkeys fed an atherogenic diet, estrogen therapy reduces lesion size by a mechanism that is independent of plasma lipoprotein concentration (32). Concurrent treatment with estrogen and MPA, however, blocks the anti-atherogenic effects of estrogen (32). Estrogen administration to female hypercholesterolemic rabbits similarly reduces cellular proliferation, as assessed by BrdU labeling, as well as aortic intimal lesion formation (27). These protective effects were completely blocked in animals receiving combined estrogen and progesterone therapy. Further, male rabbits did not benefit from estrogen treatment, which suggests gender-specific differences in the vasoprotective effects of estrogen in this model (27).

Adhesion of monocytes to the vessel wall plays an important role in the vascular response to balloon injury and atherosclerotic lesion formation. Monocyte chemotactic protein (MCP-1) is synthesized and released from monocytes, macrophages, endothelial cells and VSMCs, and promotes monocyte adhesion and migration (33). Using an in vitro assay for cell migration, Yamada and colleagues showed that estrogen inhibits monocyte migration induced by MCP-1 (33). This response was estrogen-receptor–dependent because the receptor antagonist tamoxifen restored the migratory response in estrogen-treated cells. This modulatory effect of estrogen may be mediated by inhibition of MCP-1 mRNA expression (34). Other data suggest that estrogen prevents monocyte binding/migration due to inhibition of cell surface adhesion molecule expression (35).

Estrogen and Reendothelialization

Estrogen may play an important role in maintaining the integrity of the endothelium under normal physiological conditions and by facilitating endothelial cell regrowth after injury. The Evans blue staining technique has been used to assess reendothelialization processes in injured blood vessels (36,37). Evans blue is a high molecular weight marker whose diffusion across the endothelium is restricted by the presence of an extensive network of tight junctions. Thus, reendothelialized regions of balloon injured vessels stain negatively for Evans blue, whereas endothelium-denuded regions avidly take up the dye. Estrogen clearly facilitates the anatomical reendothelialization of the balloon injured rat carotid artery. We found that estrogen supplementation in female rats promotes modest reendothelialization 2 weeks postballoon injury compared with vehicle-treated controls, but almost complete regrowth after 4 weeks (36). In this study, the plasma concentration of estrogen was within the physiological range. Another study reports extensive reendothelialization of balloon injured vessels 1 week postinjury in rats treated with supraphysiological concentrations of estrogen (37).

The functional reendothelialization of balloon injured vessels can be assessed by measurement of endothelium-dependent relaxation. We monitored acetylcholine (ACh)-mediated, NO-dependent relaxation in isolated carotid artery ring segments of female balloon injured rats that had received daily injections of estrogen or vehicle for 2 weeks prior to study (16). Isolated vessels of estrogen-supplemented rats were more sensitive to ACh, which reflects increased production of endothelial cell-derived NO. Furthermore, the enhanced relaxation of carotid rings from estrogen-supplemented females correlated positively with reduced neointima formation in this group (16). These results suggest that the estrogen facilitates the anatomical and functional reendothelialization of balloon-injured arteries and that accelerating the regrowth of the endothelium blunts neointima formation. Additional vasoprotective effects of estrogen under these conditions may be mediated indirectly via enhanced production of NO in reendothelialized blood vessels.

Endothelial cell mitogens may play an important role in vascular reendothelialization. Intravenous treatment of balloon-injured rabbits with basic-fibroblast growth factor (bFGF) results in significant reendothelialization of damaged iliac arteries compared with controls not receiving bFGF (38). The extent of neointimal thickening was not different, however, between the two groups. Functional responses of bFGF-exposed and control vessels were tested by in vitro bioassay of endothelium-dependent relaxation. Iliac arteries of rabbits receiving bFGF treatment demonstrated enhanced ACh-mediated relaxation compared with controls (38). The authors suggested that angiogenic growth factors such as bFGF may facilitate the recovery of endothelial cell function in injured vessels (38).

Local delivery of vascular endothelial growth factor (VEGF) to balloon-injured rat carotid arteries enhances reendothelialization of damaged vessels and also reduces neointima formation (39). VEGF is secreted by a variety of cell types in the vessel wall, including VSMCs and macrophages (40), and estrogen has been shown to regulate VEGF mRNA expression and protein in uterus and endometrial carcinoma cells (41,42). Thus, estrogen-mediated reendothelialization may be achieved through the enhanced synthesis of endothelial cell-specific growth factors. Secondary protective effects may be due to the enhanced formation of NO, which inhibits VSMC proliferation and macrophage/platelet adhesion.

Estrogen and Nitric Oxide Bioactivity

Estrogen stimulates the constitutive synthesis of NO in numerous tissues, including the uterine artery, heart, uterus, and skeletal muscle (14). Both pregnancy and estrogen treatment enhance neuronal and endothelial NO synthase (NOS) expression, but the inducible NOS isoform is unaffected (13). Furthermore, characterization of the gene for endothelial NOS indicates that the 5'-flanking region contains transcription factor binding sites for estrogen (43). We and others have proposed that NO acts as a secondary mediator of the protective actions of estrogen by inhibiting cell proliferation and vessel wall adhesion processes (15,25). This stimulatory effect of estrogen on NO formation and release may be achieved through both genomic and nongenomic pathways.

Estrogen-supplementation of balloon-injured female rats significantly increases plasma concentrations of the NO breakdown products nitrate and nitrite (16). Nitrate/nitrite concentration correlated inversely with neointimal area, which suggests that increased NO production is linked to a reduction in the vascular response to injury (16). Hormone replacement therapy has a similar stimulatory effect on plasma NO metabolites in postmenopausal women (8). An important protective action of estrogen on the vasculature may thus be related to enhanced production of NO. Increased production of NO attenuates vascular damage in balloon injury models (44,45). In the rat, intravenous infusion of an NO donor was shown to inhibit neointima formation and to restore the vasodilator response to ACh, which suggests that NO minimizes vessel damage and promotes the functional recovery of the endothelium (44). In vivo transfer of the gene for the endothelial NOS effectively prevents neointima formation in the balloon-injured carotid artery of the rat (46,47). Several studies show that short-term estrogen treatment results in increased coronary blood flow (10,11) and enhanced peripheral vasodilator responses in postmenopausal women (48,49). These responses occur over a time course (minutes) that is thought to be too rapid for the synthesis of new protein. This effect of estrogen on NO production, therefore, appears to be mediated by a nongenomic mechanism.

Additional benefits of estrogen-stimulated NO production may include the attenuation of inflammatory responses induced by cytokines in endothelial cells and VSMCs (50,51). Cytokines stimulate the expression of adhesion molecules such as vascular cell adhesion molecule (VCAM-1), resulting in the increased binding of inflammatory cells to the vessel wall. These processes are regulated by the NFκB transcription factor complex (52). Nitric oxide minimizes cell adhesion by stabilizing the NFκB complex, thus preventing its interaction with the gene encoding VCAM-1 (52).

Estrogen and Oxidant Stress

Antioxidant properties have also been ascribed to estrogen (12,15). Oxidants such as O_2^-, H_2O_2, and lipid peroxides are potent stimuli for VSMC growth and proliferation (18,23,24,53,54). Evidence suggests that estrogen may directly inhibit O_2^- formation (55). Estrogen supplementation of cholesterol-fed swine prevents defects in endothelium-dependent relaxation, which have been linked to the inactivation of NO by O_2^- (9,56,57). In this study, increased resistance to LDLc oxidation was positively correlated with plasma estrogen concentrations (9). Other studies show that estrogen attenuates the biochemical oxidation of LDLc in vitro and that the uptake and degradation of oxidized LDLc by macrophages is inhibited when LDLc is oxidized in the presence of estrogen (12,58–61). These studies suggest that the protective effects of estrogen on LDLc oxidation and endothelium-dependent relaxation are unrelated to shifts in plasma lipoprotein balance, but that they are due instead to direct protective effects at the level of the arterial wall (62,63).

We have reported similar effects of estrogen on the ex vivo oxidation of human LDLc (15). The mechanism underlying this activity has not been characterized, but, by analogy with other antioxidant compounds, it is likely that the hydroxy groups on the estrogen molecule bind transition metals or act as chain-breaking antioxidants by donating hydrogen atoms to peroxyl radicals (15,58,61). Similar arguments may be extended to the metabolites of membrane-associated estrogens that possess catechol functional groups (60). Additional antioxidant properties of estrogen may be exerted via the enhanced formation of NO in the vessel wall. Nitric oxide effectively inhibits lipid oxidation (64–66). This effect of NO appears to be similar to that exerted by estrogen, α-tocopherol, and other antioxidants in that it directly reacts with peroxyl radicals, thereby terminating oxidation chain propagation reactions.

Summary

Estrogen inhibits a number of inflammatory reactions at the level of the arterial wall (Fig. 15.2). With respect to balloon injury, our data suggest that

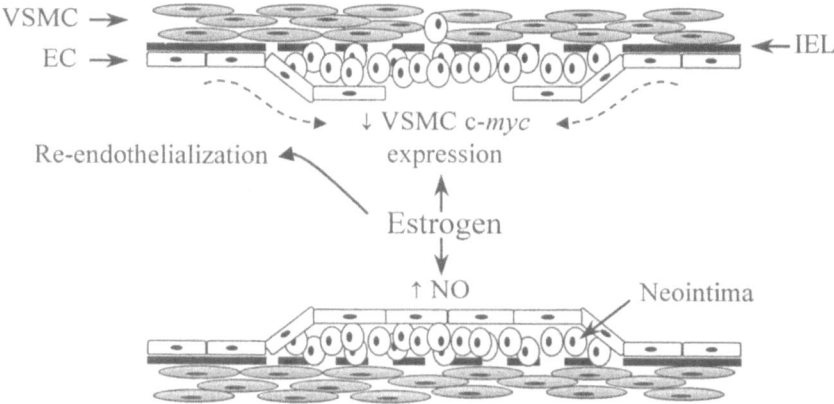

FIGURE 15.2. Mechanisms of estrogen-mediated vasoprotection. Estrogen limits cell proliferation and neointima formation in balloon injured arteries by several mechanisms. First, estrogen exerts a prominent inhibitory effect on the expression of c-myc, a proto-oncogene that plays an important role in cellular growth responses. Second, estrogen stimulates the reendothelialization of the damaged vessel by promoting the expression of endothelial cell (EC) growth factors such as VEGF. Additional benefits of estrogen may be achieved through the enhanced formation of NO in reendothelialized vessels. In this respect, NO may inhibit both VSMC proliferation and the adhesion of circulating inflammatory cells to the vessel wall. IEL denotes the internal elastic lamina.

vasoprotective effects of estrogen are exerted in two phases. During the early phase (2 weeks) postinjury, estrogen directly inhibits smooth muscle cell proliferation while also acting as a growth stimulus for endothelial cells. Evidence supports activation and migration of adventitial myofibroblasts as a novel mechanism of the early vascular injury response, and this represents another potential target for estrogen action. In the second 2 week period, when the injured vessel becomes reendothelialized, evidence suggests that estrogen indirectly limits cellular proliferation and inflammation via the stimulation of NO production.

References

1. Sullivan JM, Fowlkes LP. The clinical aspects of estrogen and the cardiovascular system. Obstet Gynecol 1996;87:S36–43.
2. Wild RA. Estrogen: effects on the cardiovascular tree. Obstet Gynecol 1996;87:S27–35.
3. Grodstein F, Stampfer M. The epidemiology of coronary heart disease and estrogen replacement in postmenopausal women. Prog Card Dis 1995;38:199–210.
4. Knopp RH, Zhu X, Bonet B. Effects of estrogens on lipoprotein metabolism and cardiovascular disease in women. Atherosclerosis 1994;110:S83–91.
5. Farhat MY, Abi-Younes S, Ramwell PW. Non-genomic effects of estrogen and the vessel wall. Biochem Pharmacol 1996;51:571–76.
6. Chen SJ, Li H, Durand J, Oparil S, Chen YF. Estrogen reduces myointimal proliferation after balloon injury of rat carotid artery. Circulation 1996;93:577–84.
7. Foegh ML, Asotra S, Howell MH, Ramwell PW. Estradiol inhibition of arterial neointimal hyperplasia after balloon injury. J Vasc Surg 1994;19:722–26.
8. Rosselli M, Imthurn B, Keller PJ, Jackson EK, Dubey RK. Circulating nitric oxide (nitrate/nitrite) levels in postmenopausal women substituted with 17β-estradiol and norethisterone acetate. A two-year follow-up study. Hypertension 1995;25:848–53.
9. Keaney JF, Shwaery GT, Xu A, Nicolosi RJ, Loscalzo J, Foxall TL, Vita JA. 17β-estradiol preserves endothelial vasodilator function and limits low-density lipoprotein oxidation in hypercholesterolemic swine. Circulation 1994;89:2251–59.
10. Herrington DM, Braden GA, Williams JK, Morgan TM. Endothelial-dependent coronary vasomotor responsiveness in postmenopausal women with and without estrogen replacement therapy. J Am Col Cardiol 1994;73:951–52.
11. Reis SE, Gloth ST, Blumenthal RS, Resar JR, Zacur HA, Gerstenblith G, et al. Ethinyl estradiol acutely attenuates abnormal coronary vasomotor responses to acetylcholine in postmenopausal women. Circulation 1994;89:52–60.
12. Nakano M, Sugioka K, Naito I, Takekoshi S, Niki E. Novel and potent biological antioxidants on membrane phospholipid peroxidation: 2-hydroxy estrone and 2-hydroxy estradiol. Biochem Biophys Res Comm 1987;142:919–24.
13. Weiner CP, Knowles RG, Moncada S. Induction of nitric oxide synthases early in pregnancy. Am J Obstet Gynecol 1994;171:838–43.
14. Weiner CP, Lizasoain I, Bayliss SA, Knowles RG, Charles IG, Moncada S. Induction of calcium-dependent nitric oxide synthases by sex hormones. Proc Natl Acad Sci USA 1994;91:5212–16.

15. White CR, Darley-Usmar V, Oparil S. Gender and cardiovascular disease: recent insights. Trends Cardio Med 1997;7:94–100.
16. White CR, Shelton J, Chen SJ, Durand J, Allen L, Darley-Usmar V, et al. Estrogen restores endothelial cell function in an experimental model of vascular injury. Circulation 1997;96:1624–30.
17. Stadius ML, Rowan R, Fleischhauer JF, Kernoff R, Billingham M, Gown AM. Time course and cellular characteristics of the iliac artery response to acute balloon injury. Arterio Thromb 1992;12:1267–73.
18. Ferns GAA, Forster L, Stewart-Lee A, Konneh M, Nourooz-Zadeh J, Anggard EE. Probucol inhibits neointimal thickening and macrophage accumulation after balloon injury in the cholesterol-fed rabbit. Proc Natl Acad Sci USA 1992;89:11312–16.
19. Wilcox JN, Scott NA. Potential role of the adventitia in arteritis and atherosclerosis. Int J Cardiol 1996;54(suppl 1):S21–35.
20. Lynch CM, Hara PS, Leonard JC, Williams JK, Dean RH, Geary RL. Adeno-associated virus vectors for vascular gene delivery. Circ Res 1997;80:497–505.
21. Shi Y, O'Brien JE, Fard A, Mannion JD, Wang D, Zalewski A. Adventitial myofibroblasts contribute to neointimal formation in injured porcine coronary arteries. Circulation 1996;94:1655–64.
22. Shi Y, Pieniek M, Fard A, O'Brien J, Mannion JD, Zalewski A. Adventitial remodeling after coronary arterial injury. Circulation 1996;93:340–48.
23. Rao GN, Berk BC. Active oxygen species stimulate vascular smooth muscle cell growth and proto-oncogene expression. Circ Res 1992;70:593–99.
24. Sundaresan M, Yu Z-X, Ferrans VJ, Irani K, Finkel T. Requirement for generation of H_2O_2 for platelet-derived growth factor signal transduction. Science 1995; 270:296–99.
25. Cooke JP, Tsao PS. Cytoprotective effects of nitric oxide. Circulation 1993;88:2451–54.
26. Oparil S, Levine RL, Chen SJ, Durand J, Chen YF. Sexually dimorphic response of the balloon injured rat carotid artery to hormone treatment. Circulation 1997; 95:1301–7.
27. Hanke H, Hanke S, Finking G, Muhic-Lohrer A, Muck AO, Schmahl FW, et al. Different effects of estrogen and progesterone on experimental atherosclerosis in female versus male rabbits. Quantification of cellular proliferation by bromo-deoxyuridine. Circulation 1996;94:175–81.
28. Bhalla RC, Toth KF, Bhatty RA, Thompson LP, Sharma RV. Estrogen reduces proliferation and agonist-induced calcium increase in coronary artery smooth muscle cells. Am J Physiol 1997;272(4 Pt 2):H1996–2003.
29. Levine RL, Chen SJ, Durand J, Chen YF, Oparil S. Medroxyprogesterone attenuates estrogen-mediated inhibition of neointima formation after balloon injury of the rat carotid artery. Circulation 1996;94:2221–27.
30. Sacks FM, Gerhard M, Walsh BW. Sex hormones, lipoproteins, and vascular reactivity. Curr Opin Lipid 1995;6:161–66.
31. Imthurn B, Rosselli M, Jaeger AW, Keller PJ, Dubey RK. Differential effects of hormone-replacement therapy on endogenous nitric oxide (nitrite/nitrate) levels in postmenopausal women substituted with 17 beta-estradiol valerate and cyproterone acetate or medroxyprogesterone acetate. J Clin Endocrinol Metab 1997;82: 388–94.

32. Adams MR, Register TC, Golden DL, Wagner JD, Williams JK. Medroxyprogesterone acetate antagonizes inhibitory effects of conjugated equine estrogens on coronary artery atherosclerosis. Arterio Thromb Vasc Biol 1997;17:217–21.

33. Yamada K, Hayashi T, Kuzuya M, Naito M, Asai K, Iguchi A. Physiological concentration of 17 beta-estradiol inhibits chemotaxis of human monocytes in response to monocyte chemotactic protein 1. Artery 1996;22:24–35.

34. Frazier-Jessen MR, Kovacs EJ. Estrogen modulation of JE/monocyte chemoattractant protein-1 mRNA expression in murine macrophages. J Immunol 1995; 154:1838–45.

35. Caulin-Glaser T, Watson CA, Pardi R, Bender JR. Effects of 17beta-estradiol on cytokine-induced endothelial cell adhesion molecule expression. J Clin Invest 1996;98:36–42.

36. Chen YF, Oparil S. Effects of sex steroids in vascular injury. In: Endocrinology of cardiovascular function. Levin ER, ed. Norwell, MA: Kluwer Academic Publishers, 1998:45–59.

37. Krasinski K, Spyridopoulos I, Asahara T, van der Zee R, Isner JM, Losordo DW. Estradiol accelerates functional endothelial recovery after arterial injury. Circulation 1997;95:1768–72.

38. Meurice T, Bauters C, Auffray JL, Vallet B, Hamon M, Valero F, et al. Basic fibroblast growth factor restores endothelium-dependent responses after balloon injury of rabbit arteries. Circulation 1996;93:18–22.

39. Asahara T, Bauters C, Pastore C, Kearney M, Rossow S, Bunting S, et al. Local delivery of vascular endothelial growth factor accelerates reendothelialization and attenuates intimal hyperplasia in balloon-injured rat carotid artery. Circulation 1995;91:2793–801.

40. Ross R. The pathogenesis of atherosclerosis: a perspective for the 1990s. Nature 1993;362:801–9.

41. Cullinan-Bove K, Koos RD. Vascular endothelial growth factor/vascular permeability factor expression in the rat uterus: rapid stimulation by estrogen correlates with estrogen-induced increases in uterine capillary permeability and growth. Endocrinology 1993;133:829–37.

42. Charnock-Jones DS, Sharkey AM, Rajput-Williams J, Burch D, Schofield JP, Fountain SA, et al. Identification and localization of alternatively spliced mRNAs for vascular endothelial growth factor in human uterus and estrogen regulation in endometrial carcinoma cell lines. Biol Reprod 1993;48:1120–28.

43. Venema RC, Nishida K, Alexander RW, Harrison DG, Murphy TJ. Organization of the bovine gene encoding the endothelial nitric oxide synthase. Biochim Biophys Acta 1994;1218:413–20.

44. Guo JP, Panday MM, Consigny PM, Lefer AM. Mechanisms of vascular preservation by a novel NO donor following carotid artery intimal injury. Am J Physiol 1995;269:H1122–31.

45. Major TC, Overhiser RW, Panek RL. Evidence for NO involvement in regulating vascular reactivity in balloon-injured rat carotid artery. Am J Physiol 1995;269: H988–96.

46. Janssens S, Flaherty D, Nong Z, Varenne O, van Pelt N, Haustermans C, et al. Human endothelial nitric oxide synthase gene transfer inhibits vascular smooth muscle cell proliferation and neointima formation after balloon injury in rats. Circulation 1998;97:1274–81.

47. Chen L, Daum G, Forough R, Clowes M, Walter U, Clowes AW. Overexpression of human endothelial nitric oxide synthase in rat vascular smooth muscle cells and in balloon-injured carotid artery. Circ Res 1998;82:862–70.
48. Lieberman EH, Gerhard MD, Uehata A, Walsh BW, Selwyn AP, Ganz P, et al. Estrogen improves endothelium-dependent, flow-mediated vasodilation in post-menopausal women. Am Coll Physicians 1994;121:936–41.
49. Gilligan DM, Badar DM, Panza JA, Quyyumi AA, Cannon RO. Effects of estrogen replacement therapy on peripheral vasomotor function in postmenopausal women. Am J Cardiol 1995;75:264–68.
50. De Caterina R, Libby P, Peng HB, Thannickal VJ, Rajavashisth TB, Gimbrone MA Jr, et al. Nitric oxide decreases cytokine-induced endothelial activation. Nitric oxide selectively reduces endothelial expression of adhesion molecules and proinflammatory cytokines. J Clin Invest 1995;96:60–68.
51. Shin WS, Hong YH, Peng HB, De Caterina R, Libby P, Liao JK. Nitric oxide attenuates vascular smooth muscle cell activation by interferon-gamma. The role of constitutive NF-kappa B activity. J Biol Chem 1996;271:11317–24.
52. Peng HB, Libby P, Liao JK. Induction and stabilization of I kappa B alpha by nitric oxide mediates inhibition of NF-kappa B. J Biol Chem 1995;70:14214–19.
53. Lafont AM, Chai YC, Cornhill JF, Whitlow PL, Howe PH, Chisholm GM. Effect of alpha-tocopherol on restenosis after angioplasty in a model of experimental atherosclerosis. J Clin Invest 1995;95:1018–25.
54. Ruef J, Hu ZY, Yin L-Y, Wu Y, Hanson SR, Kelly AB, et al. Induction of vascular endothelial growth factor in balloon-injured baboon arteries: a novel role for reactive oxygen species in atherosclerosis. Circ Res 1997;81:24–33.
55. Arnal JF, Clamens S, Pechet C, Negre-Salvayre A, Allera C, Girolami JP, et al. Ethinylestradiol does not enhance the expression of NO synthase in bovine endothelial cells but increases the release of bioactive nitric oxide by inhibiting superoxide anion production. Proc Natl Acad Sci USA 1996;93:4108–13.
56. White CR, Brock TA, Chang LY, Crapo J, Briscoe P, Ku D, et al. Superoxide and peroxynitrite in atherosclerosis. Proc Natl Acad Sci USA 1994;91:1044–48.
57. White CR, Darley-Usmar V, McAdams M, Berrington WR, Gore J, Thompson JA, et al. Circulating plasma xanthine oxidase contributes to vascular dysfunction in hypercholesterolemic rabbits. Proc Natl Acad Sci USA 1996;93:8745–49.
58. Sugioka, K, Shimosegawa Y, Nakano M. Estrogens as natural antioxidants of membrane phospholipid peroxidation. FEBS Lett 1987;210:37–39.
59. Rifici VA, Khachadurian AK. The inhibition of low-density lipoprotein oxidation by 17β-estradiol. Metabolism 1992;41:1110–14.
60. Lacort M, Leal AM, Liza M, Martin C, Martinez R, Ruiz-Larrea MB. Protective effects of estrogens and catecholestrogens against peroxidative membrane damage in vitro. Lipids 1995;30:141–46.
61. Maziere C, Auclair M, Ronveaux MF, Salmon S, Santus R, Maziere JC. Estrogens inhibit copper and cell-mediated modification of low density lipoprotein. Atherosclerosis 1991;89:175–82.
62. Williams JK, Kim YD, Adams MR, Chen MF, Myers AK, Ramwell PW. Effects of estrogen on cardiovascular responses of premenopausal monkeys. J Pharmacol Exp Ther 1994;271:671–76.
63. Giscaird V, Miller VM, Vanhoutte PM. Effect of 17β-estradiol on endothelium-dependent responses in the rabbit. J Pharmacol Exp Ther 1988;244:19–22.

64. Hogg N, Struck A, Goss SP, Santanam N, Joseph J, Parthasarathy S, et al. Inhibition of macrophage-dependent low density lipoprotein oxidation by nitric-oxide donors. J Lipid Res 1995;36:1756–62.

65. Rubbo H, Radi R, Trujillo M, Telleri R, Kalyanaraman B, Barnes S, et al. Nitric oxide regulation of superoxide and peroxynitrite-dependent lipid peroxidation. Formation of novel nitrogen-containing oxidized lipid derivatives. J Biol Chem 1994;269:26066–75.

66. Struck AT, Hogg N, Thomas JP, Kalyanaraman B. Nitric oxide donor compounds inhibit the toxicity of oxidized low-density lipoprotein to endothelial cells. FEBS Lett 1995;361:291–94.

16

Effects of Mammalian and Plant Estrogens on Lipoprotein Metabolism and Coronary Heart Disease

Janice D. Wagner, Li Zhang, Kathryn A. Greaves, and Dawn C. Schwenke

Coronary heart disease (CHD) is the leading cause of death in both pre- and postmenopausal women in Western societies. Both natural and surgical menopause are associated with increased risk of CHD (1), although estrogen replacement therapy (ERT) reduces the risk of CHD by about 50% in postmenopausal women (2–4) and the amount of atherosclerosis by about 50% in animal models (5–8). Even though there is overwhelming evidence that estrogen monotherapy markedly reduces the risk of CHD in postmenopausal women, the effects of estrogen/progestin regimens on CHD risk are less clear. A progestin is added to the ERT regimen to reduce the risk of endometrial cancer, and this may detract from estrogen's beneficial effects. For example, numerous observational studies as well as meta-analyses (2,4) have concluded that ERT reduces risk of CHD. Reports from the Nurses Health Study (9) found similar reductions in CHD risk with ERT and combined estrogen/progestin treatment (HRT); however, a report of the first randomized, blinded, secondary prevention trial—the Heart and Estrogen/progestin Replacement Study (HERS)—found no significant differences between CHD events in the placebo group compared with a group treated with combined, continuous estrogen/progestin (conjugated equine estrogens (CEE) plus medroxyprogesterone acetate (MPA) (10). As there was no estrogen-only group, it is impossible to determine whether the lack of treatment effect was due to combined HRT, the type of progestin used, or other differences between this trial and others.

The initiation and progression of coronary artery atherosclerosis and related CHD are difficult to study prospectively in human subjects because atherosclerosis develops slowly over a period of many years. In addition, there are a number of concerns in human trials that selection bias may be present (4). We have therefore used cynomolgus monkeys, a well-characterized animal model, to study the effects of sex hormone deficiency and various hormone treatments on the pathogenesis of atherosclerosis. These studies provide preclinical studies in an animal model with close homology to human beings without the constraint of selection bias, compliance problems, or ethical issues concerning the need for a progestin. Using this model, studies by Adams et al. (5) have shown that parenteral estradiol therapy with or without progesterone decreased atherosclerosis progression by about 50%. Antiatherogenic effects of hormone replacement were independent of variation in lipid, lipoprotein, and apoprotein concentrations.

A more recent study by Adams et al. (6) confirmed the cardioprotective effects of estrogens. In that study, four groups were studied: (1) no treatment (ovariectomized controls), (2) CEE at a dose equivalent to 0.625 mg/day for women, (3) MPA at a dose equivalent to 2.5 mg/day for women, or (4) combined CEE + MPA, at the preceding doses. Monkeys treated with CEE alone had a 70% decrease in atherosclerosis extent compared with controls; however, those monkeys treated with combined CEE + MPA or MPA alone were not statistically different from controls. Adjusting for plasma lipid concentrations did not affect these relationships. Consistent with the results of HERS in women, continuous combined CEE and MPA did not result in beneficial effects on CHD. Additional studies from our group show that MPA also adversely affects coronary artery vascular reactivity (11) and insulin resistance (12,13). Although the preceding findings suggest beneficial effects of ERT, mechanisms for these beneficial effects are still unclear and whether HRT confers the same benefits is questionable.

Estrogens and Lipid Metabolism

Some of the beneficial effects of ERT may be due to changes in plasma lipoproteins. The effects of exogenous estrogens on plasma lipoproteins vary with dose, route of administration, and preparation of estrogen, but generally cause a decrease in total and low density lipoprotein (LDL) cholesterol and an increase in high density lipoprotein (HDL) and triglycerides (1–4,14,15). The addition of a progestational steroid may or may not affect lipoprotein concentrations (1,14,15). Results from the Postmenopausal Estrogen-Progestin Interventions (PEPI) Trial indicate that the most favorable effect on HDL cholesterol concentrations was in women taking unopposed estrogens (14). The addition of micronized progesterone to the ERT still resulted in beneficial changes in HDL cholesterol; however, the addition of MPA blunted much of the ERT benefit. In addition, lipoprotein (LP) (a) levels are higher in postmenopausal women compared with premenopausal women and are decreased with ERT and HRT in postmenopausal women (16,17).

It is unclear whether these changes in plasma lipoproteins account for differences in risk of CHD. For example, the HERS reported an increase in

HDL and a decrease in LDL with combined HRT, yet CHD events were not reduced within the 5-year study period (10). Because there was a time trend in the HERS with more CHD events in year 1 and fewer in years 4 and 5 with HRT, a longer period of treatment may be needed for lipid lowering to reduce CHD events. Although the different ERT and HRT regimens have variable effects on plasma lipids, however, most studies still find these treatments to reduce CHD risk and atherosclerosis (2–4). This may be because only 25–50% of the beneficial effects of estrogen on CHD are believed to be due to changes in plasma HDL and LDL cholesterol concentrations (3). This suggests that estrogen use is protective largely through mechanisms other than effects on plasma lipid concentrations. Estrogens may retard atherogenesis by acting directly on the arterial wall, or by affecting such things as the LDL heterogeneity, lipoprotein oxidation, or by modifying plasma components other than lipoproteins. ERT may also improve other risk factors for CHD (e.g., obesity, hyperinsulinemia, and diabetes mellitus).

The mechanisms by which some estrogens lower plasma LDL cholesterol concentrations are unclear; however, the oral route of administration and the subsequent first-pass effect appear to be important factors (15,18). Studies in animal models and in cell culture suggest that catabolism of LDL may be increased due to upregulation of the hepatic LDL receptor by estrogen (19,20). We have shown in monkeys, however, that hepatic cholesterol content is also decreased with estrogen, which suggests the rise in LDL receptor activity may be a secondary response (21). Increased LDL catabolism has also been shown in women treated orally with CEE, but not in women given transdermal estradiol (18). This is similar to our findings in monkeys, where a number of oral estrogens increased the LDL fractional catabolic rate by 30–40%, whereas no increase was found with parenteral hormone therapy (22).

Although removal of apoB/E containing particles tends to be increased with estrogen, an important feature in preventing subsequent downregulation of LDL receptor activity is an increase in biliary cholesterol secretion (21,23). We found ERT and HRT to increase mRNA abundance for 7α-hydroxylase (the rate-limiting step in the conversion of cholesterol to bile) and to decrease hepatic cholesteryl ester content (21). Although estrogen-induced increases in biliary cholesterol secretion may represent one mechanism for decreasing plasma cholesterol concentrations, it may increase gallbladder disease (4,10,23). The decrease in hepatic cholesterol content also may be responsible for smaller LDL particles (24).

It is interesting that, whereas small LDL have been shown to be atherogenic in a number of epidemiologic studies, ERT decreases LDL size in women (15,18,25) and in monkeys (22,26,27). The decrease in LDL size may be secondary to increases in plasma triglyceride concentrations and the enrichment of very low density lipoprotein (VLDL) with triglyceride, which are exchanged for LDL cholesteryl esters (22,26). The lipolysis of LDL triglycerides then results in smaller LDL particles. This is consistent with the negative correlation between plasma triglycerides and LDL size in monkeys

(27) and in women (28). On the other hand, or in addition to this mechanism, larger lipoprotein particles, which are more apoE-enriched, may be selectively removed via either the apoB/E receptor or the lipoprotein-related protein (LRP) receptor, leaving behind smaller particles (22,26). This may also account for the lower plasma apoE concentrations associated with estrogen treatment (22,26). Estrogen was more recently shown to increase both the clearance and production of both large and small LDL, with the greatest effect on clearance of large LDL, resulting in an overall decrease in concentration of large LDL (28). The reduced CHD in both species suggests that decreased LDL size with ERT is not detrimental. The association of small dense LDL with increased risk of CHD in dyslipidemic people may be secondary to other metabolic changes (e.g., increased triglyceride concentrations associated with insulin resistance or diabetes). Furthermore, if the increased atherogenicity of smaller particles is due to their greater oxidizability, this may be less important in the presence of an antioxidant, such as estrogen (see later), or less likely if the clearance of these particles is also increased (25).

Estrogens and Arterial LDL Metabolism

Data suggest that increased arterial LDL accumulation and degradation are early events in atherogenesis. Such increases in arterial LDL metabolism occur selectively in arterial sites that are prone to atherosclerosis as well as early after onset of a hypercholesterolemic stimulus, yet before significant intimal foam cell accumulation (29,30). Using a double-labeling technique for LDL that allows separate evaluation of arterial accumulation of LDL and arterial rates of LDL degradation, we have investigated how estrogens influence these aspects of LDL metabolism (22,30). We have found that decreased arterial degradation and accumulation of LDL may be one mechanism by which ERT decreases the progression of atherosclerosis. Using combined parenteral estradiol and progesterone, coronary artery LDL degradation and accumulation decreased by 78% compared with ovariectomized controls (31). LDL metabolism varied with arterial site, but the treatments resulted in similar effects at all arterial sites.

Another study examined the effects of oral esterified estrogens on LDL metabolism (27). Again, the hormone treatment had the greatest effect on the rate of arterial LDL degradation, which was reduced an average of 73%. Despite a similar reduction in arterial LDL degradation and accumulation in the two studies, parenteral hormone replacement did not influence plasma lipids, whereas oral estrogen decreased plasma cholesterol concentrations by 30%. In a third study using oral contraceptive agents, we also found a decrease in arterial LDL degradation despite an increase in the total/HDL cholesterol ratio (22). The similar reductions in arterial degradation with varying effects on plasma lipid concentrations suggest that estrogens decrease arterial LDL accumulation, in part, independently of plasma lipids and lipoproteins. It is of interest that all three ERT/HRT regimens also reduced LDL size, which correlated with arterial LDL accumulation.

In the studies described earlier, we found that in early atherogenesis, hormone treatments decreased intracellular degradation rates more than concentrations of undegraded LDL. Our findings are consistent with the results of Hough and Zilversmit (7), who reported that 17β-estradiol cypionate reduced the arterial rate of hydrolysis of cholesteryl ester in female rabbits. These observations were independent of effect on atherosclerosis, lipids, or lipoproteins. Thus, our studies and those of Hough and Zilversmit (7), indicate that estrogen reduces arterial metabolism of both lipoprotein protein and cholesteryl ester. Other data indicate that estrogen does not reduce arterial permeability to LDL (30). This suggests that estrogen acts by reducing cellular uptake of LDL accumulated in the arterial extracellular space. In our studies, however, we found that estrogens also reduced arterial accumulation of undegraded LDL, most of which should be extracellular. In the absence of effects of estrogen on arterial permeability to LDL, this suggests that arterial retention of LDL was reduced. Reduced arterial retention of LDL could reduce potential for intraarterial oxidation of LDL, which explains the reduced arterial degradation of LDL in estrogen-treated animals.

Estrogens and Oxidative Stress

A considerable body of data suggests that oxidation of LDL could promote atherosclerosis. Structural similarities between antioxidants and estrogens have stimulated many studies of antioxidant activities of estrogens. As reviewed elsewhere (32), 11 studies investigated the influence of in vivo ERT or HRT in postmenopausal women on resistance of LDL to in vitro oxidation. In general, studies that used transdermal delivery of estrogen found estrogen to increase resistance of LDL to oxidation, whereas oral delivery of estrogen had no effect. It appears that higher plasma concentrations of estradiol are more effective and that the minimum concentration needed to increase resistance of LDL to oxidation is between 330 and 430 pmol/L. Comparison of ERT versus HRT, as well as studies in vitro, suggest that progestins do not influence resistance of LDL to oxidation (32).

Interpretation of the preceding results is complicated by the fact that most of these studies did not investigate changes of LDL composition, including antioxidant content or LDL size, all of which are known to influence resistance of LDL to oxidation (32). Overall, studies suggest that estrogens could increase LDL vitamin E relative to cholesterol, reduce average LDL size by reducing esterified and nonesterified cholesterol, and possibly reduce unsaturation of LDL fatty acids (32–35). Increased LDL vitamin E and reduced unsaturation of LDL fatty acids would be expected to reduce susceptibility of LDL to in vitro oxidation, whereas the smaller LDL size could enhance susceptibility of LDL to in vitro oxidation. It seems likely that effects on LDL composition and/or size could explain the increased resistance of LDL to in vitro oxidation reported in some studies.

The consideration of whether estrogen might inhibit intraarterial oxidation of LDL is independent of whether estrogen influences resistance of LDL to oxidation. In one study, we found esterified estrogens to reduce both aortic cholesterol and aortic thiobarbituric acid reactive material (a measure of oxidation) in monkeys (27). In a more recent study that considered a shorter period of hypercholesterolemia, we found estradiol to neither influence arterial cholesterol concentration nor arterial F2-isoprostanes, another measure of oxidation in vivo (36). Further work will be needed to clarify whether estrogens have antioxidant activity at the level of the arterial wall, and whether detection of any such antioxidant activity may require the accumulation of critical levels of arterial lipids.

Estrogens and Carbohydrate Metabolism

Some studies of ERT have reported lower fasting glucose and insulin concentrations (14) or improvement in insulin sensitivity (36,37), whereas others have not (13,38). The PEPI trial (14) reported small but significant decreases in fasting glucose and insulin concentrations in women taking primarily CEE with or without a progestin. Postchallenge glucose levels, however, tended to increase with treatment and were greatest in those treated with MPA. Likewise, we (13) and others (37) have reported decreased insulin sensitivity with MPA. Even newer progestins that lack androgenic activity, however, were found to diminish the beneficial effects of estradiol on insulin sensitivity (36). Female monkeys and women gain weight postmenopausally, and ERT tends to prevent the weight gain, primarily by reducing abdominal fat (12,39). Changes in body fat may explain, in part, some of the changes in insulin and glucose metabolism with ERT. Although previous reports of the effects of ERT on carbohydrate and insulin metabolism have yielded conflicting results, ERT users appear to have a reduced relative risk (0.80) of becoming diabetic compared with nonestrogen users (40). Furthermore, beneficial effects of ERT may be more dramatic in women with existing CHD (41) and in women who are at increased risk of CHD, such as diabetics (42).

Phytoestrogens and Atherosclerosis

An alternative to finding a progestin that does not detract from estrogen's beneficial effects is to find an estrogen that does not need a progestin. The perfect selective estrogen would result in beneficial effects in the cardiovascular system, the bones, and the brain, yet have no adverse effects in the breast or uterus. As such, our group has investigated the use of dietary estrogens, or phytoestrogens, which are found to be in high levels in soybeans (43). The two primary estrogenic compounds in soybeans are genistein and daidzein and their conjugates. These compounds bind to both estrogen receptor α and β, but with higher affinity to estrogen receptor β (44). Depending on the tissue, the phytoestrogens may exert either estrogenic or antiestrogenic effects (45). There also is evidence that ERα and ERβ differentially affect transcriptional activation, depending on the ligand and the response element (46).

Our preliminary data suggest that the phytoestrogens have beneficial effects on plasma lipoprotein concentrations, insulin sensitivity, vascular reactivity, and atherosclerosis (39,47,48), yet do not increase breast or uterine cell proliferation (49). As both ERα and ERβ have been found in coronary arteries of monkeys (50), further work will be needed to determine whether the effects of dietary soy, and phytoestrogens, are mediated by estrogen receptors, and to determine the relative importance of the ERα and β in mediating effects of both mammalian and plant estrogens.

References

1. Godsland IF, Wynn V, Crook D, Miller NE. Sex, plasma lipoproteins, and atherosclerosis: prevailing assumptions and outstanding questions. Am Heart J 1987;114:1467–503.
2. Stampfer MJ, Colditz GA. Estrogen replacement therapy and coronary heart disease: a quantitative assessment of the epidemiologic evidence. Prevent Med 1991;20:47–63.
3. Barrett-Connor E, Bush TL. Estrogen and coronary heart disease in women. JAMA 1991;265:1861–67.
4. Barrett-Connor E, Grady D. Hormone replacement therapy, heart disease, and other considerations. Annu Rev Public Health 1998;19:55–72.
5. Adams MR, Kaplan JR, Manuck SB, Koritnik DR, Parks JS, Wolfe MS, et al. Inhibition of coronary artery atherosclerosis by 17-beta estradiol in ovariectomized monkeys. Lack of an effect of adding progesterone. Arteriosclerosis 1990;10:1051–57.
6. Adams MR, Register TC, Golden DL, Wagner JD, Williams JK. Medroxyprogesterone acetate antagonizes inhibitory effects of conjugated equine estrogens on coronary artery atherosclerosis. Arterioscler Thromb Vasc Biol 1997;17:217–21.
7. Hough JL, Zilversmit DB. Effect of 17-β estradiol on cholesterol content and metabolism in cholesterol-fed rabbits. Arteriosclerosis 1986;6:57–63.
8. Haarbo J, Leth-Espensen P, Stender S, Christiansen C. Estrogen monotherapy and combined estrogen-progestogen replacement therapy attenuate aortic accumulation of cholesterol in ovariectomized cholesterol-fed rabbits. J Clin Invest 1991;87:1274–79.
9. Grodstein F, Stampfer MJ, Manson JE, Colditz MB, Willett WC, Rosner B, et al. Postmenopausal estrogen and progestin use and the risk of cardiovascular disease. N Engl J Med 1996;335:453–61.
10. Hulley S, Grady D, Bush T, Furberg C, Herrington D, Riggs B, et al., for the HERS

research group. Randomized trial of estrogen plus progestin for secondary prevention of coronary heart disease in postmenopausal women. JAMA 1998;280: 605–18.

11. Williams JK, Honoré EK, Washburn SA, Clarkson TB. Effects of hormone replacement therapy on reactivity of atherosclerotic coronary arteries in cynomolgus monkeys. J Am Coll Cardiol 1994:24:1757–61.

12. Wagner JD, Martino MA, Jayo MJ, Anthony MS, Clarkson TB, Cefalu WT. The effects of hormone replacement therapy on carbohydrate metabolism and cardiovascular risk factors in surgically postmenopausal cynomolgus monkeys. Metabolism 1996;45:1254–62.

13. Cefalu WT, Wagner JD, Bell-Farrow AD, Wang ZQ, Adams MR, Toffolo G, et al. The effects of hormonal replacement therapy on insulin sensitivity in surgically postmenopausal cynomolgus monkeys (Macaca fascicularis). Am J Obstet Gynecol 1994;171:440–45.

14. The Writing Group for the PEPI Trial. Effects of estrogen or estrogen/progestin regimens on heart disease risk factors in postmenopausal women: the Postmenopausal Estrogen/Progestin Interventions (PEPI) trial. JAMA 1995;273:199–208.

15. Seed M, Crook D. Post-menopausal hormone replacement therapy, coronary heart disease and plasma lipoproteins. Curr Opinions Lipidol 1994;5:48–58.

16. Espeland MA, Marcovina SM, Miller V, Wood PD, Wasilauskas C, Sherwin R, et al. Effect of postmenopausal hormone therapy on lipoprotein(a) concentration. Circulation 1998;97;979–86.

17. Kim CJ, Ryu WS, Kwak JW, Park CT, Ryoo UH. Changes in Lp(a) lipoprotein and lipid levels after cessation of female sex hormone production and estrogen replacement therapy. Arch Intern Med 1996;156:500–4.

18. Walsh BW, Schiff I, Rosner B, Greenberg L, Ravnikar V, Sacks FM. Effects of postmenopausal estrogen replacement on the concentrations and metabolism of plasma lipoproteins. N Engl J Med 1991;325:1196–204.

19. Kovanen PT, Brown MS, Goldstein JL. Increased binding of low density lipoprotein to liver membranes from rats treated with 17α-ethinyl estradiol. J Biol Chem 1979;254:11367–73.

20. Ma PT, Yamamoto T, Goldstein JL, Brown MS. Increased mRNA for low density lipoprotein receptors in livers of rabbits treated with 17α-ethinyl estradiol. Proc Natl Acad Sci USA 1986;83:792–96.

21. Colvin PL, Wagner JD, Adams MR, Sorci-Thomas MG. Sex steroids increase cholesterol 7 alpha-hydroxylase mRNA in nonhuman primates. Metabolism Clin Exp 1998;47:391–95.

22. Wagner JD, Zhang L, Adams MR. Effects of estrogens on arterial LDL metabolism. In: Forte T, ed. Hormonal, metabolic, and cellular influences on cardiovascular disease in women. Armonk, NY: Futura Publishing Co., 1997:153–74.

23. Everson GT, McKinley C, Kern F, Jr. Mechanisms of gallstone formation in women. Effects of exogenous estrogen (Premarin) and dietary cholesterol on hepatic lipid metabolism. J Clin Invest 1991;87:237–46.

24. Parks JS, Wilson MD, Johnson FL, Rudel LL. Fish oil decreases hepatic cholesteryl ester secretion but not apoB secretion in African green monkeys. J Lipid Res 1989;30:1535–44.

25. Campos H, Walsh BW, Judge H, Sacks FM. Effect of estrogen on very low density lipoprotein and low density lipoprotein subclass metabolism in postmenopausal women. J Clin Endocrinol Metab 1997;82:3955–63.

26. Manning JM, Campos G, Edwards IJ, Wagner WD, Wagner JD, Adams MR, et al. Effects of hormone replacement modalities on low density lipoprotein composition and distribution in ovariectomized cynomolgus monkeys. Atherosclerosis 1996;121:217–30.
27. Wagner JD, Zhang L, Williams JK, Register TC, Ackerman DM, Wiita B, et al. Esterified estrogens with and without methyltestosterone decrease arterial LDL metabolism in cynomolgus monkeys. Arterioscler Thromb Vasc Biol 1996;16:1473–80.
28. Wakatsuke A, Ikenoue N, Sagara Y. Estrogen-induced small low-density lipoprotein particles in postmenopausal women. Obstet Gynecol 1998;91:234–40.
29. Schwenke DC, Carew TE. Initiation of atherosclerotic lesions in cholesterol-fed rabbits. I. Focal increases in arterial LDL concentration precede development of fatty streak lesions. Arteriosclerosis 1989;9:895–907.
30. Wagner JD, Schwenke DC. Lipoprotein metabolism in the vessel wall. In: Rubanyi G, ed. Estrogens and the vessel wall. Berkshire, U.K.: Harwood Academic Publishers, 1998:107–20.
31. Wagner JD, St. Clair RW, Schwenke DC, Shively CA, Adams MR, Clarkson TB. Regional differences in arterial low density lipoprotein metabolism in surgically postmenopausal cynomolgus monkeys: effects of estrogen and progesterone therapy. Arterioscler Thromb 1992;12:717–26.
32. Schwenke DC. Aging, menopause and free radicals. In: Murphy AA, Parthasarathy S, eds. Seminars in reproductive endocrinology, free radicals in gynecology and obstetrics. Semin Reprod Endocrinol 1998;16:281–308.
33. Clemente C, Caruso MG, Berloco P, Buonsante A, Giannandrea B, DiLeo A. Alpha-tocopherol and beta-carotene serum levels in post-menopausal women treated with transdermal estradiol and oral medroxyprogesterone acetate. Horm Metab Res 1996;28:558–61.
34. Mcmanus J, Mceneny J, Thompson W, Young IS. The effect of hormone replacement therapy on the oxidation of low density lipoprotein in postmenopausal women. Atherosclerosis 1997;135:73–81.
35. Nenseter MS, Volden V, Berg T, Drevon CA, Ose L, Tonstad S. Effect of hormone replacement therapy on the susceptibility of low-density lipoprotein to oxidation among postmenopausal hypercholesterolaemic women. Eur J Clin Invest 1996;26:1062–68.
36. Wagner JD, Thomas MJ, Williams JK, Zhang L, Greaves KA, Cefalu WT. Insulin sensitivity and cardiovascular risk factors in ovariectomized monkeys with estradiol alone or combined with nomegestrol acetate. J Clin Endocrinol 1998;83:896–901.
37. Lindheim SR, Presser SC, Ditkoff EC. A possible bimodal effect of estrogen on insulin sensitivity in postmenopausal women and the attenuating effect of added progestin. Fertil Steril 1993;60:664–67.
38. Godsland IF, Gangar K, Walton C, et al. Insulin resistance, secretion, and elimination in postmenopausal women receiving oral or transdermal hormone replacement therapy. Metabolism 1993;42:846–53.
39. Wagner JD, Cefalu WT, Anthony MS, Litwak KN, Zhang L, Clarkson TB. Dietary soy protein and estrogen replacement therapy improve cardiovascular risk factors and decrease aortic cholesteryl ester content in ovariectomized cynomolgus monkeys. Metabolism Clin Exp 1997;46:698–705.
40. Manson JE, Rimm EB, Colditz GA, Willett WC, Nathan DM, Arky RA, et al. A

prospective study of postmenopausal estrogen therapy and subsequent incidence of non-insulin-dependent diabetes mellitus. Ann Epidemiol 1992;2:665–73.

41. Sullivan JM, Vander Zwaag R, Hughes JP, Maddock V, Kroetz FW, Ramanathan KB, et al. Estrogen replacement and coronary artery disease. Arch Intern Med 1990;150:2557–62.

42. Grodstein F, Stampfer MJ, Colditz GA, Willett WC, Manson JE, Joffe M, Rosner B, et al. Postmenopausal hormone therapy and mortality. N Engl J Med 1997; 336:1769–75.

43. Hughes CL, Cline JM, Williams JK, Anthony MS, Wagner JD, Clarkson TB. Dietary soy phytoestrogens and the health of menopausal women: overview and evidence of cardioprotection from studies in non-human primates. In: Wren BG, ed. Progress in the management of the menopause. New York: Parthenon Publishing, 1997:30–39.

44. Kuiper GGJM, Carlson B, Grandien K, Enmark E, Häggblad J, Nilsson S, et al. Comparison of the ligand binding specificity and transcript tissue distribution of estrogen receptors α and β. Endocrinology 1997;138:863–70.

45. Setchell KDR. Naturally occurring non-steroidal estrogens of dietary origin, and discussion. In: McLachlan JA, ed. Estrogens in the environment. II. Influences on development. New York: Elsevier, 1985:69–85.

46. Paech K, Webb P, Kuiper GGJM, Nilsson S, Gustafsson JA, Kushner PJ, et al. Differential ligand activation of estrogen receptors ERα and ERβ at AP1 sites. Science 1997;277:1508–10.

47. Anthony MS, Clarkson TB, Bullock BC, Wagner JD. Soy protein versus soy phytoestrogens in the prevention of diet-induced coronary artery atherosclerosis of male cynomolgus monkeys. Arterioscler Thromb Vasc Biol 1997;17:2524–31.

48. Honoré EK, Williams JK, Anthony MS, Clarkson TB. Soy isoflavones enhance coronary vascular reactivity in atherosclerotic female macaques. Fertil Steril 1997;67:148–54.

49. Foth D, Cline JM. Effects of mammalian and plant estrogens on mammary glands and uteri of macaques. Am J Clin Nutr 1998;68:14135–75.

50. Register TC, Adams MR. Coronary artery and cultured aortic smooth muscle cells express mRNA for both the classical estrogen receptor and the newly described estrogen receptor beta. J Steroid Biochem Mol Biol 1998;64:187–91.

17

Potential Cardioprotective Effects of Estrogen Involving Ion Channels, Endothelium, and Coronary Artery Reactivity

Joseph E. Brayden, Adrian Bonev, Karen Lounsbury, Harm Knot, George C. Wellman, and Mark T. Nelson

Introduction

The protective effects of estrogen have been studied extensively in the cardiovascular system, where reproductive hormones are known to have major influence on morbidity and mortality. Epidemiological data indicate that women in their reproductive years have a much lower incidence of coronary disease than men of similar age, an advantage that diminishes rapidly with the onset of menopause (1,2). These studies demonstrate a direct correlation between plasma estrogen levels and coronary disease among these populations. A component of the cardioprotective effect of estrogen appears to be related to favorable effects on lipid profiles, which results in less atherosclerotic disease. Estrogen may also have antiproliferative effects by interfering with fibroblast activity that may account for part of the antiatherogenic activity of this hormone (3). Direct or indirect effects of estrogen may also lead to decreased platelet and monocyte adhesion, and a lower likelihood of thrombosis. A number of studies, however, suggest that effects on coronary artery contractility might also account for a substantial portion of the cardioprotective action of estrogen (1,2). Estrogen appears to reduce coronary vasoconstrictor activity, which may increase coronary blood flow, and/or decrease the likelihood or severity of an ischemic event in the coronary circulation.

Estrogen appears to have effects on several cell types within the vascular wall that could reduce contractility. Possible targets include nerves, connective tissue, smooth muscle cells, and endothelial cells. The latter two cell types have been the focus of many studies.

Estrogen and Coronary Artery Reactivity

Direct inhibitory effects of estrogen on smooth muscle contractility have been reported for several vascular preparations, including coronary artery smooth muscle. In many arterial smooth muscle preparations, particularly those involved in regulation of vascular resistance, control of calcium entry through voltage-dependent calcium channels is a major mechanism of vasodilation and vasoconstriction (4). Estrogen relaxes arteries in vitro, apparently through antagonism of voltage-dependent calcium channels on the smooth muscle cells (5,6). The direct inhibitory effect of estrogen on calcium channel function, however, does require concentrations of estrogen far in excess of even the highest normal circulating levels, so the physiological significance of such an effect of estrogen is ambiguous. Another direct smooth muscle action of estrogen involving ion channels has been described (7). Estrogen, apparently in concentrations approaching physiological levels, has been found to activate coronary artery smooth muscle potassium channels, specifically large-conductance, calcium-activated potassium channels. This would lead to membrane hyperpolarization, closure of voltage-dependent calcium channels, and decreased calcium entry, resulting in vasodilation.

Although the studies described earlier suggest smooth muscle targets for the action of estrogen, the effects of estrogen on the vasculature may be mediated in large part through actions on endothelial cells. The vascular endothelium produces several potent vasodilator substances (e.g., prostacyclin, nitric oxide, endothelium-derived hyperpolarizing factor), and alteration in the synthesis or activity of these endogenous factors can have substantial effects on vascular tone. In vitro production of prostacyclin is decreased in uterine arteries from postmenopausal compared with premenopausal women (8). Estrogen-stimulated prostacyclin production has been found in cell cultures from both vascular smooth muscle (9) and endothelium (10). Thus, estrogen may exert some vasodilator action through enhanced prostacyclin formation.

Nitric oxide (NO) is the most widely described relaxing factor produced by the vascular endothelium, and is produced by the conversion of L-arginine to citrulline by the enzyme nitric oxide synthase (NOS) (11). A growing body of evidence indicates that estrogen increases the formation of nitric oxide by endothelial cells and this reduces vascular tone. The following observations are consistent with this proposal. A positive correlation has been found between plasma 17β-estradiol and stable metabolites of NO (nitrite/nitrate) during follicular development in women (12). Agonist-induced, endothelium-dependent coronary artery vasodilation mediated by NO is increased following 17β-estradiol treatment in ovariectomized monkeys (13). Endothelium-dependent coronary artery vasodilation defined by angiography is enhanced following 17β-estradiol in postmenopausal women (14). In vitro studies examining isometric tension development also report enhanced agonist-induced, endothelium-dependent relaxations in arteries from animals with elevated estrogen levels (15–17). Twenty-four-hour incubation with

17β-estradiol in vitro also potentiated endothelium-dependent relaxations in porcine coronary arteries (18).

Other data suggest that tonic NO production may be related to plasma estrogen status. A number of studies indicate that NO release is elevated in arteries from females (7,12,19–24) and estrogen appears to be responsible for this observed gender difference (17,19,21,24,25). Inhibition of NOS produced a greater increase in tension in aorta from female rabbits compared with male or ovariectomized animals (20). An enhanced production of NO, but not of prostacyclin, in aortas from female compared with male rats has also been described using a bioassay system (26). Sensitivity to exogenously applied NO is similar in endothelium-denuded arteries from males and females (20), which suggests that a differential production of NO may exist between sexes.

The effects of estrogen on NO activity appear to be due to a genomic action. Through interaction with its receptor, estrogen may increase the transcription of endothelial constitutive NOS (ecNOS). Estrogen has been shown to increase the level of ecNOS mRNA in cultured pulmonary artery endothelial cells (27) and in uterine artery (28). A study by Kleinert et al. provides direct evidence for a 1.8-fold increase in ecNOS protein expression by physiological levels of estrogen in cultured human endothelial cells (29). mRNA levels for ecNOS were similarly increased in aorta from estradiol treated rats (30). The preceding evidence suggests that estrogen promotes vascular smooth muscle relaxation through induction of ecNOS and increased production of nitric oxide from the vascular endothelium.

Our work supports this concept and demonstrates that estrogen has important effects on tonic regulation of intrinsic (myogenic) tone in coronary resistance arteries (24). Within the circulatory system, blood pressure primarily drops through small arteries less than 200 μm in diameter that are termed *resistance arteries*. In vivo, these resistance arteries are maintained in a partially constricted state, which allows for additional constriction or dilation to occur which can increase or decrease blood flow to target organs, respectively. The majority of this underlying constriction is due to the effects of intravascular pressure and has been termed *myogenic tone* (31). This myogenic tone in coronary arteries isolated from female rats with physiological levels of plasma estrogen is about 50% of that observed in arteries from male or estrogen-deficient (ovariectomized) females (24). Myogenic tone in arteries from OVX rats with estrogen replaced to physiological levels is identical to that observed in untreated females. Removal of the endothelium or chemical inhibition of NOS activity increases myogenic tone in arteries from females to the level found in arteries from male or OVX animals. This suggests that estrogen enhances NO production and this reduces myogenic tone.

We have also investigated the possible *mechanisms* by which NO causes coronary artery dilation. Nitrovasodilators (e.g., nitric oxide, nitroprusside) cause vasodilation by elevating intracellular levels of guanosine-3',5'-cyclic monophosphate (cGMP) (32,33) and stimulating cGMP-dependent protein kinase (G-kinase) activity. G-kinase may relax vascular smooth muscle by reducing the calcium sensitivity of the contractile proteins (34) and also by uncoupling myosin light chain phosphorylation and force generation (35). Nitrovasodilators and G-kinase also relax vascular smooth muscle by decreasing intracellular calcium concentrations through increased extrusion via Ca^{2+}–ATPase activity (36) or increased sequestration by the sarcoplasmic reticulum (37). One of the most important physiological actions of NO, however, may be to indirectly decrease $(Ca^{2+})_i$ through activation of large conductance calcium-activated potassium (K_{Ca}) channels.

Dilations of coronary arteries from female and OVX animals to sodium nitroprusside, a nitrovasodilator that generates nitric oxide, are reduced by >50% by iberiotoxin (IBTX), which is an inhibitor of K_{Ca} channels (24). Sodium nitroprusside (10 µM) hyperpolarizes coronary arteries by 10–15 mV, an effect that is greatly diminished (~80%) by IBTX. Coronary arteries isolated from female rats produce greater constrictions in response to IBTX and KT 5823, which is an inhibitor of G-kinase, compared with coronary arteries from OVX animals. G-kinase increases the activity of K_{Ca} channels about 16-fold in excised membrane patches from vascular smooth muscle cells enzymatically isolated from these small coronary arteries. This finding indicates that enhanced endothelial NO production is responsible for the increased diameter of coronary arteries obtained from animals with physiological levels of circulating estrogen. NO and G-kinase dilate these arteries in part by stimulation of K_{Ca} channels in coronary vascular smooth muscle.

Additional studies in our laboratory have focused on the possible mechanisms by which estrogen may regulate NO activity in the coronary endothelium (Knot, Lounsbury, Brayden, Nelson, unpublished). Endothelial intracellular Ca^{2+} ($[Ca^{2+}]_i$) and NOS activity were measured in coronary arteries isolated from male and female rats. Reduced myogenic tone in females is associated with substantial increases in coronary endothelial $[Ca^{2+}]_i$ compared with males (female endothelial Ca^{2+}: 174 nM; male endothelial Ca^{2+}: 90 nM). Due to the calcium sensitivity of NOS, the Ca^{2+} increase alone should increase NOS activity in female endothelial cells. In addition, maximal ecNOS activity in arterial lysates is elevated (1.4-fold) in females compared with males. Together, the increased endothelial $[Ca^{2+}]_i$ and maximal NOS activity are calculated to increase tonic ecNOS activity in intact arteries from females by nearly threefold. This elevated ecNOS activity is likely to contribute to the overall decrease in myogenic activity observed in coronary arteries exposed to circulating estrogen in vivo.

Summary and Conclusion

Current evidence suggests the following general mechanism by which estrogen may decrease coronary vascular contractility (Fig. 17.1): (1) Estrogen increases endothelial $[Ca^{2+}]_i$ and induces increased expression of ecNOS; (2) the combined effects of increased $[Ca^{2+}]_i$ and NOS lead to elevated NOS

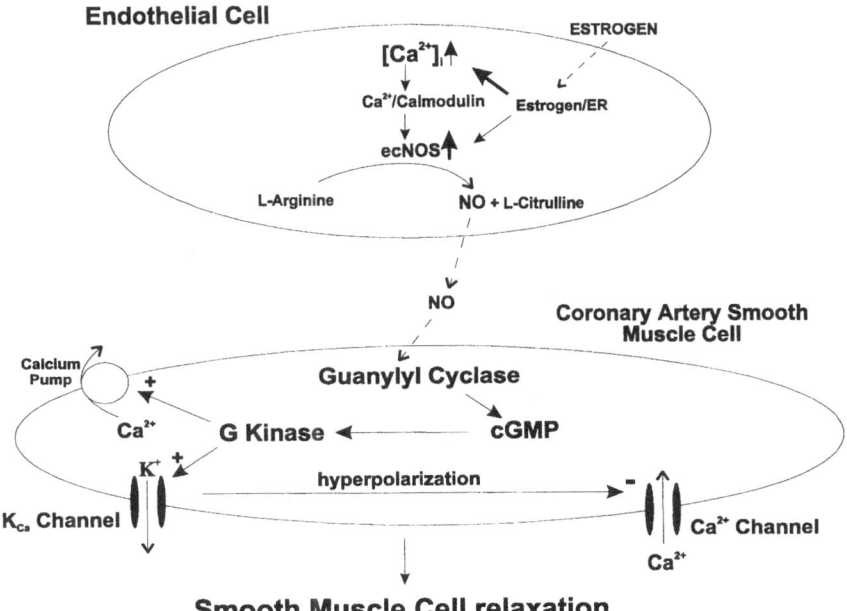

FIGURE 17.1. A proposed mechanism for the gender differences in the modulation of coronary artery diameter by the actions of estrogen on the coronary endothelium. Both increased levels of endothelial Ca^{2+} and ecNOS protein levels in female coronary endothelium contribute to an increase in nitric oxide (NO) production. The increase in basal NO production exerts a tonic relaxing effect on the arterial wall smooth muscle by increasing calcium extrusion and decreasing calcium entry, leading to coronary vasodilation.

activity and NO formation in the presence of estrogen; (3) increased basal NO production stimulates cGMP formation and G-kinase activity in the coronary vascular smooth muscle cells; (4) G-kinase activates K_{Ca} channels, which hyperpolarizes the smooth muscle cells, leading to closure of voltage-dependent calcium channels, less calcium entry, and reduced smooth muscle $[Ca^{2+}]_i$. G-kinase may also increase calcium extrusion. The net effect of decreased smooth muscle $[Ca^{2+}]_i$ is vasodilation. The coronary vasodilator effects of estrogen may provide several advantages in vivo. First, oxygen delivery will be increased. In addition, when the smooth muscle membrane is hyperpolarized (i.e., in the presence of greater amounts of NO) and myogenic tone is reduced, sensitivity to endogenous vasoconstrictors (e.g., angiotensin II) is decreased (38). Thus, hyperpolarization and reduced myogenic tone may lower the incidence of arterial spasm in response to vasoactive stimuli. Enhanced NO activity may also reduce thrombosis and atherosclerosis by interfering with the activity of vascular cells involved in these pathologies.

The effects of estrogen on coronary artery function appear to be due to a sustained, genomic effect because they have been observed in isolated arteries that were no longer exposed to estrogen. The exact mechanisms by which estrogen influences endothelial cell Ca^{2+} (i.e., increased calcium influx, decreased calcium extrusion or sequestration) or NOS activity (increased expression or specific activity) remain to be determined. The present observations, however, stress the importance of coronary endothelial cell Ca^{+2} regulation, and may promote new perspectives for basic and clinical research into vascular changes in estrogen deficiency states and diseases associated with impaired endothelial function.

References

1. Bush TL, Barrett-Connor E, Cowan LD, et al. Cardiovascular mortality and noncontraceptive use of estrogen in women: results from the Lipid Research Clinics Program Follow-up Study. Circulation 1987;75:1102–9.
2. Barrett-Connor E, Bush TL. Estrogen and coronary heart disease in women [see comments]. JAMA 1991;265:1861–67.
3. White CR, Shelton J, Chen SJ, et al. Estrogen restores endothelial cell function in an experimental model of vascular injury. Circulation 1997;96:1624–30.
4. Nelson MT, Patlak JB, Worley JF, Standen NB. Calcium channels, potassium channels, and voltage dependence of arterial smooth muscle tone. Am J Physiol 1990;259:C3–18.
5. Zhang F, Ram JL, Standley PR, Sowers JR. 17 beta-Estradiol attenuates voltage-dependent Ca2+ currents in A7r5 vascular smooth muscle cell line. Am J Physiol 1994;266:C975–80.
6. Shan J, Resnick LM, Liu QY, Wu XC, Barbagallo M, Pang PK. Vascular effects of 17 beta-estradiol in male Sprague-Dawley rats. Am J Physiol 1994;266:H967–73.
7. Darkow DJ, Lu L, White RE. Estrogen relaxation of coronary artery smooth muscle is mediated by nitric oxide and cGMP. Am J Physiol 1997;272:H2765–73.
8. Steinleitner A, Stanczyk FZ, Levin JH, et al. Decreased in vitro production of 6-keto-prostaglandin F1 alpha by uterine arteries from postmenopausal women. Am J Obstet Gynecol 1989;161:1677–81.
9. Chang WC, Nakao J, Orimo H, Murota SI. Stimulation of prostaglandin cyclooxygenase and prostacyclin synthetase activities by estradiol in rat aortic smooth muscle cells. Biochim Biophys Acta 1980;620:472–82.
10. Seillan C, Ody C, Russo-Marie F, Duval D. Differential effects of sex steroids on prostaglandin secretion by male and female cultured piglet endothelial cells. Prostaglandins 1983;26:3–12.
11. Palmer RM, Ferrige AG, Moncada S. Nitric oxide release accounts for the biological activity of endothelium-derived relaxing factor. Nature 1987;327:524–26.
12. Rosselli M, Imthurm B, Macas E, Keller PJ, Dubey RK. Circulating nitrite/nitrate levels increase with follicular development: indirect evidence for estradiol mediated NO release. Biochem Biophys Res Commun 1994;202:1543–52.
13. Williams JK, Adams MR, Klopfenstein HS. Estrogen modulates responses of atherosclerotic coronary arteries. Circulation 1990;81:1680–87.
14. Gilligan DM, Quyyumi AA, Cannon RO. Effects of physiological levels of estro-

gen on coronary vasomotor function in postmenopausal women. Circulation 1994;89:2545–51.

15. Gisclard V, Miller VM, Vanhoutte PM. Effect of 17 beta-estradiol on endothelium-dependent responses in the rabbit. J Pharmacol Exp Ther 1988;244:19–22.

16. Williams SP, Shackelford DP, Iams SG, Mustafa SJ: Endothelium-dependent relaxation in estrogen-treated spontaneously hypertensive rats. Eur J Pharmacol 1988;145:205–7.

17. Kim TH, Weiner CP, Thompson LP. Effect of pregnancy on contraction and endothelium-mediated relaxation of renal and mesenteric arteries. Am J Physiol 1994;267: H41–47.

18. Bell DR, Rensberger HJ, Koritnik DR, Koshy A. Estrogen pretreatment directly potentiates endothelium-dependent vasorelaxation of porcine coronary arteries. Am J Physiol 1995;268:H377–83.

19. Guetta V, Quyyumi AA, Prasad A, Panza JA, Waclawiw M, Cannon RO. The role of nitric oxide in coronary vascular effects of estrogen in postmenopausal women. Circulation 1997;96:2795–801.

20. Hayashi T, Fukuto JM, Ignarro LJ, Chaudhuri G. Basal release of nitric oxide from aortic rings is greater in female rabbits than in male rabbits: implications for atherosclerosis. Proc Natl Acad Sci USA 1992;89:11259–63.

21. Huang A, Sun D, Koller A, Kaley G: Gender difference in myogenic tone of rat arterioles is due to estrogen-induced, enhanced release of NO. Am J Physiol 1997;272:H1804–9.

22. Kawano H, Motoyama T, Kugiyama K, et al. Gender difference in improvement of endothelium-dependent vasodilation after estrogen supplementation. J Am Coll Cardiol 1997;30:914–19.

23. Ma L, Robinson CP, Thadani U, Patterson E. Effect of 17-beta estradiol in the rabbit: endothelium-dependent and -independent mechanisms of vascular relaxation. J Cardiovasc Pharmacol 1997;30:130–35.

24. Wellman GC, Bonev AD, Nelson MT, Brayden JE. Gender differences in coronary artery diameter involve estrogen, nitric oxide, and Ca(2+)-dependent K+ channels. Circ Res 1996;79:1024–30.

25. Skarsgard P, van Breemen C, Laher I. Estrogen regulates myogenic tone in pressurized cerebral arteries by enhanced basal release of nitric oxide. Am J Physiol 1997;273:H2248–56.

26. Kauser K, Rubanyi GM. Gender difference in bioassayable endothelium-derived nitric oxide from isolated rat aortae. Am J Physiol 1994;267:H2311–17.

27. MacRitchie AN, Jun SS, Chen Z, et al. Estrogen upregulates endothelial nitric oxide synthase gene expression in fetal pulmonary artery endothelium. Circ Res 1997;81:355–62.

28. Weiner CP, Lizasoain I, Baylis SA, Knowles RG, Charles IG, Moncada S. Induction of calcium-dependent nitric oxide synthases by sex hormones. Proc Natl Acad Sci USA 1994;91:5212–16.

29. Kleinert H, Wallerath T, Euchenhofer C, Ihrig-Biedert I, Li H, Forstermann U. Estrogens increase transcription of the human endothelial NO synthase gene: analysis of the transcription factors involved. Hypertension 1998;31: 582–88.

30. Goetz RM, Morano I, Calovini T, Studer R, Holtz J. Increased expression of endothelial constitutive nitric oxide synthase in rat aorta during pregnancy. Biochem Biophys Res Commun 1994;205:905–10.

31. Bayliss WM. On the local reactions of the arterial wall to changes of internal pressure. J Physiol (Lond) 1902;28:220–31.
32. Arnold WP, Mittal CK, Katsuki S, Murad F. Nitric oxide activates guanylate cyclase and increases guanosine 3':5'- cyclic monophosphate levels in various tissue preparations. Proc Natl Acad Sci USA 1977;74:3203–7.
33. Furchgott RF, Cherry PD, Zawadzki JV, Jothianandan D. Endothelial cells as mediators of vasodilation of arteries. J Cardiovasc Pharmacol 1984;6(suppl 2):S336–43.
34. Pfitzer G, Ruegg JC, Flockerzi V, Hofmann F. cGMP-dependent protein kinase decreases calcium sensitivity of skinned cardiac fibers. FEBS Lett 1982;149:171–75.
35. McDaniel NL, Chen XL, Singer HA, Murphy RA, Rembold CM. Nitrovasodilators relax arterial smooth muscle by decreasing [Ca2+]i and uncoupling stress from myosin phosphorylation. Am J Physiol 1992;263:C461–67.
36. Rashatwar SS, Cornwell TL, Lincoln TM. Effects of 8-bromo-cGMP on Ca2+ levels in vascular smooth muscle cells: possible regulation of Ca2+-ATPase by cGMP-dependent protein kinase. Proc Natl Acad Sci USA 1987;84:5685–89.
37. Lincoln TM, Cornwell TL. Towards an understanding of the mechanism of action of cyclic AMP and cyclic GMP in smooth muscle relaxation. Blood Vessels 1991;28:129–37.
38. Dunn WR, Wellman GC, Bevan JA. Enhanced resistance artery sensitivity to agonists under isobaric compared with isometric conditions. Am J Physiol 1994;266:H147–55.

18

Role of the Membrane Estrogen Receptor in Vascular Cell Physiology

ELLIS R. LEVIN

Previous Studies of the Membrane Effects of Estrogen

Estrogen's cellular actions are felt to be mediated through response elements on the promoters of target genes, or through modifying transcription via protein–protein interactions (1–4). There is increasing evidence, however, that ligands for various members of the steroid receptor superfamily modulate cell functions via nongenomic actions. Some of these actions appear to originate through plasma membrane protein interactions (5–8). As examples of nongenomic functions, progesterone quickly stimulates increased $[Ca^{2+}]_i$ in sperm (6), and aldosterone rapidly activates inositol1,4,5-trisphosphate generation in several cell types (8), or stimulates hemodynamic changes quickly (9). 17β-Estradiol (17βE2) can induce various signal transduction events in seconds to a few minutes. These events include the stimulation of calcium flux (10), cAMP (11), phospholipase C activation, and inositol phosphate generation (12,13), as well as the rapid release of prolactin (14). Many of the rapid actions of 17β-E2 have been attributed to interactions at the cell membrane. These could include 17β-E2 indirectly activating tyrosine kinase growth factor receptors (e.g., epidermal growth factor receptor, EGFR), with subsequent signal transduction initiated through these receptors (15). On the other hand, the existence of a cell membrane estrogen receptor (ER) was reported more than 20 years ago (16). This putative receptor has been investigated by several laboratories more recently (17,18), and it appears capable of enacting signal transduction.

Estrogen Inhibits Vascular Smooth Muscle Proliferation

To begin to understand the participation of a putative membrane ER in cell biologic actions, we examined the role of ER in vascular cell physiology. Acute and chronic responses to vascular insult lead to vascular pathology, including atherosclerosis. These processes are in part mediated through the migration to and the proliferation of vascular smooth muscle cells (VSMC) at the neointimal lining of the blood vessel. Many vascular growth factors stimulate this process, upon release from platelets and macrophages attracted to the site of vessel injury, or from the vascular wall itself. One important such protein, endothelin-1 (ET-1), is made in the endothelial cells and is a potent vasoconstrictor and mitogen (reviewed in Ref. 19). We examined the pathway by which endothelin stimulates VSMC proliferation, and found that this was dependent on signal transduction to the activation of the MAP kinase, ERK. ERK activation, therefore, resulted in the stimulation of c-fos production and c-myc phosphorylation, steps shown to be crucial for VSMC proliferation both *in vivo* and *in vitro*. Estrogen or progesterone were capable of inhibiting an upstream signal transduction pathway through Ras-Raf and MEK, leading to the inhibition of ERK activation by ET-1 (20). We also found that E2 conjugated to BSA, a bulky complex that could not enter the cell, also inhibited ET-1-activated ERK and VSMC proliferation. E2-BSA was incapable of activating a transfected ERE-Luciferase reporter at 24 hours in these cells, which indicates that the complex did not internalize/bind the nuclear ER, nor was E2 dislodging from its BSA conjugate. We then showed that fluorinating the E2-BSA yielded a probe that labeled a putative cell membrane receptor on the VSMC. Unlabeled E2, tamoxifen, or H222, an antibody to the ligand-binding domain of the classical ER reduced probe binding to this putative receptor. In total, these data support the ability of a putative membrane ER to signal negatively and to disrupt growth factor signaling to the nuclear growth program in VSMC. This identified a novel potential mechanism by which ER could protect the vasculature against the development of disease.

Endothelin Transcription Is Stimulated by Angiotensin II: Inhibition by Sex Steroids

We then asked the question whether E2 (or P) could interfere with ET-1 synthesis. We chose the model of angiotensin II (Ang II)-induced ET-1 synthesis because this relationship is felt to be important for understanding the cardiac hypertrophic and vasoconstrictive effects of Ang II. We found that Ang II induced transcription of ET-1 via an AP-1 site at -102nt of the proximal ET-1 gene promoter. This occurred because Ang II stimulated ERK activation, which was essential to induce the transcription of c-fos, increasing AP-1 binding activity in the endothelial cell (EC) nucleus, and stimulating ET-1 transcription (21). E2 or P was capable of substantially inhibiting all of these events. If we transfected the EC with a constituitively active ERK2, we were able to reverse the E2 inhibitory effects by at least 60%. This indicated that the dominant mechanism of E2 action stemmed from inhibiting

ERK activation. This also identified a novel mechanism by which steroids can negatively modulate transcription. This is not a direct genomic effect; rather, it originates from the interruption of cytoplasmic signaling to the nuclear transcriptional program.

Membrane and Nuclear Estrogen Receptors Originate from a Single Transcript

What is the nature of the putative membrane ER that appears to play such an important role in the vascular cell models that we have investigated? Both VSMC and EC are known to express ER, and probably express both ERα and ERβ. To begin to understand the nature and participation of each membrane ER, we transfected CHO cells with cDNAs for either of the two receptors. CHO-ERα expressed a single transcript, and both membrane and nuclear receptors. The receptors had identical Kds of about 0.1 nM, but vastly different population densities, whereby the nuclear pool was 40-fold in excess of the membrane pool. After cross-linking the receptors from membrane or nuclear compartments to labeled E2, we found that membrane and nuclear ERα were of very similar size. We also found very comparable results when we expressed the ERβ receptor in CHO cells, yielding both membrane and nuclear receptors. The membrane receptors for either isoform of ER were capable of enacting signal transduction and cell proliferation in response to both E2 and E2-BSA. Labeling of the cells with a fluorinated E2-BSA probe supported the existence of a membrane ER (22).

These results indicate that mammalian cells are capable of producing a functional cell membrane ER, which is likely to participate in the cell biologic actions of the sex steroid. Similar functional findings in primary cells that express native ER support the role of the putative membrane receptor. Understanding the independent and interdependent functions of the membrane and nuclear ER will ultimately lead to a fuller understanding of the physiologic actions of steroids and will afford unique opportunities for therapeutic intervention.

References

1. Halachmi S, Marden E, Martin G, MacKay H, Abbondanza C, Brown M. Estrogen receptor-associated proteins: possible mediators of hormone-induced transcription. Science 1994;264:1455–58.
2. Glass CK, Rose DW, Rosenfeld MG. Nuclear receptor coactivators. Curr Opin Cell Biol 1997;9(2):222–32.
3. Budhram-Mahadeo V, Parker M, Latchman DS. POU transcription factors Brn-3a and Brn-3b interact with the estrogen receptor and differentially regulate transcriptional activity via an estrogen response element. Mol Cell Biol 1998; 18(2):1029–41.

4. Pfahl M. Nuclear receptor/AP-1 interactions. Endocrine Rev 1993;14(6):651–58.
5. Gametchu B. Glucocorticoid receptor-like antigen in lymphoma cell membranes: correlation to cell lysis. Science 1987;236:456–61.
6. Blackmore PF, Neulen J, Lattanzio F, Beebe SJ. Cell surface-binding sites for progesterone mediate calcium uptake in human sperm. J Biol Chem 1991;266:18655–59.
7. Nemere I, Dormanen MC, Hammond MW, Okamura WH, Norman AW. Identification of a specific binding protein for 1α,25-dihydroxyvitamin D3 in basal-lateral membranes of chick intestinal epithelium and relationship to transcaltachia. J Biol Chem 1994;269:23750–56.
8. Wehling M. Nongenomic aldosterone effects: the cell membrane as a specific target of mineralocorticoid action. Steroids 1995;60:153–56.
9. Wehling M, Spes CH, Win N, Janson CP, Schmidt BM, Theisen K, et al. Rapid cardiovascular actions of aldosterone. J Clin Endocrinol Metab 1998;83:3517–22.
10. Tesarik J, Mendoza C. Nongenomic effects of 17 beta-estradiol on maturing human oocytes: relationship to oocyte developmental potential. J Clin Endocrinol Metab 1995;80(4):1438–43.
11. Aronica SM, Kraus WL, Katznellenbogen BS. Estrogen action via the cAMP signaling pathway: stimulation of adenylate cyclase and cAMP-regulated gene transcription. Proc Natl Acad Sci USA 1994;91:8517–21.
12. Lieberherr M, Grosse B, Kachkache M, Balsan S. Cell signaling and estrogens in female rat osteoblasts: a possible involvement of unconventional non-nuclear receptors. J Bone Miner Res 1993;8(11):1365–76.
13. Le Mellay V, Grosse B, Lieberherr M. Phospholipase C beta and membrane action of calcitriol and estradiol. J Biol Chem 1997;272(18):11902–7.
14. Pappas TC, Gametchu B, Yannariello-Brown J, Collins TJ, Watson CS. Membrane estrogen receptors in GH3/B6 cells are associated with rapid estrogen-induced release of prolactin. Endocrine 1994;2:813–22.
15. Nelson KG, Takahashi T, Bossert NL, Walmer DK, McLachlan JA. Epidermal growth factor replaces estrogen in the stimulation of female genital-tract growth and differentiation. Proc Natl Acad Sci USA 1991;88:21–25.
16. Pietras R, Szego CM. Specific binding sites for oestrogen at the outer surfaces of isolated endometrial cells. Nature 1977;265:69–72.
17. Berthois Y, Pourreau-Schneider N, Gaudilhon P, Mittre H, Tubiana N, Martin PM. Estradiol membrane binding sites on human breast cancer cell lines. Use of fluorescent estradiol conjugate to demonstrate plasma membrane binding systems. J Steroid Biochem 1986;25:963–72.
18. Pappas TC, Gametchu B, Watson CS. Membrane estrogen receptors identified by multiple antibody labeling and impeded-ligand binding. Faseb J 1995;9(5):404–10.
19. Levin ER. Endothelins. N Engl J Med 1995;333(6):356–63.
20. Morey AK, Pedram A, Razandi M, Prins BA, Hu R-M, Biesiada E, et al. Estrogen and progesterone inhibit human vascular smooth muscle proliferation. Endocrinology 1997;138(8):3330–39.
21. Morey AK, Razandi M, Pedram A, Hu R-M, Prins B, Levin ER. Estrogen and progesterone inhibit the stimulated production of endothelin-1: differential positive and negative regulatory mechanisms. Biochem J 1998;330(3):1097–105.
22. Razandi M, Pedram A, Greene GL, Levin ER. Cell membrane and nuclear estrogen receptors derive from a single transcript: studies of ERα and ERβ expressed in CHO cells. Mol Endocrinol 1999;13:307–19.

19

Role of Nitric Oxide in the Antiatherosclerotic Effect of 17β-Estradiol

KATALIN KAUSER AND GABOR M. RUBANYI

Introduction

Women before menopause appear to be protected against cardiovascular diseases compared with men at similar age (1). This protection is lost after menopause, which indicates the role of female sexual steroid hormones in the modulation of cardiovascular risk factors and clinical events. Epidemiological studies show that estrogen replacement therapy after menopause reduces the morbidity and mortality of coronary artery diseases (2).

The underlying cause of coronary artery disease is the development of atherosclerotic plaques in the coronary circulation. The types of human atherosclerotic lesions are characterized from fatty streaks to advanced complicated lesions with plaque instability and rupture (3). Endothelial dysfunction, loss of nitric oxide (NO)-mediated endothelium-dependent vasodilation, is an early event and a potential progression factor in the development of atherosclerosis. Estrogen replacement therapy of postmenopausal women showed improvement in endothelial function (4). Direct vascular effects of estrogens may account for more than 50% of the reduction in cardiovascular events, potentially through the improvement of endothelial function (5).

Marked reduction of nitric oxide synthase-III (NOS-III) protein expression is reported in the endothelium at later stages of atherosclerosis (6), and the expression of the inducible isoform, NOS-II, has been demonstrated (7). The role of peroxynitrite, generated by the interaction of large amounts of NO (produced by NOS-II) and superoxide anion free radicals, has been implicated in apoptosis (8), matrix metalloproteinase (MMP) activation (9), and, subsequently, in plaque destabilization leading to plaque rupture. 17β-estradiol inhibits excessive production of NO in endotoxemia (10) and downregulates NOS-II expression in rat aorta (11). It is conceivable that the cardioprotective benefit of 17β-estradiol treatment in later stages of atherosclerosis involves the reciprocal regulation of NO production by the two isoforms of NOS.

Vasculoprotective Mechanisms Shared by 17β-Estradiol and Endothelial Nitric Oxide

Hypercholesterolemia is associated with the incidence of coronary heart disease. Clinical trials with lipid-lowering agents (i.e., HMG CoA reductase inhibitors) demonstrated that the reduction in serum cholesterol level is correlated with improved survival of patients with coronary artery disease (12). Estrogens have favorable effects on lipid profile: lowering serum levels of total cholesterol and low-density lipoprotein (LDL) cholesterol and increasing the level of high-density lipoprotein (HDL) cholesterol (13). Cholesterol-lowering properties of estrogens account at least in part for their clinical benefit (14). In most studies, however, inhibition of lesion development by 17β-estradiol could not be explained by changes in serum cholesterol or lipoprotein composition alone. Many of the uncovered vasculoprotective mechanisms of estrogen directly target the vascular wall.

The documented vasculoprotective effects of 17β-estradiol show striking similarities to those attributed to endothelial NO (EDNO).

Inhibition of Endothelial Cell Activation and Leukocyte Recruitment

The earliest event in the development of atherosclerotic lesions is the increased adherence of circulating monocytes to the vascular endothelium. Endothelial leukocyte adhesion molecules [e.g., vascular cell adhesion molecule-1 (VCAM-1), intercellular adhesion molecule-1 (ICAM-1) and endothelial-leukocyte adhesion molecule-1 (ELAM-1 or E-selectin)] facilitate the recruitment of monocytes and lymphocytes to sites of lesion formation (15). 17β-estradiol inhibited cytokine-mediated induction of adhesion molecules in cultured human umbilical vein endothelial cells (HUVEC) (16). Similar to 17β-estradiol, NO has also been shown to decrease cytokine-induced endothelial cell activation and adhesion molecule expression (17).

Monocyte chemoattractant protein 1 (MCP-1) represents a potent stimulus for monocyte migration into the atherosclerotic lesions (18). Physiological levels of 17β-estradiol inhibited lipopolysaccharide (LPS)-induced expression of the murine homologue (JE) of MCP-1 in murine macrophages in vitro (19). 17β-estradiol decreased aortic plaque formation in apolipoproteinE-deficient (apoE-KO) mice concomitant to significant reduction in MCP-1 expression in the vessel wall (20). Inhibition of EDNO synthesis resulted in increases in MCP-1 expression in cultured endothelial cells, and it was postulated that part of the NO in vivo antiatherogenic effect is due to the continuous suppression of MCP-1 expression (21).

Effect on Vascular Smooth Muscle Cell Proliferation

Proliferation of vascular smooth muscle cells (VSMC) contributes to the progression of atherosclerotic lesions and causes vessel narrowing. Suppression of intimal hyperplasia by estrogen administration was among the earliest observations that reported cardiovascular protection by the female sexual steroid hormone (22). Other investigators have since confirmed the effect of estrogens on VSMC proliferation (23,24). The inhibitory effect of estrogen on intimal proliferation was not abolished in the estrogen receptor α-deficient (ERKO) mouse (25), indicating that the estrogen receptor beta isoform (ERß) (26,27) may mediate the effect of 17β-estradiol on VSMC. NO has also been shown to inhibit VSMC proliferation by both cyclic GMP-mediated (28) and cyclic GMP-independent mechanisms, possibly via upregulation of cell cycle inhibitory proteins (29). Studies using NOS-III-deficient mice elegantly illustrate the role of EDNO as a negative regulator of VSMC proliferation by demonstrating significantly greater intimal hyperplasia and aberrant vascular remodeling in the balloon injured vessels of the NOS-III-deficient compared with vessels of the wild type mice (30).

Effect on Endothelial Cell Proliferation and Angiogenesis

Both estrogen and EDNO were reported to stimulate endothelial cell growth, in contrast to their inhibitory effect on VSMC proliferation (see earlier). 17β-estradiol exerts direct angiogenic effects on endothelial cells (31). Studies from our laboratory demonstrated that potentiation of angiogenesis by 17β-estradiol was blunted in ERKO mice, indicating the potential role of ERα mediating the angiogenic effect of estrogen (32). Estradiol has also been shown to accelerate reendothelialization after arterial injury (33). The potential cellular mechanism of estrogen-induced angiogenesis may involve the upregulation of vascular endothelial growth factor (VEGF) (34). Several reports indicate the angiogenic effect of EDNO (35–37). EDNO seems to be the mediator of VEGF's angiogenic effect in vitro (38). Lack of VEGF-induced neovascularization in NOS-III-KO mice can serve as a proof for this hypothesis (39). VEGF has been reported to increase NOS-III mRNA stability in cultured human endothelial cells (40).

Inhibition of Early Lesion Development in Cholesterol-Fed Rabbits

Cholesterol-fed rabbits are one of the most frequently used models of early atherosclerosis. Several studies documented the antiatherosclerotic effect of 17β-estradiol in hypercholesterolemic rabbits, which was associated with significant reduction of aortic plaque area or lipid accumulation independent of changes in plasma cholesterol or lipoprotein concentration (41–43). The hypercholesterolemic rabbit was the first model of atherosclerosis in which impairment of EDNO-mediated relaxation was reported (44). The antiatherosclerotic effect of EDNO was demonstrated either by increasing EDNO production via L-arginine supplementation (45) or by decreasing EDNO via chronic inhibition of NOS with N^{ω}-nitro-L-arginine methylester (L-NAME) (46,47). Studies from our laboratory reported that estrogen reduced atherosclerotic plaque development in ovariectomized hypercholesterolemic rabbits, which coincided with a significant improvement in EDNO-mediated endothelium-dependent relaxation to acetylcholine (48).

Does EDNO Mediate the Vasculoprotective Effect of 17β-Estradiol?

The possible influence of sexual steroid hormones on EDNO production is indicated by studies that demonstrate gender difference in endothelium-dependent modulation of vascular tone (49–52). These studies showed that the vascular endothelium of females is capable to produce larger amount of EDNO. Results from our laboratory, which utilized a superfusion bioassay system, demonstrated that more EDNO is released from the perfused aortas of female rats than from that of the males (49). The hypothesis that estrogen regulates the production of EDNO in females was also supported by studies investigating the effect of ovariectomy and/or 17β-estradiol treatment on endothelium-dependent vascular responses (52–54). The original observation by Gisclard et al. (54) described increased endothelium-dependent relaxation in isolated femoral arteries of rabbits after 4 days treatment with 17β-estradiol. Similar potentiation of EDNO-mediated responses were found in coronary arteries isolated from 17β-estradiol treated monkeys (55).

The significant correlation between the antiatherosclerotic effect of estrogen and its stimulation of EDNO mediated endothelial function (56) suggest that increased EDNO release could potentially play a role in the antiatherosclerotic effect of 17β-estradiol. The numerous shared vasculoprotective mechanisms between estrogen and NO make EDNO a likely candidate to mediate the antiatherosclerotic effect of the sexual steroid hormone (57).

The obligatory role of an intact endothelium in the antiatherosclerotic effect of estrogen has been demonstrated in cholesterol-clamped rabbits (58). In this experiment removal of the endothelium by balloon catheter abolished the inhibitory effect of estrogen on aortic cholesterol accumulation. A similar study was conducted in which rabbits were chronically treated with the NOS inhibitor L-NAME to confirm that the role of EDNO is responsible for the antiatherogenic effect of estrogen in the presence of an intact endothelium (59). L-NAME significantly attenuated the vasculoprotective effect of estrogen, indicating that it was mediated (at least in part) by EDNO (59).

The hypothesis that EDNO is responsible for the vasculoprotective effect of estrogen was also investigated in our laboratory by analyzing the effect of the inhibition of EDNO production on the antiatherosclerotic effect of 17β-estradiol in ovariectomized, hypercholesterolemic rabbits. In agreement with earlier observations (46,47), our data showed a strong inverse relationship between the extent of vascular lesion development and changes in basal EDNO production (Fig. 19.1). Basal EDNO production was significantly enhanced by 17β-estradiol even in the presence of L-NAME. In our experiment the reduction in plaque area by 17β-estradiol was not significantly affected by L-NAME treatment despite the significant inhibition of EDNO production in these animals (Fig. 19.1). The fact that the antiatherosclerotic effect of 17β-estradiol could not be dissociated from increased EDNO production, however, suggests that EDNO may play an important role in mediating the vasculoprotective effect of estrogen (60).

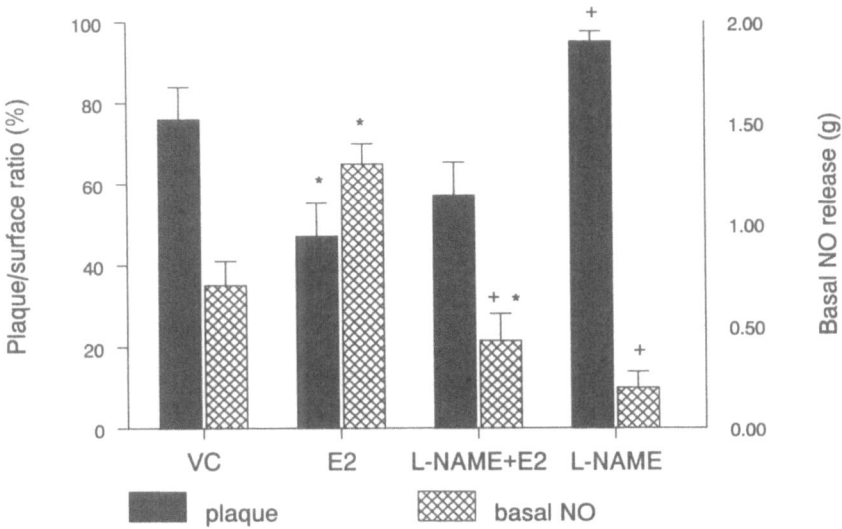

FIGURE 19.1. Effect of chronic NOS inhibition on the antiatherosclerotic effect of 17β-estradiol in hypercholesterolemic rabbits. Filled bars illustrate plaque/surface area ratio (left y axis) of thoracic aortae isolated from control, 17β-estradiol treated, 17β-estradiol + L-NAME and L-NAME treated cholesterol fed rabbits. six-week treatment of the rabbits with 17β-estradiol resulted in a significant (*p < 0.05) reduction in the plaque/ surface area, which was diminished by chronic NOS inhibition. Chronic NOS inhibition in vivo by L-NAME resulted in a significant acceleration of atherosclerosis (+p < 0.05). Cross-hatched bars represent endothelium-dependent contraction to N^G-nitro-L-arginine (L-NA) (right y-axis). 17β-estradiol significantly (*p < 0.05) augmented the endothelium-mediated contraction to L-NA (estimation of basal NO production), whereas L-NAME treatment was accompanied by significantly (+p < 0.05) attenuated basal NO production. Basal EDNO production was significantly enhanced by 17β-estradiol, even in the presence of L-NA. In each group n = 8 rabbits were studied. Bars represent mean ± SEM plaque/surface area (left y-axis) and NO mediated contraction (right y-axis). VC = vehicle control.

L-NAME treatment of ovariectomized apoE-KO mice did not modify the vasculoprotective effect of estrogen (61). In this study increase in the oxidative modification of LDL [malondialdehyde (MDA)-LDL activity] was used to assess the inhibitory effect of L-NAME on EDNO production. L-NAME treatment enhanced MDA-LDL activity, suggesting the role of EDNO in the inhibition of LDL oxidation, but it had no effect on atherosclerotic plaque development (61). On the other hand, estrogen treatment had no measurable effect on the oxidative modification of LDL, which contradicts earlier observations in other species (62,63). The lack of correlation between lipid peroxidation and plaque development in this study questions the conclusion that EDNO is not a mediator of the vasculoprotective effect of estrogen (64).

In most studies, specifically those investigating the question of EDNO mediating the vasculoprotective effect of estrogen, the effect of the female sexual steroid hormone could not be dissociated from increases in EDNO production. Our conclusion, therefore, is that although other factors may contribute to it, increased EDNO production is a key contributor to the vasculoprotective and antiatherosclerotic effect of 17β-estradiol.

Potential Mechanisms Involved in the Regulation of NOS-III by 17β-Estradiol

NOS-III Gene Expression

One of the mechanisms by which 17β-estradiol enhances endothelial NO synthesis is the upregulation of NOS-III gene expression. Transcriptional regulation of the NOS-III gene by 17β-estradiol has been studied in cultured endothelial cells. In early passages of human aortic and ovine fetal pulmonary endothelial cells, physiological concentrations of 17β-estradiol caused a significant increase in NOS-III protein levels and consequently augmented NO production (65,66). Bovine aortic endothelial cells (BAEC) in culture contain functional estrogen receptors (67), which mediate 17β-estradiol-induced transactivation of a luciferase reporter gene with an estrogen response element (ERE) as a transcriptional enhancer (67). The existence of the two half-palindromic sites of ERE in the promoter region of human NOS-III gene supports a potential receptor-mediated effect of estrogen on NOS-III gene expression in endothelial cells (68). Studies using a 1.6-κB human NOS-III promoter construct transfected into EA.hy926 cells demonstrated increased transcriptional activity in response to estrogen treatment mediated by the transcription factor Sp1 (69).

In vivo data from several species suggest that 17β-estradiol can facilitate NOS-III gene expression and increase the amount of NOS-III enzyme. Pregnancy has been associated with a fourfold increase in NOS-III activity in the guinea pig uterine artery (70). Increases of NOS-III mRNA have similarly been demonstrated in pregnant rat aortas using semiquantitative polymerase chain reaction (PCR) (71).

NOS-III Enzyme Activity

Increased synthesis of NO can be a result of elevated enzyme activity without changes in gene expression or NOS-III protein level. Activity of NOS-III can be augmented by increased availability of the substrate: L-arginine, or cofactors: tetrahydrobiopterin (BH_4), calmodulin, or intracellular free cytosolic calcium level. Studies have demonstrated that the administration of exogenous L-arginine restores endothelium-dependent relaxation in hypercholesterolemic humans (72) and decreases aortic lesion formation in cholesterol-fed rabbits (45). BH_4 substitution has also been shown to restore impaired endothelium-dependent relaxation in hypercholesterolemic patients (73) and in diabetic rats (74). These findings indicate that under certain pathological conditions substrate and cofactor availability of NOS-III may be limited, and a treatment that results in the increase in these molecules can provide vasculoprotective benefit. A report by Hayashi et al. (75) argues against the possibility that 17β-estradiol would regulate the level of circulating L-arginine, but there is no data available describing the effect of estrogen on BH_4 synthesis. 17β-estradiol may facilitate the activity of NOS-III via calmodulin as it has been described in rabbit myometrium (76), but it has not yet been investigated in vascular endothelial cells. On the other hand, 17β-estradiol treatment of HUVEC resulted in a rapid increase in NO release independent of changes in intracellular calcium concentration (77). Although the mechanism of this calcium-independent activation of NOS-III by estrogen is not completely understood, it seems to involve the activation of ER because the ER antagonist ICI 164,384 inhibited this response.

NOS-III activity has most recently been shown to be regulated by association with other intracellular proteins [e.g, caveolin-1 (78) and heat shock protein 90 (Hsp90) (79)]. It is possible that estrogen influences NOS-III activity by modulating these protein–protein interactions. Hsp90 associates with NOS-III in a stimulus-dependent manner and contributes to the subsequent activation of NOS-III to physiological stimuli, such as shear stress or histamine (79). Hsp90 is a molecular chaperone that has been shown to participate actively in steroid-induced signal transduction (80). It is possible that the rapid ER-dependent effect of 17β-estradiol on NOS-III activity is mediated by the intracellular redistribution of Hsp90, permitting association with NOS-III.

Bioactivity of NO

In addition to its rate of synthesis, the level of bioavailable NO can be regulated by degradation or by inhibition of its inactivation. Increased amount of cGMP, as a readout of augmented NO release, has been documented in ethinyl estradiol–treated BAEC without seeing any increase in NOS-III mRNA and protein amount or enzyme activity (81). Measurement of superoxide anion radical production from the same cells revealed that estrogen treatment significantly reduced the generation of superoxides. Superoxide anion is an effective inactivator of NO (82). Decreasing the level of superoxide anions could lead to increased amount of bioavailable NO without changing its synthetic rate. It is also conceivable that the antioxidant properties of 17β-estradiol itself may contribute to the increased bioactivity of EDNO, in addition to suppressing superoxide radical production in endothelial cells.

Role of Estrogen Receptor (ER) in the Regulation of NOS-III

The presence of functional ER has been demonstrated in endothelial cells (66, 67, 83–85). The role of ER in mediating the effect of 17β-estradiol on NOS-III regulation has been investigated using the ER antagonists ICI 182780 and tamoxifen. Both compounds inhibited the dose-dependent increase in NOS-III mRNA and protein as well as enzyme activity in response to estrogen in cultured HUVEC and BAEC (66,67,85).

Studies from our laboratory using the homozygous ERKO mouse provided the first in vivo proof of the role of ERα in mediating the effect of estrogen on the release of EDNO (86). We found that the lack of expression of ERα led to significant impairment in basal endothelial NO release, without affecting the amount of NOS-III enzyme (Fig. 19.2). This finding demonstrated that ERα is involved in the regulation of NOS-III enzyme activity in vivo, probably by affecting the level of free cytosolic calcium and/or participating in post-translational modifications of NOS-III, resulting in increased enzyme activity.

Paradoxical Role of NO in a Therosclerosis

In addition to the numerous antiatherogenic properties of EDNO generated by NOS-III described earlier, large amounts of NO produced by NOS-II, the inducible isoform of NOS, can potentially be proatherogenic. NO reacts with equimolar concentrations of superoxide radical resulting in peroxynitrite formation, which in large amount acts as a strong oxidant damaging intracellular proteins by nitration of the tyrosine residues (87). Extensive protein nitration has been demonstrated in human atherosclerotic lesions (87,88), although the effect of nitration on protein function in vivo is not known (87). Peroxynitrite and its decomposition products, hydroxyl radicals, are able to induce lipid peroxidation (89). Via the alteration of the cellular redox state, it can also induce the expression of redox-sensitive genes that participate in the recruitment of inflammatory cells to the endothelial surface, such as VCAM-1 (90).

Large quantities of NO have been reported to cause apoptosis in macro-

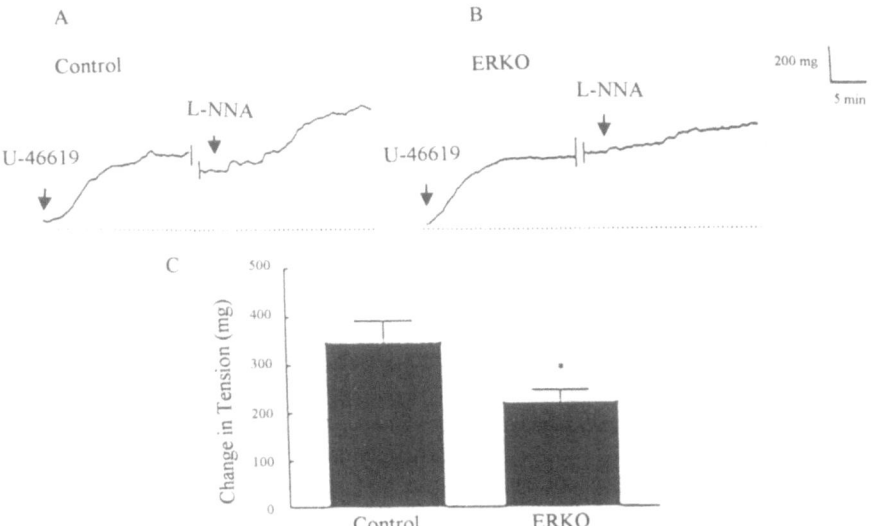

FIGURE 19.2. Decreased basal endothelial NO production in estrogen receptor deficient (ERKO) mouse aorta. Endothelium-dependent contraction of the thoracic aorta
to L-NA was significantly ($*p < 0.05$) diminished in rings isolated from ERKO mice
compared with age matched wild type (WT) controls. The upper two panels of the
graph illustrate original tracing of the organ chamber experiments and the lower
panel demonstrates bar graph of the average data (mean ± SEM) obtained from $n = 5$
animals. The bar graphs represent total contraction of the isolated mouse thoracic
aortic rings to the thromboxane receptor agonist U-46619 and L-NA. Reproduced from
J Clin Invest 1997;99:2429–37 by copyright permission of The American Society of
Clinical Investigation (Ref. 86).

phages and smooth muscle cells (91,92). The role of apoptosis in atherosclerosis is not understood completely. Increases in apoptotic cell death can be
beneficial by resulting in plaque regression, or it could contribute to plaque
instability by playing a role in the development of large necrotic core and
thinning of the fibrous cap. Unstable plaques are at high risk to rupture, and the
exposed extracellular matrix provides a surface for thrombotic complications.
Large amounts of NO may also promote plaque destabilization by the activation
of matrix metalloproteinases (MMPs) (9) or by the inhibition of the tissue inhibitor of metalloproteinase-1 (TIMP-1) (93). Expression of NOS-II has been reported
in human coronary atherosclerosis (7) and more extensively in transplant atherosclerosis in animal models (94) and in humans (87). Intimal vascular smooth
muscle cells can be a significant source of NOS-II expression in atherosclerotic
arteries (95). The generation of large amounts of peroxynitrite consequent to the
increased levels of NO and superoxide in the atherosclerotic lesion may be a
factor contributing to the progression of the disease (87).

Inhibition of NOS-II by 17β-Estradiol

The effect of estrogen has been first investigated on the spontaneous NO production by NOS-II expressed in cultured hepatocytes (96) and later in rat peritoneal macrophages (97) in vitro. Studies from our laboratory reported for the first time the effect of 17β-estradiol on cytokine-induced NOS-II expression in aortic rings isolated from rats. We found that, like the NOS-II inhibitor, aminoguanidine, co-incubation of the aortic rings with the cytokines and 17β-estradiol significantly reduced nitrite accumulation and increased contractile reactivity to phenylephrine. This effect of 17β-estradiol was accompanied by significant reduction in NOS-II protein expression and mRNA levels determined by semiquantitative RT-PCR (11). In contrast with the augmentation of NOS-III expression/activity in endothelial cells, 17β-estradiol can inhibit cytokine-induced excessive nitrite/nitrate generation and is able to restore contractile responsiveness of vascular smooth muscle in isolated rat aortic rings (Fig. 19.3).

ER is present in the rat aorta (98,99). Based on our data, the effect of 17β-estradiol is mediated by the activation of ER because the partial ER agonist, 4-OH-tamoxifen, also inhibited nitrite production and attenuated the suppression of phenylephrine-induced contraction. In agreement with our finding, Hayashi et al. (97) reported that 17β-estradiol inhibited NOS-II protein expression and NO production in murine macrophages. This effect of the hormone was abolished by simultaneous treatment with the ER receptor antagonist, ICI 182780.

We also demonstrated for the first time that physiological substitution dose of 17β-estradiol reduced nitrite/nitrate accumulation in the plasma of lipopolysaccharide (LPS)-treated ovariectomized rats (10). Ovariectomized and sham-operated female rats were injected with LPS, and excessive NO generation was estimated by measuring plasma nitrite levels 12 hours after LPS injection. LPS-induced increase in circulating plasma nitrite level was significantly higher in the ovariectomized animals compared with sham-operated females, which suggests that endogenous ovarian sex steroid hormones may also suppress excessive NO production by NOS-II in this model. Our finding suggests that estrogens may be also beneficial in pathological conditions associated with excessive NO generation via the inducible NOS isoform (NOS-II).

Summary

Although gender difference in morbidity and mortality due to cardiovascular diseases and the cardiovascular benefits of estrogen replacement therapy in postmenopausal women have been known for a long time, the exact mechanism of vasculoprotection by estrogens remained elusive. Significant progress has been made in this field. One of the most exciting developments was the discovery of a

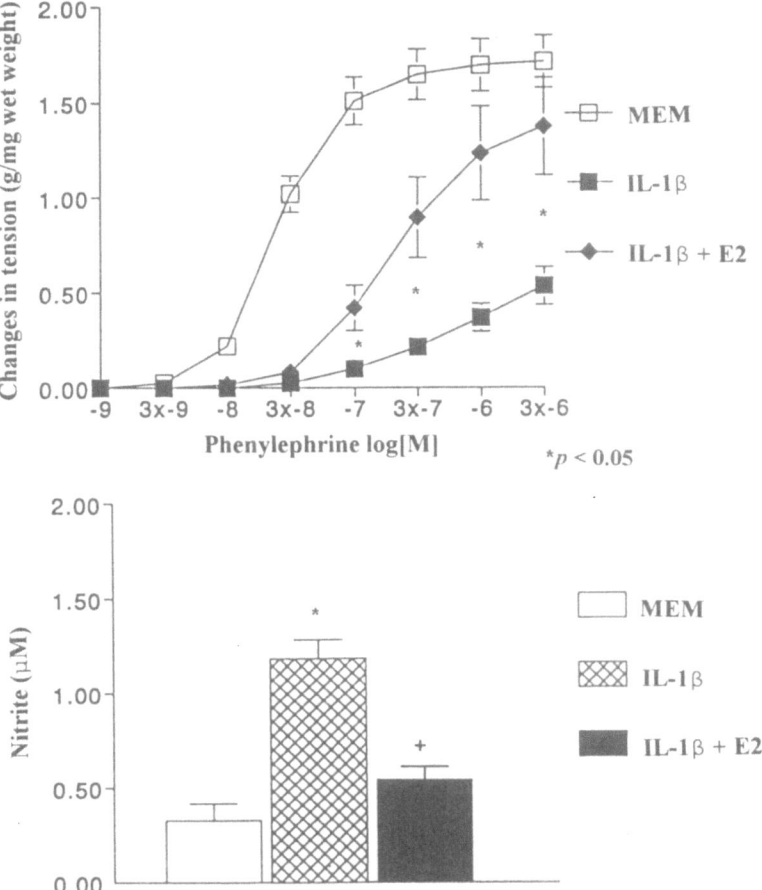

FIGURE 19.3. Effect of 17β-estradiol on IL-1β-induced suppression of contractility and nitrite production by isolated rat thoracic aortic rings. The upper panel of the figure illustrates dose-response curves to phenylephrine in control (open square), IL-1β (filled square), and IL-1β+17β-estradiol (filled diamond) incubated de-endothelialized rat aortic rings. IL-1β significantly attenuated smooth muscle contraction. Treatment with 17β-estradiol significantly (*$p < 0.05$) reversed the inhibition of contractility by IL-1β. The lower panel demonstrates nitrite accumulation in the incubation media of the vessels. IL-1β induced significant (*$p < 0.05$) nitrite production (cross-hatched bar) by the aortic rings in the organ culture, which was significantly inhibited (+$p < 0.05$) by 17β-estradiol treatment of the cytokine incubated rings (filled bar).

potential link between vasculoprotection by estrogen and NO produced by the endothelium (57,100). In this chapter we listed evidence demonstrating that the upregulation of EDNO production/activity plays an important role in the antiatherosclerotic action of 17β-estradiol. The other interesting discovery has been the inhibition of excessive (destructive) NO production via NOS-II by the same doses of 17β-estradiol that upregulate vasculoprotective EDNO. This unique property of the female sexual steroid hormone may explain (and mediate at least in part) the remarkable cardiovascular protection by estrogens.

References

1. Isles CG, Hole DJ, Hawthorne VM, Lever AF. Relation between coronary risk and coronary mortality in women of the Renfrew and Paisely survey: comparison with men. Lancet 1992;339:702–6.
2. Stampfer MJ, Colditz, GA. Estrogen replacement therapy and coronary heart disease: a quantitative assessment of the epidemiological evidence. Prev Med 1991;20:47–63.
3. Fuster V, Badimon JJ, Chesebro JH, Fallon JT. Plaque rupture, thrombosis, and therapeutic implications. Haemostasis 1996;26(suppl 4):269–84.
4. Bell DM, Johns TE, Lopez LM. Endothelial dysfunction: implications for therapy of cardiovascular diseases. Ann Pharmacother 1998;32:459–70.
5. Vogel RA, Corretti MC. Estrogens, progestins, and heart disease: can endothelial function divine the benefit? Circulation 1998;97:1223–26.
6. Oemar BS, Tschudi MR, Godoy N, Brovkovich V, Malinski T, Lüscher TF. Reduced endothelial nitric oxide synthase expression and production in human atherosclerosis. Circulation 1998;97:2494–98.
7. Buttery LDK, Springall DR, Chester AH, Evans TJ, Standfield N, Parums DV, et al. Inducible nitric oxide synthase is present within human atherosclerotic lesions and promotes the formation and activity of peroxynitrite. Lab Invest 1996;75: 77–85.
8. Fukuo K, Hata S, Suhara T, Nakahashi T, Shinto Y, Tsujimoto Y, et al. Nitric oxide induces upregulation of fas and apoptosis in vascular smooth muscle. Hypertension 1996;27:823–26.
9. Murrell GAC, Jang D, Williams RJ. Nitric oxide activates metalloprotease enzymes in articular cartilage. Biochem Biophys Res Commun 1995;206:15–21.
10. Kauser K, Sonnenberg D, Tse J, Rubanyi GM. 17β-Estradiol attenuates endotoxin-induced excessive nitric oxide production in ovariectomized rats in vivo. Am J Physiol 1997;273:H506–9.
11. Kauser K, Sonnenberg D, Diel P, Rubanyi GM. 17β-Estradiol inhibits cytokine-induced nitric oxide production in isolated rat aorta. Br J Pharmacol 1998;123: 1089–96.
12. Shepherd J, Cobbe SM, Ford I, Isles CG, Lorimer AR, MacFarlane PW, et al. Prevention of coronary heart disease with pravastatin in men with hypercholesterolemia: West of Scotland Coronary Prevention Study Group. N Engl J Med 1995;333:1301–7.
13. Walsh BW, Schiff I, Rosner B, Greenberg L, Ravnikar V, Sacks FM. Effects of postmenopausal estrogen replacement on the concentrations and metabolism of plasma lipoproteins. N Engl J Med 1991;325:1196–204.
14. Knopp RH, Zhu X, Bonet B. Effects of estrogens on lipoprotein metabolism and cardiovascular disease in women. Atherosclerosis 1994;110:583–91.
15. Cybulsky MI, Gimbrone MA Jr. Endothelial expression of a mononuclear leukocyte adhesion molecule during atherogenesis. Science 1991;251:788–91.

16. Caulin-Glaser T, Watson CA, Pardi R, Bender JR. Effects of 17β-estradiol on cytokine-induced endothelial cell adhesion molecule expression. J Clin Invest 1996;98:36–42.
17. DeCaterina R, Libby P, Peng H-B, Thannickal VJ, Rajavashisth TB, Gimbrone MA Jr., et al. Nitric oxide decreases cytokine-induced endothelial activation. J Clin Invest 1995;96:60–68.
18. Yla-Herttuala S, Palinski W, Rosenfeld M, Parthasarathy S, Carew TE, et al. Evidence for the presence of oxidatively modified low density lipoprotein in atherosclerotic lesions of rabbit and man. J Clin Invest 1989;84:1086–95.
19. Frazier-Jessen MR, Kovacs EJ. Estrogen modulation of je/monocyte chemoattractant protein-1 mRNA expression in murine macrophages. J Immunol 1995;154: 1838–45.
20. Bourassa P, Milos P, Lira M, Aiello RJ. Estrogen's effects on atherosclerotic lesion formation in the apolipoprotein e deficient mouse. J Vascular Res 1996;33(suppl 1): 36.
21. Zeiher AH, Fisslthaler B, Schray-Utz B, Busse R. Nitric oxide modulates the expression of monocyte chemoattractant protein 1 in cultured human endothelial cells. Circ Res 1995;76:980–86.
22. Rhee CY, Spaet TH, Stemerman MB, Lajam F, Shiang HH. Estrogen suppression of surgically induced vascular intimal hyperplasia in rabbits. J Lab Clin Med 1977;90:77–84.
23. Foegh ML, Asotra S, Howell MH, Ramwell PW. Estradiol inhibition of arterial neointimal hyperplasia after balloon injury. J Vasc Surg 1994;19:722–26.
24. Vargas R, Wroblewska B, Rego A, Hatch J, Ramwell PW. Oestradiol inhibits smooth muscle cell proliferation of pig coronary artery. Br J Pharmacol 1993;109:612–17.
25. Iafrati MD, Karas RH, Aronovitz M, Kim S, Sullivan TR Jr, Lubahn DB, et al. Estrogen inhibits the vascular injury response in estrogen receptor α-deficient mice. Nature Med 1997;3:545–48.
26. Kuiper GGJM, Enmark E, Pelto-Huikko M, Nilsson S, Gustafsson J-A. Cloning of a novel estrogen receptor expressed in rat prostate and ovary. PNAS 1996;93: 5925–30.
27. Mosselman S, Polman J, Dijkema R. ERβ: identification and characterization of a novel human estrogen receptor. FEBS Lett 1996;392:49–53.
28. Garg UC, Hassid A. Nitric oxide-generating vasodilators and 8-bromocyclic guanosine monophosphate inhibit mitogenesis and proliferation of cultured rat vascular smooth muscle cells. J Clin Invest 1989;83:1774–77.
29. Kibbe MR, Billiar TR, Tzeng E. Adenoviral iNOS gene transfer inhibits smooth muscle cell proliferation through the upregulation of p27. Nitric Oxide 1998;2:80.
30. Rudic RD, Shesely EG, Maeda N, Smithies O, Segal SS, Sessa WC. Direct evidence for the importance of endothelium-derived nitric oxide in vascular remodeling. J Clin Invest 1998;101:731–36.
31. Morales DE, McGowan KA, Grant DS, Maheshwari S, Bhartiya D, Cid MC, et al. Estrogen promotes angiogenic activity in human umbilical vein endothelial cells in vitro and in a murine model. Circulation 1995;91:755–63.
32. Johns A, Freay AD, Fraser W, Korach KS, Rubanyi GM. Disruption of estrogen receptor gene prevents 17β estradiol-induced angiogenesis in transgenic mice. Endocrinology 1996;137:4511–13.
33. Krasinski K, Spyridopoulos I, Asahara T, van der Zee R, Isner JM, Losordo DW.

Estradiol accelerates functional endothelial recovery after arterial injury. Circulation 1997;95:1768–72.

34. Karas RH, Bieber HE, Baur WE, Mendelsohn ME. Estrogen enhances vascular endothelial growth factor (VEGF) gene expression in human vascular smooth muscle cells. Circulation 1996;94(suppl 1):1–595 (abstract).

35. Ziche ML, Morbidelli E, Masini S, Amerini HJ, Granger CA, Maggi P, et al. Nitric oxide mediates angiogenesis in vivo and endothelial cell growth and migration in vitro promoted by substance P. J Clin Invest 1994;94:2036–44.

36. Pipili-Synetos E, Sakkoula E, Haralabopoulos G, Andriopoulou P, Peristeris P, Maragoudakis ME. Evidence that nitric oxide is an endogenous antiangiogenic mediator. Br J Pharmacol 1994;111:894–902.

37. Noiri E, Hu Y, Bahou WF, Kesse CR, Giaever I, Goligorsky MS. Permissive role of nitric oxide in endothelin-induced migration of endothelial cells. J Biol Chem 1997;272:1747–52.

38. Papapetropoulos A, García-Cardeña G, Madri JA, Sessa WC. Nitric oxide production contributes to the angiogenic properties of vascular endothelial growth factor in human endothelial cells. J Clin Invest 1997;100:3131–39.

39. Murohara T, Asahara T, Silver M, Bauters C, Masuda H, Kalka C, et al. Nitric oxide synthase modulates angiogenesis in response to tissue ischemia. J Clin Invest 1998;101:2567–78.

40. Bouloumie A, Schini-Kerth VB, Busse R. Vascular endothelial growth factor upregulates the expression of nitric oxide synthase III in native and cultured endothelial cells. Circulation 1997;96 (suppl 1):1–550.

41. Kushwaha RS, Hazzard WR. Exogenous estrogens attenuate dietary hypercholesterolemia and atherosclerosis in the rabbit. Metabolism 1981;30:359–66.

42. Hough JL, Zilversmit DB. Effect of 17 beta estradiol on aortic cholesterol content and metabolism in cholesterol-fed rabbits. Arteriosclerosis 1986;6:57–63.

43. Haarbo J, Leth-Espensen P, Stender S, Christiansen C. Estrogen monotherapy and combined estrogen-progestogen replacement therapy attenuate aortic accumulation of cholesterol in ovariectomized cholesterol-fed rabbits. J Clin Invest 1991; 87:1274–79.

44. Verbeuren TJ, Jordaens FH, Zonnekeyn LL, Van Hove CE, Coene M-C, Herman AG. Effect of hypercholesterolemia on vascular reactivity in the rabbit: 1. Endothelium-dependent and endothelium-independent contractions and relaxations in isolated arteries of control and hypercholesterolemic rabbits. Circ Res 1986;58: 552–64.

45. Cooke JP, Singer AH, Tsao P, Zera P, Rowan RA, Billingham ME. Antiatherogenic effects of L-arginine in the hypercholesterolemic rabbit. J Clin Invest 1992;90: 168–72.

46. Naruse KM, Shimizu M, Muramatsu M, Toki Y, Miyazaki Y, Okurama K, et al. Long-term inhibition of NO synthesis promotes atherosclerosis in the hypercholesterolemic rabbit thoracic aorta. Arterioscler Thromb 1994;14:746–52.

47. Cayatte AJ, Palacino JJ, Horten K, Cohen RA. Chronic inhibition of nitric oxide production accelerates neointima formation and impairs endothelial function in hypercholesterolemic rabbits. Arterioscler Thromb 1994;14:753–59.

48. Kauser K, Rubanyi GM. Vasculoprotection by estrogen contributes to gender difference in cardiovascular diseases; potential mechanism and role of endothelium. In:

Rubanyi GM, Dzau, VJ, eds. The endothelium in clinical practice. New York: Marcel Dekker, 1997:439–67.

49. Kauser K, Rubanyi GM. Gender difference in bioassayable endothelium-derived nitric oxide release from isolated rat aortae. Am J Physiol 1994;267:H2311–17.

50. Kauser K, Rubanyi GM. Gender difference in endothelial dysfunction in the aorta of spontaneously hypertensive rats. Hypertension 1995;25:517–23.

51. Stallone JN, Crofton JT, Share L. Sexual dimorphism in vasopressin-induced contraction of rat aorta. Am J Physiol 1991;260:H453–58.

52. Hayashi T, Fukuto JM, Ignarro LJ, Chaudhuri G. Basal release of nitric oxide from aortic rings is greater in female rabbits than in male rabbits: implications for atherosclerosis. PNAS 1992;89:11259–63.

53. Kauser K, Rubanyi GM. Effect of 17β-estradiol on endothelial dysfunction in the aorta of spontaneously hypertensive rats. Endothelium 1995;25:517–23.

54. Gisclard V, Miller VM, Vanhoutte PM. Effect of 17β-estradiol on endothelium-dependent responses in the rabbit. J Pharmacol Exp Therap 1988;244:19–22.

55. Nascimento C, Kauser K, Rubanyi GM. Effect of 17β-estradiol in hypercholesterolemic rabbits with severe endothelial dysfunction. Am J Physiol 1999;276: H1788–94.

56. Williams JK, Adams MR, Klopfenstein HS. Estrogen modulates responses of atherosclerotic coronary arteries. Circulation 1990;81:1680–87.

57. Kauser K, Rubanyi GM. 17β-estradiol and endothelial nitric oxide synthase. Endothelium 1994;2:203–8.

58. Holm P, Stender S, Andersen HO, Hansen BF, Nordestgaard BG. Antiatherogenic effect of estrogen abolished by balloon catheter injury in cholesterol-clamped rabbits. Arterioscler Thromb Vasc Biol 1997;53:45–54.

59. Holm P, Korsgaard N, Shalmi M, Andersen HL, Hougaard P, Skouby SO. Significant reduction of the antiatherogenic effect of estrogen by long-term inhibition of nitric oxide synthesis in cholesterol-clamped rabbits. J Clin Invest 1997;100: 821–28.

60. Nascimento, CA, Kauser K, Rubanyi GM. Effect of 17β-estradiol in hypercholesterolemic rabbits with severe endothelial dysfunction. Am J Physiol 1999; 276:H1788–94.

61. Elhage R, Bayard F, Richard V, Holvoet P, Duverger N, Fiévet C, et al. Prevention of fatty streak formation of 17β-estradiol is not mediated by the production of nitric oxide in apolipoprotein e-deficient mice. Circulation 1997;96:3048–52.

62. Maziere C, Auclair M, Ronveaux MF, Salmon S, Santus RL, et al. Estrogens inhibit copper and cell-mediated modification of low density lipoprotein. Arteriosclerosis 1991;89:175–82.

63. Schröder J, Dören M, Schneider B, Oettel M. Are the antioxidative effects of 17β-estradiol modified by concomitant administration of a progestin? Maturitas 1996;25:133–39.

64. White CR, Darley-Usmar V, Oparil S. NO in estrogen-mediated vasoprotection? Circulation 1997;96:2769–71.

65. Hishikawa K, Makaki T, Marumo T, Suzuki H, Kato R, Saruta T. Up-regulation of nitric oxide synthase by estradiol in human aortic endothelial cells. FEBS Lett 1995;36:291–93.

66. MacRitchie AN, Jun SS, Chen Z, German Z, Yuhanna IS, Sherman TS, Shaul PW.

Estrogen upregulates endothelial nitric oxide synthase gene expression in fetal pulmonary artery endothelium. Circ Res 1997;81:355–62.

67. Bayard F, Clamens S, Delson G, Blaes N, Maret A, Faye J-C. Oestrogen synthesis, oestrogen metabolism and functional oestrogen receptors in bovine aortic endothelial cells. In: Non-Reproductive Actions of Sex Steroids Ciba Foundation Symposium. Chichester: John Wiley and Sons, 1995;191:122–38.

68. Venema RC, Nishida K, Alexander RW, Harrison DG, Murphy TJ. Organization of the bovine gene encoding the endothelial nitric oxide synthase. Biochem Biophys Acta 1994;1218:413–20.

69. Kleinert H, Wallerath T, Euchenhofer C, Ihrig-Biedert I, Li H, Förstermann U. Estrogens increase transcription of the human endothelial NO synthase gene: analysis of the transcription factors involved. Hypertension 1998;31:582–88.

70. Weiner CP, Knowles RG, Moncada S. Induction of nitric oxide synthases early in pregnancy. Am J Obstet Gynecol 1994;71(suppl 3):838–43.

71. Goetz RM, Morano I, Calvoni T, Studer R, Holtz J. Increased expression of endothelial constitutive nitric oxide synthase in rat aorta during pregnancy. Biochem Biophys Res Commun 1994;205:905–10.

72. Creager MA, Girerd XJ, Gallagher SH, Coleman S, Dzau VJ, Cooke JP. L-arginine improves endothelium-dependent vasodilation in hypercholesterolemic humans. J Clin Invest 1992;90:1248–53.

73. Stroes E, Kastelein J, Cosentino F, Erkelens W, Wever R, Koomans H, et al. Tetrahydrobiopterin restores endothelial function in hypercholesterolemia. J Clin Invest 1997;99:41–46.

74. Pieper GM. Acute amelioration of diabetic endothelial dysfunction with a derivative of the nitric oxide synthase cofactor, tetrahydrobiopterin. J Cardiovasc Pharmacol 1997;29:8–15.

75. Hayashi T, Fukuto JM, Ignarro LJ, Chaudhuri G. Gender differences in atherosclerosis: possible role of nitric oxide. J Cardiovasc Pharmacol 1995;26:792–802.

76. Matsui K, Higashi K, Fukunga K, Miyazaki K, Maeyama M, Miyamoto E. Hormone treatments and pregnancy alter myosin light chain kinase and calmodulin levels in rabbit myometrium. J Endocrinol 1983;97:11–16.

77. Caulin-Glaser T, García-Cardeña G, Sarrel P, Sessa WC, Bender JR. 17β-estradiol regulation of human endothelial cell basal nitric oxide release, independent of cytosolic Ca^{2+} mobilization. Circ Res 1997;81:885–92.

78. García-Cardeña G, Martasek P, Masters BSS, Skidd PM, Couet J, Li S, et al. Dissecting the interaction between nitric oxide synthase (NOS) and caveolin. J Biol Chem 1997;272:25437–40.

79. García-Cardeña G, Fan R, Shah V, Sorrentino R, Cirino G, Papapetropoulos A, et al. Dynamic activation of endothelial nitric oxide synthase by Hsp90. Nature 1998;392:821.

80. Segnitz B, Gehring U. The function of steroid hormone receptors is inhibited by the hsp90-specific compound geldanamycin. J Biol Chem 1997;272(30): 18694–701.

81. Arnal JF, Clamens S, Pechet C, Negre-Salvayre A, Allera C, Girolami J-P, et al. Ethinylestradiol does not enhance the expression of nitric oxide synthase in bovine endothelial cells but increases the release of bioactive nitric oxide by inhibiting superoxide anion production. PNAS 1996;93:4108–13.

82. Rubanyi GM, Vanhoutte PM. Oxygen-derived free radicals, endothelium, and responsiveness of vascular smooth muscle. Am J Physiol 1986;250:H815–21.

83. Kim-Schulze S, McGowan KA, Hubchak SC, Cid MC, Martin MB, Kleinman HK, et al. Expression of an estrogen receptor by human coronary artery and umbilical vein endothelial cells. Circulation 1996;94:1402–7.

84. Venkov CD, Rankin AB, Vaughan DE. Identification of authentic estrogen receptor in cultured endothelial cells. A potential mechanism for steroid hormone regulation of endothelial function. Circulation 1996;94:727–33.

85. Hayashi TK, Yamada K, Esaki T, Kuzuya M, Satake S, Ishikawa T, et al. Estrogen increases endothelial nitric oxide by a receptor-mediated system. Biochem Biophys Res Commun 1995;214:847–55.

86. Rubanyi GM, Freay AD, Kauser K, Sukovich D, Burton G, Lubahn DB, et al. Vascular estrogen receptors and endothelium-derived nitric oxide production in the mouse aorta gender difference and effect of estrogen receptor gene disruption. J Clin Invest 1997;99:2429–37.

87. Beckman JS, Ye YZ, Anderson PG, Chen J, Accavitti MA, Tarpey MM, et al. Extensive nitration of protein tyrosines in human atherosclerosis detected by immunohistochemistry. Biol Chem Hoppe Seyler 1994;375:81–88.

88. Ravalli S, Albala A, Ming M, Szabolcs M, Barbone A, Michler RE, et al. Inducible nitric oxide synthase expression in smooth muscle cells and macrophages of human transplant coronary artery disease. Circulation 1998;97:2338–45.

89. Darley-Usmar VM, Hogg N, O'Leary VJ, Moncada S. The simultaneous generation of superoxide and nitric oxide can initiate lipid peroxidation in human low density lipoprotein. Free Radic Res Commun 1992;17:9–20.

90. Marui N, Offerman MK, Swerlick R, Kunsch C, Rosen CA, Ahmad M, et al. Vascular cell adhesion molecule-1 (VCAM-1) gene transcription and expression are regulated through an antioxidant-sensitive mechanism in human vascular endothelial cells. J Clin Invest 1993;92:1866–74.

91. Sarih M, Souvannavong V, Adam A. Nitric oxide synthase induces macrophage death by apoptosis. Biochem Biophys Res Commun 1993;191:503–8.

92. Albina JE, Cui S, Mateo RB, Reichner JS. Nitric oxide-mediated apoptosis in murine peritoneal macrophages. J Immunol 1993;150:5080–85.

93. Frears ER, Zhang Z, Blake DR, O'Connell JP, Winyard PG. Inactivation of tissue inhibitor of metalloproteinase-1 by peroxynitrite. FEBS Lett 1996;381:21–24.

94. Russell ME, Wallace AF, Wyner LR, Newell JB, Karnovsky MJ. Upregulation and modulation of inducible nitric oxide synthase in rat cardiac allografts with chronic rejection and transplant arteriosclerosis. Circulation 1995;92:457–64.

95. Sobey CG, Brooks RM II, Heistad DD. Evidence that expression of inducible nitric oxide synthase in response to endotoxin is augmented in atherosclerotic rabbits. Circ Res 1995;77:536–43.

96. Pittner RA, Spitzer JA. Steroid hormones inhibit induction of spontaneous nitric oxide production in cultured hepatocytes without changes in arginase activity or urea production. PSEBM 1993;202:499–504.

97. Hayashi T, Yamada K, Esaki T, Muto E, Chaudhuri G, Iguchi A. Physiological concentrations of 17β-estradiol inhibit the synthesis of nitric oxide synthase in macrophages via a receptor-mediated system. J Cardiovasc Pharmacol 1998;31:292–98.

98. Lin AL, Shain SA, Gonzales R. Sexual dimorphism characterizes steroid hormone

modulation of rat aortic steroid hormone receptors. Endocrinology 1986;119: 296–302.

99. Knauthe R, Diel P, Hegele-Hartung CH, Engelhaupt A, Fritzemeier K-H. Sexual dimorphism of steroid hormone receptor messenger ribonucleic acid expression and hormonal regulation in rat vascular tissue. Endocrinology 1996;137:3220–27.

100. Kauser K, Rubanyi GM. Potential cellular signaling mechanisms mediating upregulation of endothelial nitric oxide production by estrogen. J Vasc Res 1997/b;34:229–36.

20

Biological and Psychosocial Factors Affecting Sexual Functioning During the Menopausal Transition

Lorraine Dennerstein, Philippe Lehert, Henry Burger, and Emma Dudley

Introduction

There has been much controversy about whether women experience changes in sexual functioning during the menopausal transition. Studies of women attending menopause clinics report a high prevalence of sexual difficulties and marital problems (1). Clinicians and scientists are therefore concerned about whether aspects of sexual functioning diminish with the menopausal transition. If changes do occur, then does this reflect particular hormonal changes, or concomitant aging?

Clinical experience, however, is known to be based on a small proportion of self-selecting, predominantly ill women, and it may not be representative of most women's experience of the menopause (2,3). Population-based studies are needed to determine the range of experiences associated with menopause and to provide us with the opportunity to investigate the relative role of hormones in a naturalistic setting. Relatively few of the population-based studies of the menopausal transition, however, have inquired about sexual functioning. Differing measures of sexual functioning have been used, with studies often failing to offer any data on the validity or reliability of these measures in their local population. Some of the assessments have been very vague constructs indeed. For example, Osborn et al. (4) asked a general practice population if they suffered from "sexual dysfunction," whereas Koster and Garde (5) asked women whether sexual desire was decreased or infrequent compared with 11 years previously. Even fewer studies have undertaken any hormonal determinations, and these have either been of small sample size (6), followed the sample for only a short time, or are still in progress. Studies of the role of menopausal status seem to hypothesize that any changes in health outcomes reflecting the menopause will be evident in postmenopausal women having different levels of the variable under study to that of women who have not reached menopause (12 months of amenorrhoea). Endocrine change occurs for some years prior to cessation of menopause, however, so it would seem important also to examine for changes in health outcomes some years before menses cease. Exogenous hormones may mask the effects of changing ovarian function so that women taking the oral contraceptive pill or any sort of hormone therapy must be excluded when the objective is to examine the role of the natural menopausal transition. Population-based studies often suffer from being cross-sectional in design (7) rather than having the power of longitudinal analysis of the same women through the menopausal transition (6). Cross-sectional studies can only indicate whether associations exist and are unable to determine the direction of causality. Most studies have only utilized univariate analysis and have thus been unable to take into account the role of confounding or interacting factors. The findings of these studies are thus contradictory, with some studies (8,6,9) finding a link between declining sexual functioning and menopausal status and others finding no such association (7,4,5,10).

The Melbourne Women's Midlife Health Project set out to overcome many of the methodological limitations of previous research by utilizing a community-derived sample of women aged 45–55 years and following them through the menopausal transition with annual assessments. We have previously reported (11) the initial baseline cross-sectional findings of a significant association between self-reports of decline in sexual interest with advancing menopausal status. Logistic regression also found that reports of decline in sexual interest were significantly associated with decreased well-being, hormone therapy use, less than full-time paid employment, and presence of bothersome symptoms. Increasing years spent in education were associated with less decrease in sexual interest. The baseline cross-sectional analysis did not use a detailed or validated questionnaire, nor were any hormonal measures available. These measures were introduced into the longitudinal study described later. This chapter presents the first results of longitudinal analysis of data from the initial 5 years of follow-up study with regard to changes in female sexual functioning during the natural menopausal transition.

Sample

The Melbourne Women's Midlife Health Project involves a community based cohort of middle-aged women. The study began in 1991 with population sampling by random telephone digital dialing to find 2,001 Australian-born women aged between 45 and 55 years and resident in Melbourne. There was a 71% response rate to participation in a telephone interview used to acquire baseline data (11,12). The study was approved by the Human Research Ethics Committee of the University of Melbourne.

All of those women who had experienced menses in the prior 3 months, and who were not taking the oral contraceptive pill or hormone therapy, were invited to participate in a longitudinal study. Of those eligible to enter the longitudinal study, only 56% chose to do so ($n = 438$). Volunteers for the longitudinal study were more likely than nonparticipants to report: better self-rated health, paid employment, more than 12 years of education, having ever had a pap smear, exercising at least once a week, and having undergone dilatation and curettage (13).

The retention rate of the 438 women who volunteered for the longitudinal study was 97% for year 2, 94% for year 3, and 91% for years 4 and 5. Women were excluded from analysis for the following reasons: drop-outs ($n = 39$), surgical menopause during the study ($n = 26$), oral contraceptive use at any year ($n = 4$), hormone therapy users ($n = 110$), and incomplete data ($n = 4$), leaving a sample size of 255 women who were in the natural menopausal transition and for whom there was sufficient data.

Methods

The longitudinal study involves annual assessments in the women's own homes. Quantitative information was collected on moods (14), symptoms (12), sexual functioning (Personal Experiences Questionnaire, PEQ) (15), and lifestyle. Blood samples for hormone assays and measurement of cholesterol were taken between days 4 and 8 for those still cycling, or after 3 months of amenorrhoea. Estradiol (E_2), follicle stimulating hormone (FSH), and immunoreactive inhibin (I) were determined by radioimmunoassay (13).

PEQ

The PEQ, derived from the McCoy Female Sexual Questionnaire (16), was handed to women to self-complete and returned to field workers in a closed envelope. The PEQ focuses on current sexual experience over the previous 2 weeks and uses a five-point Likert scale. Women were also asked to indicate their sexual preference and whether or not they have a sexual partner.

Using cross-sectional data from the third year of study, the psychometric properties of the PEQ were assessed (15). Six factors were extracted by principal components factor analysis, and these explained 70% of the variance. Two of these factors related to possible determinants of sexual functioning: feelings for partner and partner problems. The other four factors were considered to be indicators of different aspects of sexual functioning. These were: sexual responsivity, sexual frequency, libido, and vaginal dryness/ dyspareunia. Internal consistency was considered adequate with a Cronbach's alpha of 0.71.

This chapter presents the results of longitudinal analysis of data from the first 5 years of study. For this very complex analysis we chose to investigate only one factor: Factor 4 (libido), which highly loads on the following items:

- How many times have you had sexual thoughts or fantasies during the last month?
- Do you currently masturbate?

Thus, this factor is largely a measure of autoeroticism or women's own sexual interest and behavior. Factor 4 was chosen for this preliminary analysis because there was evidence from primate and laboratory and observational studies of humans that female initiation of sexual activity and autoerotic activity were more dependent on hormonal phase than were receptivity and arousal (17).

Aims

The present analysis aimed to determine:

1. What happens to women's autoeroticism (Factor 4: libido) during the natural menopausal transition?
2. Whether any changes were related to aging per se or to specific hormonal change.
3. Whether changes in women's autoeroticism (Factor 4: libido) were affected by feelings for partner (Factor 1).

Hypotheses

1. Autoeroticism as measured by Factor 4: libido will be reduced by the hormonal changes accompanying menopause rather than by aging.
2. Feelings for partner (Factor 1) will also affect female sexual functioning.

Statistical Analysis

All the analyses in this report were carried out on 255 women. Missing values were calculated using pairwise correlations between values in time.

Menopausal status is traditionally assigned by knowledge of menstrual status. A preliminary discriminant analysis found that a binary variable based on whether or not the woman had menstruated in the last 3 months best divided the sample, preserving good balance among sizes and the best discriminant effect size. The menstruating group (M) included both those still menstruating regularly (premenopausal) and those who reported a change in menstrual regularity (early perimenopausal). The nonmenstruating group (NM) comprised all those with more than 3 months amenorrhoea, including both late perimenopausal and postmenopausal women.

In order to take into account the combined influence of age and menstrual status, a lagged time variable was defined in which the origin (0 value) was placed at the first report of more than 3 months amenorrhoea. All hormonal data were first log transformed, which restored normal distributions (Kolmogorov Smirnov, $p > 0.05$).

As this first analysis focused on only one PEQ factor as an outcome (i.e., Factor 4: libido), we sequentially estimated several models instead of systematically using structural modeling. Repeated measures multivariate analyses of variance (MANOVA) were performed on data from the first 5 years of annual assessments. To account for possible nonlinear longitudinal influence, we used polynomial regression coefficients up to the fourth order as contrasts.

These analyses were first carried out on each hormone in order to determine how hormone levels were related to age and to menstrual status. ANOVA was used in separate partial models adding each hormone to menstrual status and age to explain libido.

The relationships among hormones are complex and their variations cannot be modeled on an annual measurement basis. A long-term trend can only be derived from the 5 yearly measurements. Strong correlations were observed at each time between FSH, E_2, and I. A nonobservable latent variable λ was estimated by the first factor in an exploratory factor analysis based on the evolution of the three hormones, FSH, E_2, and I. The final model used MANOVA to investigate the effects on libido of age, menstrual status, hormones, and feelings for partner (Factor 1).

Results

Sample Profile

At the first year of the longitudinal study, the 255 women had a mean (SD) age of 49.4 (2.4) years (range, 45.7–56.9). Forty percent had more than 12 years of education and the median parity was 3 (range, 0–9). At the fifth year, 86% had a current sexual partner and five women reported their sexual preference as homosexual/bisexual. The frequency of sexual activities was reported as never (21%), less than once a week (27%), once or twice a week (42%), several times a week (9%), and at least once a day (1%).

Hormone Levels During the Menopausal Transition

FSH Studied by Age and Menopausal Status

Between menstruating (M) and nonmenstruating (NM) women the effect size is big, as shown by the following means and standard deviation (SD): M 1.18 ± 0.40, NM 1.89 ± 0.50. ANOVA detected a highly significant age effect ($T = 5.2722, p = 0.0000$), a highly significant menstrual status effect ($T = 13.0520, p = 0.0000$), and no interaction. In other words, this seems to indicate a linear increase with age, but an additional very fast sudden rise of FSH around the time prolonged amenorrhoea was first reported (see Fig. 20.1).

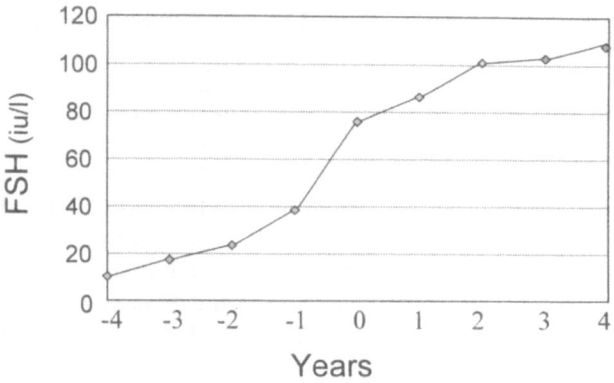

FIGURE 20.1. Mean FSH levels across the menopausal transition. On the horizontal axis, the origin (0) represents the time of the first report of more than 3 months amenorrhoea. Negative (positive) numbers represent years before (after) time "0."

Estradiol Studied by Age and Menopausal Status

Mean ± SD values are 2.34 ± 0.45 for menstruating women compared with 1.74 ± 0.50 for nonmenstruating women. ANOVA found a highly significant age effect ($T = 4.3187$, $p = 0.0000$), a highly significant menstrual status effect ($T = 8.1578$, $p = 0.0000$), and no interaction. This seems to indicate a linear decrease with age and additional effect of menstrual status (see Fig. 20.2).

Inhibin Studied by Age and Menopausal Status

Mean ± SD values are M 2.21 ± 0.36, NM 1.86 ± 0.34. ANOVA found a significant age effect ($T = 3.3493$, $p < 0.0011$) and a highly significant men-

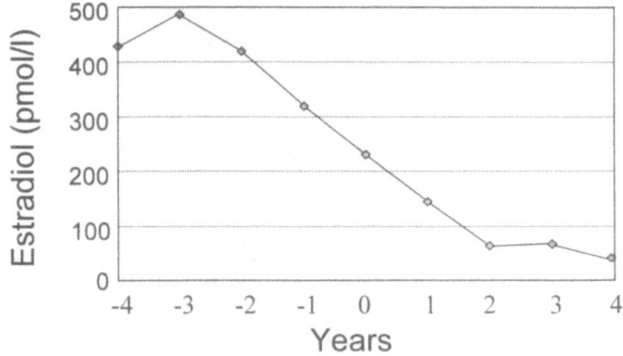

FIGURE 20.2. Mean estradiol levels across the menopausal transition. On the horizontal axis, the origin (0) represents the time of the first report of more than 3 months amenorrhoea. Negative (positive) numbers represent years before (after) time "0."

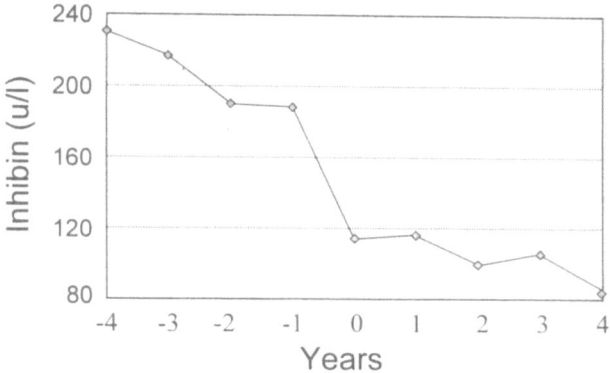

FIGURE 20.3. Mean inhibin levels across the menopausal transition. On the horizontal axis, the origin (0) represents the time of the first report of more than 3 months amenorrhoea. Negative (positive) numbers represent years before (after) time "0."

strual status effect ($T = 5.9581$, $p = 0.0000$), and no interaction. ANOVA found a linear decrease with age, but an additional effect of menstrual status (see Fig. 20.3).

Libido (Factor 4) Studied by Age and Menopausal Status

ANOVA found that libido, as the main endpoint of this analysis, had a nonsignificant decrease with age and a significant decrease associated with menstrual status ($T = 3.8129$, $p < 0.0002$). Mean ± SD effect sizes were: M, 1.85 ± 1.86; NM, 1.03 ± 1.22. There was no significant interaction of menstrual status and age. Figure 20.4 shows that a major fall in libido occurs around 3

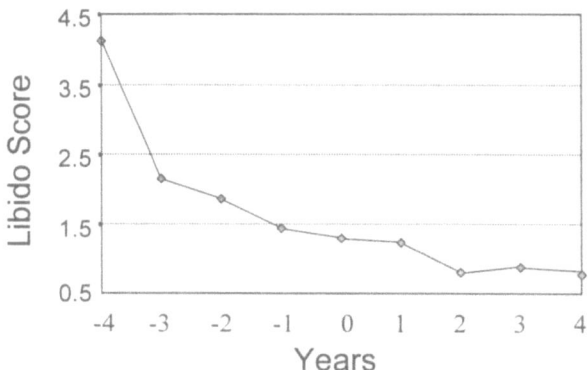

FIGURE 20.4. Mean factor score for libido (Factor 4) across the menopausal transition. On the horizontal axis, the origin (0) represents the time of the first report of more than 3 months amenorrhoea. Negative (positive) numbers represent years before (after) time "0."

years before amenorrhoea was reported. The linear goodness of fit remained small (0.045), suggesting that these variables (age and menstrual status) were not the most explanatory of libido or that they only influenced it indirectly.

Effect of Each Hormone on Libido FSH

MANOVA found that FSH introduced in the model had a highly significant effect ($T = 2.84485$, $p = 0.0047$) and significantly increased the total coefficient of determination ($R = 0.085$). Age and menstrual status were then found to be nonsignificant. This result suggests that menopausal status (menstrually determined) has significant effects on libido through FSH.

Effect of Each Hormone on Libido: Estradiol

E_2 introduced in the model had a separate significant contribution ($T = 2.3023$, $p < 0.0208$) and age was not significant, but menstrual status remained significant ($T = 2.5032$, $p = 0.0123$). E_2 did not produce a larger coefficient of determination ($R = 0.057$).

Effect of Each Hormone on Libido: Inhibin

MANOVA found that inhibin did not have a separate significant contribution and did not contribute to a larger coefficient of determination. Age remained not significant and menstrual status remained highly significant ($T = 2.9752$, $p = 0.0033$). The global goodness of fit was 0.050.

FSH levels and, to less extent, E_2 levels therefore appear to influence libido, whereas I appears to have no detectable effect. Accounting for a priori known interrelationships of FSH, E_2, and I, factor analysis using the centroid algorithm confirmed that the three variables, FSH, E_2, and I, could not be dissociated as separate in the model, as they were highly intercorrelated, but could be aggregated into the latent variable λ. The model fitted exceptionally well ($R^2 = 0.75$). λ may be thought of as a global measurement of ovarian functioning.

Simultaneous Effect on Libido of FSH, E_2, and I, and Feelings for Partner

The final model used MANOVA to investigate the effects on libido of age and menstrual status, with the addition of the new covariates of l and Factor 1 (feelings for partner). MANOVA found a highly significant linear effect and no higher-order significant effects. Feelings for partner were found to be a very important criterion ($T = 3.0261$, $p = 0.003$), with a confirmation of significant hormonal (λ) effect ($T = 2.2457$, $p = 0.02$). The other variables—menstrual status ($T = 1.8513$, $p = 0.06$) and age ($T = 0.0602$, $p = 0.95$)—were no longer significant. The global goodness of fit was 0.080.

Discussion

This study confirms the findings from our first cross-sectional study of 2,001 middle-aged women that sexual interest was decreased by the menopausal transition rather than by aging. Sexual interest, as measured within Factor 4: libido, also showed a trend ($p = 0.07$) toward decline with the menopausal transition in the cross-sectional analysis of data from the fourth year of study (15). The repeated measures analyses used in this chapter are a more powerful way to deal with the data than was a single cross-sectional analysis. The analysis by menopausal group was also carried out differently in this chapter where data were divided according to whether women were menstruating or had more than 3 months of amenorrhoea. Published data from The Melbourne Women's Midlife Health Project provides further endocrinological justification for such a dichotomy (18).

Our conclusions are interesting to compare with those of Cawood and Bancroft (19), who found no relationship between hormone measures and measures of sexuality. These authors followed 141 women recruited from the community aged 40–60 years for 4 weeks, obtaining blood at weekly intervals. This small time interval would not allow for the change in hormone levels and libido evident over the 5 years of data collection in the Melbourne Women's Midlife Health Project, and it may account for the failure of Cawood and Bancroft to find any effect of estrone or estradiol on measures of sexuality. Their results that the quality of the relationship with the partner was a most powerful indicator of sexual outcomes are confirmed by our findings.

Hawton et al. (10) noted a trend for more postmenopausal than premenopausal women to report lower interest in sex compared with the previous year. This may have reflected the small number of women in the age-matched groups and the type of measure used.

An earlier small ($n = 16$) longitudinal study of the natural menopausal transition by McCoy and Davidson (6) also found a significant reduction in the frequency of sexual thoughts or fantasies reported by women over the previous month when the means were compared before or after the last cycle. This item showed the most change of any item in their questionnaire. Fourteen of the 16 women showed a decrease from pre- to postmenopause. There was little change in frequency of masturbation in their study using the same analysis technique.

The results reported in this chapter demonstrate that both age and menstrual status influence hormone levels. Further analysis of changes in hormone levels with age and menopausal status has been carried out using nonlinear modeling on the data from all women who reached final menstrual period during the first 6 years of study (20, unpublished observations). The results are confirmatory of those reported here using MANOVA. Androgens were not included in either modeling because we are awaiting the results of other androgen assays that were carried out in 1999 at the conclusion of this section of data collection.

Our study found that decreasing libido (Factor 4) was significantly linked to declining ovarian functioning, as indicated by increasing FSH and decreasing estradiol and inhibin. The hormone most significantly associated with declining libido was FSH. Only a single blood sample for each woman was collected on an annual basis. Given the variability of menstrual cycles at this time and the associated variability of hormonal levels during the perimenopause (13), a great deal of variance in hormone levels would be expected. This may be why estradiol (which may show more variation than FSH) was less strongly associated with libido.

These findings provide further evidence that women's sexual interest is significantly associated with active ovarian follicular functioning. In a previous study of the menstrual cycle, women collected daily ratings of sexual interest on a five-point scale and 24-hour urines for an entire menstrual cycle. Sexual interest was significantly higher in the follicular and ovulatory phase than it was in the luteal and menstrual phases (21).

Conclusions

This study found that declining ovarian functioning produced a decline in autoeroticism, an aspect of sexuality that is also profoundly affected by feelings for the partner. Whether the hormonal changes of the menopausal transition also affect sexual responsivity, frequency of sexual activities, and vaginal dryness/dyspareunia will require further complex modeling in order to include these parameters along with other variables that may influence sexual functioning, such as HRT use, education, and paid employment. Structural modeling will allow all of these possible relationships to be examined simultaneously.

Acknowledgments. The Melbourne Women's Midlife Health Study is funded by the Victorian Health Promotion Foundation, the Public Health Research and Development Committee of NH&MRC, and has received grants from: Percy Baxter Trust, H & L Hecht Trust, Estate of Late Daniel Scott, Ian Potter Foundation, Smorgon Family Trust, Leigh & Marjorie Bronwen Murray Charitable Trust, and Helen M. Schutt Trust. The Prince Henry's Institute for Medical Research received a grant in aid from Organon (Aust) Pty. Ltd. toward the costs of hormone assays for which the assistance of Mr. N. Balazs and his staff is gratefully acknowledged.

The research team would like to thank all those women who participated in this study and the Community Advisory Committee. We are also indebted to Professor Norma McCoy for her contribution to the development of the Personal Experiences Questionnaire.

References

1. Sarrel PM, Whitehead MI. Sex and menopause: defining the issues. Maturitas 1985;7:217–24.
2. McKinlay JB, McKinlay SM, Brambilla DJ. Health status and utilisation behaviour associated with menopause. Am J Epidemiol 1987;125(1):110–21.
3. Morse CA, Smith A, Dennerstein L, Green A, Hopper J, Burger H. The treatment-seeking woman at menopause. Maturitas 1994;18(3):161–73.
4. Osborn M, Hawton K, Gath D. Sexual dysfunction among middle aged women in the community. Br Med J 1988;296:959–62.
5. Køster A, Garde K. Sexual desire and menopausal development: a prospective study of Danish women born in 1936. Maturitas 1993;16:49–60.
6. McCoy N, Davidson JM. A longitudinal study of the effects of menopause on sexuality. Maturitas 1985;7:203–10.
7. Van Keep PA, Kellerhals JM. The ageing woman. Acta Obstet Gynaecol Scand 1976;51(suppl):17–27.
8. Hallstrom T. Sexuality in the climacteric. Clin Obstet Gynaecol 1977;4:227–39.
9. Hunter MS. Emotional well-being, sexual behaviour and hormone replacement therapy. Maturitas 1990;12:299–314.
10. Hawton K, Gath D, Day A. Sexual function in a community sample of middle-aged women with partners: effects of age, marital, socioeconomic, psychiatric, gynecological and menopausal factors. Arch Sex Behav 1994;23(4):375–95.
11. Dennerstein L, Smith AMA, Morse CA, Burger HG. Sexuality and the menopause. J Psychosom Obstet Gynecol 1994;15:59–66.
12. Dennerstein L, Smith A, Morse C, Burger H, Green A, Hopper J, et al. Menopauseal symptomatology: the experience of Australian women. Med J Aust 1993;159: 232–36.
13. Burger HG, Dudley EC, Hopper JL, Shelley JM, Green A, Smith A, et al. The endocrinology of the menopausal transition: a cross-sectional study of a population-based sample. J Clin Endocrinol Metab 1995;80(12):3537–45.
14. Kamman R, Flett R. Affectometer 2: a scale to measure current level of general happiness. Aust J Psychol 1983;35:259–65.
15. Dennerstein L, Dudley EC, Hopper JL, Burger H. Sexuality, hormones and the menopausal transition. Maturitas 1997;26:83–93.
16. McCoy NL, Matyas JR. Oral contraceptives and sexuality in university women. Arch Sex Behav 1996;25:73–79.
17. Dennerstein L, Burrows GD. Hormones and female sexuality. In: Dennerstein L, De Senarclens M, eds. The young woman: psychosomatic aspects of obstetrics and gynaecology. International Congress Series 618, Amsterdam: Excerpta Medica, 1983:201–14.
18. Dudley EC, Hopper JL, Taffe J, Guthrie JR, Burger HG, Dennerstein L. Using longitudinal data to define the perimenopause by menstrual cycle characteristics. Climacteric 1998;1:18–25.
19. Cawood EHH, Bancroft J. Steroid hormones, the menopause, sexuality and well-being of women. Psychol Med 1996;26:925–36.
20. Burger HG, Dudley EC, Hopper JL, Groome N, Guthrie JR, Green A, Dennerstein L. Prospectively measured levels of serum follicle-stimulating hormone, estradiol, and the dimeric inhibins during the menopausal transition in a population-based cohort of women. J Clin Endocrinol Metab 1999;84:4025–30.

21. Dennerstein L, Gotts G, Brown JB, Morse CA, Farley TMM, Pinol A. The relationship between the menstrual cycle and female sexual interest in women with PMS complaints and volunteers. Psychoneuroendocrinology 1994;19(3):293–304.

Part III

New Methods and Emerging Areas

21

Transgenic Models to Study Reproduction, Oncogenesis, and Development

Julia A. Elvin and Martin M. Matzuk

In mammals, there are approximately 100,000 genes that govern the development of an organism. For development to proceed normally, there must be coordinate interaction of thousands of these gene products in any given cell of the organism. Beginning with fertilization, precise expression of these gene products is required during embryonic, fetal, postnatal, and adult development. Aberrant synthesis of even one of these gene products can be disastrous: Birth defects, cancer, infertility, and even death are all possible consequences when this developmental program is altered. To understand these processes in humans fully, it is necessary to have physiological models that closely mimic developmental events that occur during the creation of a human being.

It is now possible to manipulate the mammalian germline to generate transgenic mice that either overexpress a wild-type or mutant gene or lack a functional copy of an endogenous gene (1,2). In particular, the rapid advances in embryonic stem (ES) cell technology have enabled the generation of mice functionally deficient in specific gene products (i.e., knockout mice) and has allowed investigators to address the unique and essential functions of these gene products in vivo. Several genes essential for sex determination and the establishment and maintenance of the reproductive system have been investigated using transgenic mouse technology (3–5). Because the hypothalamic–pituitary–gonadal axis is a complex network of interacting feedback loops between multiple tissues, in vivo analysis using the knockout mouse approach has contributed to the definition and understanding of its components and has advanced research on mammalian reproduction. In addition, developmental roles for genes originally discovered for their role in reproduction have been uncovered through gene targeting, emphasizing the power of this technique.

For the past 6 years, studies in this laboratory have been directed at elucidating some of these critical gene products involved in both normal and abnormal mammalian development. Use of knockout technology has led our laboratory and those of our collaborators to create a number of models for birth defects (Table 21.1). Among the models we have created are several lines of mice that die at birth. For example, mice with mutations in the activin βA and follistatin genes have cleft palate, a common birth defect of un- known etiology in humans (6,7). In addition, a fraction of the mice with mutations in another related gene, the activin receptor type II (ActRII) gene, were discovered to have skeletal and facial abnormalities that mimic the human Pierre-Robin syndrome (7). Human newborns with this syndrome have hypoplasia of the mandible, leading to respiratory distress, that must be cor- rected immediately with surgery. We have also shown that 100% of mice lacking the FK506 binding protein, FKBP12, have dilated cardiomyopathy

TABLE 21.1. Knockout models that have birth defects or reproductive findings created by Matzuk and colleagues.

Knockout mouse model	Phenotype	References
Activin/inhibin βA	Neonatal lethal; craniofacial defects (cleft palate, lack of whiskers, and tooth defects)	7, 32
Activin/inhibin βB	Large litters but delayed parturition; nursing defects; eyelid closure defects at birth	33
Follistatin	Neonatal lethal; craniofacial defects, growth retardation, and skin defects	6
Activin receptor type II	Infertility in females due to folliculogenesis defect; delayed fertility in males; small gonads; 25% of mice die neonatally secondary to mandible defects	7
α-inhibin	Infertility in females; secondary infertility in males; granulosa/ Sertoli cell tumours; cachexialike syndrome	10–12, 34
Growth differentiation factor 9	Infertile; defect in folliculogenesis at one-layer follicle stage	13
Follicle stimulating hormone β subunit	Female infertility; folliculogenesis block prior to antral follicle stage; males fertile but decreased testis size	14
Oxytocin	Nursing defect	35
Mouse atonal homolog 1 (Math1)	Neonatal lethal; absence of external germinal layer in cerebellum	9
FKBP12	Majority of mice die between E14.5 and birth due to cardiomyopathy and neural tube defects	8
Superoxide dismutase I	Reduced fertility; increase in embryonic lethality	36

and ventricular septal defects that mimic a human congenital heart disorder, noncompaction of the left ventricular myocardium (8). A smaller percentage of the FKBP12 mutants also have neural tube defects that lead to exencephaly. Mice lacking the mouse atonal homolog I also die at birth and have defects in cerebellum development (9). Last, mice lacking the α inhibin gene develop ovarian and testicular tumors that resemble juvenile granulosa cell tumors, which arise in young human females (10). The development of these tumors leads to a cancer cachexialike (wasting) syndrome that we have shown is due to activins secreted from the tumors signaling directly through ActRII in the liver and glandular stomach (11,12).

In addition to these developmental and cancer models, we have generated several mouse models that have reproductive defects. Female mice deficient in growth differentiation factor 9 (GDF-9), follicle stimulating hormone (FSH), and activin receptor type II (ActRII) are infertile due to complete blocks at specific stages of folliculogenesis (Table 21.1, Fig. 21.1) (7,13,14). As shown, absence of GDF-9 results in a block in folliculogenesis at the one-layer primary follicle stage (13), whereas absence of FSH blocks folliculogenesis at the multilayer preantral follicle stage (14). Ovaries of mice lacking ActRII contain follicles slightly more developed than do those of the FSH knockouts, but are still defective in antral follicle development (7).

The function of growth-differentiation factor-9 (GDF-9), a novel, oocyte-expressed member of the TGF-β superfamily of secreted growth factors, was first defined as a key regulator of early preantral follicle growth by our gene-targeting approach. GDF-9 is first expressed by oocytes in type 3a follicles and its expression persists in the oocyte through ovulation (15). The GDF-9-deficient female mouse shows a block in follicular development at the type 3b stage I-layer follicle. In these mice, the oocytes continue to grow in the one-layer follicles and eventually die, leaving behind a ribbon of zona pellucida. The granulosa cells, however, fail to proliferate normally, and they persist after the oocyte degenerates, eventually taking on a steroidogenic phenotype (13). In addition, no identifiable or functional thecal layer ever forms around these follicles, which emphasizes the disruption of follicle development that occurs as a result of GDF-9 absence. We have observed a twofold increase in luteinizing hormone (LH) serum levels and a threefold increase in FSH serum levels, which is likely a result that is secondary to the ovarian failure. As the GDF-9 homozygotes age, we have also observed a high frequency of ovarian cyst formation. By 6 months of age, approximately 40% develop single, large, fluid-filled follicular cysts that can be greater than 1 cm in diameter. We have not seen an increase, however, in the rate of tumor formation in this model.

It is of interest that Steel[panda] and Steel[t] mice, which have altered expression of the granulosa cell–produced growth factor, kit ligand, have a block in folliculogenesis similar to that of the GDF-9 knockouts (16–18). In a reciprocal signaling scenario to the one presumed for GDF-9, the receptor

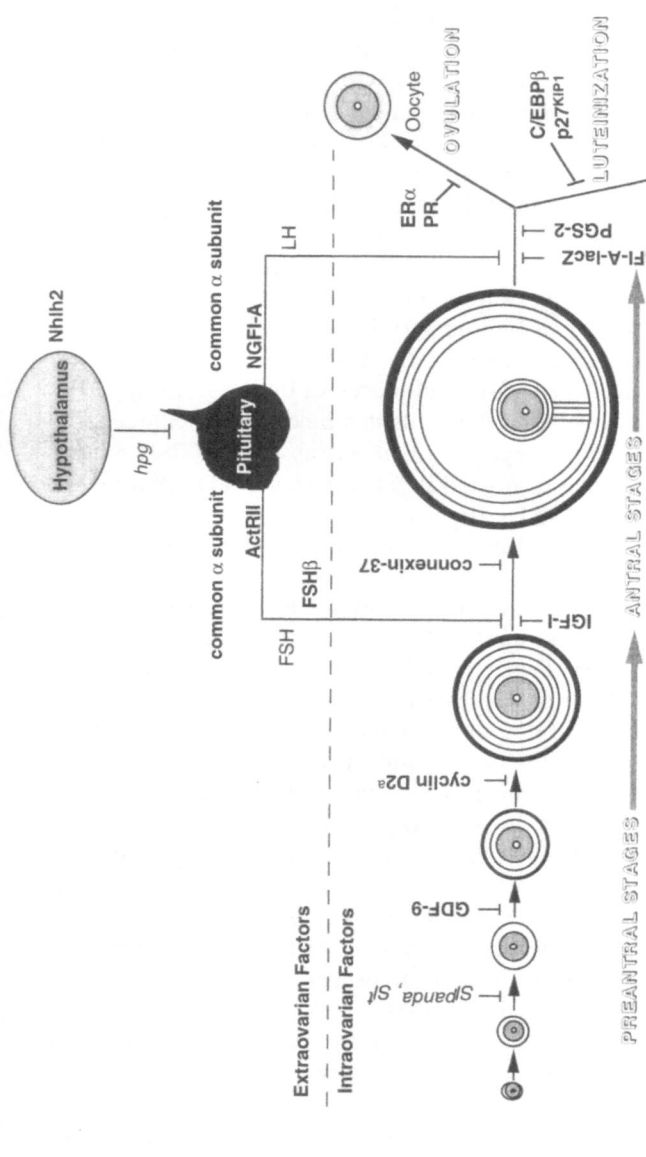

FIGURE 21.1. Mouse models of altered ovarian function. This schematic illustrates the stage of folliculogenesis blocked in selected mouse models. Mouse models of intraovarian factors are listed below the dashed line, whereas those altering ovarian function secondary to hypothalamic or pituitary defects are listed next to the affected organ above the dashed line. Knockout models are in **bold** type, spontaneous mutations in *italics*, and blocks are denoted by —|. Abbreviations: growth differentiation factor-9 (GDF-9); insulinlike growth factor I (IGF-I); glycoprotein hormone common α subunit (common α subunit); activin type II receptor (ActRII); estrogen receptor α (ERα); progesterone receptor (PR); CCAAT/enhancer binding protein β (C/EBPβ); prostaglandin synthase 2 (PGS-2); neuronal helix–loop–helix 2 (Nhlh2); nerve growth factor induced A (NGFI-A). aCyclin D2-deficient follicles develop antra, but granulosa cell proliferation is essentially blocked at this stage. Adapted from Elvin JA, Matzuk MM, Rev Reprod 1998 (Ref. 3).

for kit ligand, c-kit, is expressed by oocyte and thecal cells. From these models, it is clear that paracrine signaling from the oocyte to the granulosa cells, as well as from the granulosa cells to the oocyte, regulates preantral follicle growth.

The two pituitary gonadotroph–derived glycoprotein hormones, FSH and LH, are composed of a common α subunit, which is shared with thyroid-stimulating hormone and chorionic gonadotropin, and a unique β subunit (19) that confers hormone-specific receptor binding and hence biological activity. In the FSH-β knockout mouse generated in our laboratory (14), preantral follicle growth proceeds normally until the multilayer follicle stage, confirming the gonadotropin independence of preantral follicle development, but antrum formation is completely blocked. A block in folliculogenesis at the same stage is seen in the hypogonadal (hpg) mouse, in which a naturally occurring mutation in the gonadotropin releasing hormone (GnRH) gene drastically reduces release of FSH and LH from the pituitary (20,21). These models confirm the pivotal role of FSH in antral follicle development.

To determine whether interspecies FSH hybrids can function properly in vivo, we have rescued the mutant phenotype of the FSH-deficient mice using two different genetic strategies. In one case, a pituitary-expressed human FSH-β transgene was introduced into the mouse FSH-β–deficient genetic background (22). In the pituitary gonadotrophs of these mice, the mouse α subunit dimerized with the human β subunit, generating a mouse–human "hybrid" FSH that was able to restore follicular development and fertility completely to all of the females. These females were even able to produce normal size litters repeatedly. In the second type of rescue, human FSH was ectopically produced by multiple tissues by introduction of mouse metallothionein I-driven human FSH-α and human FSH-β transgenes. Only 3 out of 10 females producing ectopic human FSH were fertile and able to carry their pregnancies to term and parturition, reinforcing the critical role of FSH regulation by the feedback loops in the hypothalamic–pituitary–gonadal axis for fertility. These studies clearly demonstrate that interspecies FSH is biologically active and equivalent to endogenous mouse FSH if expressed correctly, and that the FSH-deficient mouse is likely a useful in vivo bioassay for testing bioactivity of human FSH isoforms and analogs.

The insulinlike growth factor I (IGF-I) knockout mice have a very similar ovarian phenotype to that of FSH knockout mice (23,24). IGF-I and IGF-I receptor are expressed by follicles that appear healthy; thus, they may be markers of follicular selection. In situ hybridization analysis of ovaries from FSH-deficient mice proved that IGF-I expression is not dependent upon gonadotropins because IGF-I is still expressed in the FSH-deficient follicles (24). IGF-I expression is instead thought to be initiated by paracrine signaling from the oocyte. Expression of the FSH receptor (FSH-R), however, is stimulated by both FSH and IGF-I, as evidenced by a 50% reduction in

FSH-R concentration in the IGF-I-deficient ovaries or after elimination of gonadotropins by hypophysectomy. A decrease in FSH-R expression leading to relative insensitivity in the IGF-I-deficient follicles is proposed as a mechanism for follicular arrest in the IGF-I-deficient model.

Activins are dimeric members of the TGF-β superfamily originally isolated for their ability to stimulate FSH synthesis and secretion by the pituitary. Mice lacking one of the activin receptors, ActRII, exhibit markedly suppressed serum and pituitary concentrations of FSH (7). This indicates that ActRII is the major pituitary receptor through which activins signal to regulate FSH levels. The block in folliculogenesis in the ActRII-deficient females occurs at a similar but slightly later developmental stage than it does in the FSH-deficient mice, which suggests that the phenotype is secondary to decreased FSH concentrations and is due to lack of paracrine signaling through the ActRII in the pituitary. Consistent with this primary reproductive role of ActRII in the pituitary, ActRII-deficient ovaries transplanted into ovariectomized female mice with normal FSH levels resumed normal folliculogenesis (Coerver KA, Guo Q, and Matzuk MM, unpublished).

Mice overexpressing follistatin, the activin-binding protein, under the control of the metallothionein promoter have been generated to understand further the importance of activins and other members of the TGF-β superfamily in reproduction and development (25). Female mice from the two lines with the highest levels of follistatin transgene expression often had blocks in folliculogenesis at the preantral and antral follicle stage, resulting in infertility in the most severely affected mice. These results suggest a role for follistatin in the local regulation of activin and possibly other TGF-β family members in the ovary. Due to the perinatal lethality of the activin βA-deficient mice, it is unfortunately still unknown whether activin A is required for folliculogenesis.

Other models that have reproductive defects in females only or in both males and females are summarized in Tables 21.2 and 21.3. The stage at which folliculogenesis is blocked gives critical information about the function of the gene product (Fig. 21.1) (3). In particular, spontaneous mutations in the c-kit (White-spotting mutants) and kit ligand (Steel mutants) genes have demonstrated roles for c-kit signaling in both gonad formation and early stages of folliculogenesis. Disruptions of later stages of follicle development are seen in mice deficient for cyclin D2, IGF-I, and connexin-37. Defects in ovulation and/or corpus luteum formation occur in mice carrying mutations in NGFI-A, PGS-2, ERα, PR, C/EBPβ, and p27[Kip1]. Future studies using these transgenic mice as in vivo tools and new technology, such as gene substitution through "knockin transgenics" (26,27), the Cre-loxP system for spatiotemporal knockouts and large deletions (28,29), and tagged sequence mutagenesis (30,31), will allow investigators to understand more fully the relationship of these many proteins in reproductive development and function.

TABLE 21.2. Knockout models with reproductive defects only in females.

Knockout mouse model	Major reproductive findings	References
c-mos	Decreased fertility in females only; ovarian cysts and teratomas	37, 38
Leukemia inhibitory factor (LIF)	Infertility; implantation defect	39
Progesterone receptor	Infertility; defects in all reproductive tissues	40
activin/inhibin βB	Large litters but delayed parturition; nursing defects	33, 41
α-lactalbumin	Normal fertility but inability to nurse offspring	42
Growth differentiation factor 9	Infertility; defect in folliculogenesis at one-layer follicle stage	13
Oxytocin	Nursing defect	35, 43
Transcription factor NGFI-A	Infertility; luteinizing hormone suppression causing no corpora lutea (and/or ovulation)	44, 45
Zona protein 3 (ZP3)	Infertile; no zone pellucida	46, 47
Steroid 5α-reductase type I	Reduced litter size; parturition defects (fetal death due to excess estrogens)	48, 49
Mf3	Nursing defect (also embryonic and post-natal defects)	50
Connexin 37	Infertile; defect in folliculogenesis at the Graafian follicle stage	51
Cyclooxygenase II synthase-2) (prostaglandin endoperoxide synthase-2)	Largely infertile; absence of corpora lutea due to apparent ovulation defect	52
C/EBPb	Infertile; ovulation and corpora lutea defects	53
Interleukin II receptor α	Infertile; implantation defect	54
Prostaglandin F receptor	Infertile due to lack of induction of oxytocin receptor	55
Prolactin	Infertile; irregular estrous cycles	56
Hmx3	Normal preimplantation development, but implantation failure of embryos	57
Superoxide dismutase I	Reduced fertility; increase in embryonic lethality	36, 58
Caspase-2	Excess number of germ cells in the ovaries; oocytes resistant to cell death following exposure to chemotherapeutic drugs	59
Stat5a/Stat5b double mutants	Infertile; absence of corpora lutea leading to implantation defect	60

TABLE 21.3. Knockout models with reproductive defects in both sexes.

Knockout mouse model	Major reproductive findings	References
α-inhibin	Infertility in females; secondary infertility in males; granulosa/Sertoli cell tumours; in males; granulosa/Sertoli cell tumours; cachexia-like syndrome	10–12, 34
Activin receptor type II	Infertility in females; delayed fertility in males; small gonads	7
Estrogen receptor α	Uterine/ovarian defects in females; small testes, reduced number of spermatozoa in males	61, 62
Glycoprotein hormone α-subunit	Infertile; hypogonadal and hypothyroid	63
Hoxa10	Variable infertility in males and females due to cryptorchidism and preimplantation embryonic loss, respectively	64
Insulin-like growth factor (IGF-1)	Hypogonadal and infertile; pre-antral block in folliculogenesis in females	23, 24
Neuronal helix-loop-helix 2 (Nhlh2)	Males infertile; females fertile only in presence of males; hypothalamic defect	65
Zfx	Reduced germ cell number in both sexes due to defective proliferation	66
Follicle stimulating hormone β subunit	Female infertility; folliculogenesis block prior to antral follicle stage; males fertile but decreased testis size	14
p27^{Kip1} CDK inhibitory protein	Female infertility; corpus luteum defects; males fertile and increased testis size	67, 68
MLH1 DNA mismatch repair enzyme	Male and female infertility; Defective meiosis at pachytene stage (males) and failure to complete meiosis II (females)	69, 70
Ataxia telangiectasia (Atm)	Male and female infertility; complete absence of germ cells	71, 72
Cyclin D2	Female infertility secondary to a block in folliculogenesis; males fertile but decreased testis size	73
Prolactin receptor	Female infertility due to multiple abnormalities, including irregular estrous cycles and implantation defects; males infertile or subfertile of unknown origin	74
Dazla	Male and female infertility; loss of germ cells and complete absence of gamete production	75
β 1,4-Galactogyltransferase	Male and female infertility due to abnormal glycoprotein hormone glycosylation	76
A-myb	Male infertility; pachytene stage arrest of germ cells; nursing defects in females due to underdevelopment of mammary glands	77
Emx2	Accelerated degeneration of Wolffian duct and mesonephric tubules without the formation of the Müllerian duct	78
Hoxa11	Partial homeotic transformation of vas deferens to epididymis; failure of testicular descent; absence of uterine stromal, decidual, and glandular cells in females	79
TIAR	Infertility; complete absence of primordial germ cells by E13.5 leading to absence of spermatogonia and oogonia	80
Dmc1	Arrest of spermatogenesis at zygotene stage in males; no oocytes in the adult ovary	81, 82
Telomerase	Progressive infertility in males and females; increased apoptosis in testicular germ cells, and reduced testis size; decreased number of oocytes and uterine abnormalities	83

Acknowledgments. Studies in the Matzuk laboratory on ovarian development and ovarian cancer have been supported in part by a sponsored research grant from Genetics Institute and National Institutes of Health grants HD33438, CA60651, HD37231, and HD07495 as part of the specialized cooperative centers program in reproductive research. Dr. Julia A. Elvin is a student in the Medical Scientist Training Program supported in part by NIH grants GM-07330 and GM-08307 and the Baylor Research Advocates for Student Scientists (BRASS) organizaton.

References

1. Camper SA, Saunders TL, Kendall SK, Keri RA, Seasholtz AF, Gordon DF, et al. Implementing transgenic and embryonic stem cell technology to study gene expression, cell-cell interactions and gene function. Biol Reprod 1995;52:246–57.
2. Capecchi MR. Targeted gene replacement. Sci Am 1994;270:52–59.
3. Elvin JA, Matzuk MM. Mouse models of ovarian failure. Rev Reprod 1998;3: 183–95.
4. Nishimori K, Matzuk MM. Transgenic mice in the analysis of reproductive development and function. Rev Reprod 1996;1:203–12.
5. Sassone-Corsi P. Transcriptional checkpoints determining the fate of male germ cells. Cell 1997;88:163–66.
6. Matzuk MM, Lu H, Vogel H, Sellheyer K, Roop DR, Bradley A. Multiple defects and perinatal death in mice deficient in follistatin. Nature 1995;372:360–63.
7. Matzuk MM, Kumar TR, Bradley A. Different phenotypes for mice deficient in either activins or activin receptor type II. Nature 1995;374:356–60.
8. Shou W, Aghdasi B, Armstrong DL, Guo Q, Bao S, Charng M-J, et al. FKBP12-deficient mice display cardiac defects and altered ryanodine receptor function. Nature 1998;391:489–92.
9. Ben-Arie N, Bellen HJ, Armstrong DL, McCall AE, Gordadze PR, Guo Q, et al. *Math1* is essential for genesis of cerebellar granule neurons. Nature 1997;390: 169–72.
10. Matzuk MM, Finegold MJ, Su J-GJ, Hsueh AJW, Bradley A. α-Inhibin is a tumor-suppressor gene with gonadal specificity in mice. Nature 1992;360:313–19.
11. Coerver KA, Woodruff TK, Finegold MJ, Mather J, Bradley A, Matzuk MM. Activin signaling through activin receptor type II causes the cachexia-like symptoms in inhibin-deficient mice. Mol Endocrinol 1996;10:534–43.
12. Matzuk MM, Finegold MJ, Mather JP, Krummen L, Lu H, Bradley A. Development of cancer cachexia-like syndrome and adrenal tumors in inhibin-deficient mice. Proc Natl Acad Sci USA 1994;91:8817–21.
13. Dong J, Albertini DF, Nishimori K, Kumar TR, Lu N, Matzuk MM. Growth differentiation factor-9 is required during early ovarian folliculogenesis. Nature 1996;383:531–35.
14. Kumar TR, Wang Y, Lu N, Matzuk MM. Follicle stimulating hormone is required for ovarian follicle maturation but not male fertility. Nature Gen 1997;15:201–4.
15. McGrath SA, Esquela AF, Lee S-J. Oocyte-specific expression of growth/differentiation factor-9. Mol Endocrinol 1995;9:131–36.
16. Beechey CV, Loutit JF, Searle AG. Panda, a new steel allele. Mouse News Lett 1986;74:92.

17. Huang EJ, Manova K, Packer AI, Sanchez S, Bachvarova RF, Besmer P. The murine steel panda mutation affects kit ligand expression and growth of early ovarian follicles. Dev Biol 1993;157:100–9.

18. Kuroda H, Terada N, Nakayama H, Matsumoto K, Kitamura Y. Infertility due to growth arrest of ovarian follicles in the *Sl/Slᵗ* mice. Dev Biol 1988;126:71–79.

19. Bonsfield GR, Perry WW, Ward DN. Gonadotropins: chemistry and biosynthesis. In: Knobil E, Neill JD, eds. The physiology of reproduction. New York: Raven Press, 1994.

20. Cattanach BM, Iddon CA, Charlton HM, Chiappa SA, Fink G. Gonadotrophin-releasing hormone deficiency in a mutant mouse with hypogonadism. Nature 1977;269:338–40.

21. Mason AJ, Hayflick JS, Zoeller RT, Phillips HS, Nikolics K, et al. A deletion truncating the gonadotropin-releasing hormone gene is responsible for hypogonadism in the *hpg* mouse. Science 1986;234:1366–71.

22. Kumar TR, Low MJ, Matzuk MM. Genetic rescue of follicle-stimulating hormone β-deficient mice. Endocrinology 1998;139:3289–95.

23. Baker J, Hardy MP, Zhou J, Bondy C, Lupu F, Bellvé AR, et al. Effects of an *Igf1* gene null mutation on mouse reproduction. Mol Endocrinol 1996;10:903–18.

24. Zhou J, Kumar TR, Matzuk MM, Bondy C. Insulin-like growth factor I regulates gonadotropin responsiveness in the murine ovary. Mol Endocrinol 1997;11:1924–33.

25. Guo Q, Kumar TR, Woodruff T, Hadsell LA, DeMayo FJ, Matzuk MM. Overexpression of mouse follistatin causes reproductive defects in transgenic mice. Mol Endocrinol 1998;12:96–106.

26. Hanks M, Wurst W, Anson-Cartwright L, Auerbach A, Joyner A. Rescue of the En-1 mutant phenotype by replacement of En-1 with En-2. Science 1995;269:679–82.

27. Wang Y, Schnegelsberg PNJ, Dausman J, Jaenisch R. Functional redundancy of the muscle-specific transcription factor Myf5 and myogenin. Nature 1996;379:823–25.

28. Gu H, Zou Y-R, Rajewsky K. Independent control of immunoglobulin switch recombination at individual switch regions evidenced through Cre-loxP-mediated gene targeting. Cell 1993;73:1155–64.

29. Ramirez-Solis R, Liu P, Bradley A. Chromosome engineering in mice. Nature 1995;378:720–24.

30. Hicks GG, Shi E-G, Li X-M, Li C-H, Pawlak M, Ruley HE. Functional genomics in mice by tagged sequence mutagenesis. Nat Gen 1997;16:338–44.

31. Zambrowicz BP, Friedrich GA, Buxton EC, Lilleberg SL, Person C, Sands AT. Disruption and sequence identification of 2,000 genes in mouse embryonic stem cells. Nature 1998;392:608–11.

32. Ferguson CA, Tucker AS, Christensen L, Lau AL, Matzuk MM, Sharpe PT. Activin is an essential early mesenchymal signal in tooth development that is required for patterning of the murine dentition. Genes Dev 1998;12:2636–49.

33. Vassalli A, Matzuk MM, Gardner HAR, Lee K-F, Jaenisch R. Activin-inhibin bB subunit gene disruption leads to defects in eyelid development and female reproduction. Genes Dev 1994;8:414–27.

34. Matzuk MM, Kumar TR, Shou W, Coerver KA, Lau AL, Behringer RR, et al. Transgenic models to study the roles of inhibins and activins in reproduction, oncogenesis, and development. Rec Prog Hormone Res 1996;51:123–57.

35. Nishimori K, Young LJ, Guo Q, Wang Z, Insel TR, Matzuk MM. Oxytocin is required for nursing but is not essential for parturition or reproductive behavior. Proc Natl Acad Sci USA 1996;93:11699–704.

36. Matzuk MM, Dionne L, Guo Q, Kumar TR, Lebovitz RM. Analysis of superoxide dismutase 1 and 2 in ovarian function using knockout mice. Endocrinology 1998;139:4008–11.

37. Colledge WH, Carlton MB, Udy GB, Evans MJ. Disruption of c-*mos* causes parthenogenetic development of unfertilized mouse eggs. Nature 1994;370:65–68.

38. Hashimoto N, Watanabe N, Furuta Y, Tamemoto H, Sagata N, Yokoyama M, et al. Parthenogenetic activation of oocytes in c-*mos*-deficient mice. Nature 1994;370: 68–71.

39. Stewart CL, Kaspar P, Brunet LJ, Bhatt H, Gadi I, Kontgen F, et al. Blastocyst implantation depends on maternal expression of leukemia inhibitory-function. Nature 1992;359:76–79.

40. Lydon JP, Demayo FJ, Funk CR, Mani SK, Hughes AR, Montgomery CA, et al. Mice lacking progesterone receptor exhibit pleiotropic reproductive abnormalities. Genes Dev 1995;9:2266–78.

41. Schrewe H, Gendron-Maguire M, Harbison ML, Gridley T. Mice homozygous for a null mutation of activin bB are viable and fertile. Mech Dev 1994;47: 43–51.

42. Stinnakre MG, Vilotte JL, Soulier S, Mercier JC. Creation and phenotypic analysis of a-lactalbumin-deficient mice. Proc Natl Acad Sci USA 1994;91:6544–48.

43. Young WS, Shepard E, Amico J, Hennighausen L, Wagner KU, LaMarca ME, et al. Deficiency in mouse oxytocin prevents milk ejection, but not fertility or parturition. J Neuroendocrinol 1996;8:847–53.

44. Lee SL, Sadovsky Y, Swirnoff AH, Polish JA, Goda P, Gavrilina G, et al. Luteinizing hormone deficiency and female infertility in mice lacking the transcription factor NGFI-A (Egr-1). Science 1996;273:1219–21.

45. Topilko P, Schneider-Maunory S, Levi G, Trembleau A, Gourdji D, Driancourt M-A, et al. Multiple pituitary and ovarian defects in *Krox-24* (*NGFI-A, EGR-1*)-targeted mice. Mol Endocrinol 1997;12:107–22.

46. Liu C, Litscher ES, Mortillo S, Sakai Y, Kinloch RA, Stewart CL, et al. Targeted disruption of the *mZP3* gene results in production of eggs lacking a zona pellucida and infertility in male mice. Proc Natl Acad Sci USA 1996;93:5431–36.

47. Rankin T, Familari M, Lee E, Ginsberg A, Dwyer N, Blanchette-Mackie J, et al. Mice homozygous for an insertional mutation in the *Zp3* gene lack a zona pellucida and are infertile. Development 1996;122:2903–10.

48. Mahendroo MS, Cala KM, Russell DW. 5α-reduced androgens play a key role in murine parturition. Mol Endocrinol 1996;10:380–92.

49. Mahendroo MS, Cala KM, Landrum CP, Russell DW. Fetal death in mice lacking 5α-reductase type I caused by estrogen excess. Mol Endocrinol 1997;11:1–11.

50. Labosky PA, Winnier GE, Jetton TL, Hargett L, Ryan AK, Rosenfeld MG, et al. The winged helix gene, *Mf3*, is required for normal development of the diencephalon and midbrain, postnatal growth and the milk-ejection reflex. Development 1997;124:1263–74.

51. Simon AM, Goodenough DA, Li E, Paul DL. Female infertility in mice lacking connexin 37. Nature 1997;385:525–29.

52. Dinchuk JE, Car BD, Focht RJ, Johnston JJ, Jaffee BD, Covington MB, et al. Renal

abnormalities and an altered inflammatory response in mice lacking cyclooxygenase II. Nature 1995;378:406–9.

53. Sterneck E, Tessarollo L, Johnson PF. An essential role for C/EBPβ in female reproduction. Genes Dev 1997;11:2153–62.

54. Robb L, Li R, Hartley L, Nandurkar HH, Koentgen F, Begley CG. Infertility in female mice lacking the receptor for interleukin II is due to a defective uterine response to implantation. Nat Med 1998;4:303–8.

55. Sugimoto Y, Yamasaki A, Segi E, Tsuboi K, Aze Y, Nishimura T, et al. Failure of parturition in mice lacking the prostaglandin F receptor. Science 1997;277: 681–83.

56. Horseman ND, Zhao W, Montecino-Rodriguez E, Tanaka M, Nakashima K, Engle SJ, et al. Defective mammopoiesis, but normal hematopoiesis, in mice with a targeted disruption of the prolactin gene. EMBO J 1997;16:6926–35.

57. Wang W, Water TVD, Lufkin T. Inner ear and maternal reproductive defects in mice lacking the Hmx3 homeobox gene. Development 1998;125:621–34.

58. Ho Y-S, Gargano M, Cao J, Bronson RT, Heimler I, Hutz RJ. Reduced fertility in female mice lacking copper-zinc superoxide dismutase. J Biol Chem 1998; 273:7765–69.

59. Bergeron L, Perez GI, Macdonald G, Shi L, Sun Y, Jurisisova A, et al. Defects in regulation of apoptosis in caspase-2-deficient mice. Genes Dev 1998;12: 1304–14.

60. Teglund S, McKay C, Schuetz E, van Deursen JM, Stravopodis D, Wang D, et al. Stat5a and Stat5b proteins have essential and nonessential, or redundant, roles in cytokine responses. Cell 1998;93:841–50.

61. Lubahn DB, Moyer JS, Golding TS, Couse JF, Korach KS, Smithies O. Alteration of reproductive function but not prenatal sexual development after insertional disruption of the mouse estrogen receptor gene. Proc Natl Acad Sci USA 1993; 90:11162–66.

62. Couse JF, Curtis SW, Washburn TF, Lindzey J, Golding TS, Lubahn DB, et al. Analysis of transcription and estrogen insensitivity in the female mouse after targeted disruption of the estrogen receptor gene. Mol Endocrinol 1995;9:1441–54.

63. Kendall SK, Samuelson LC, Saunders TL, Wood RI, Camper SA. Targeted disruption of the pituitary glycoprotein hormone α-subunit produces hypogonadal and hypothyroid mice. Genes Dev 1995;9:2007–19.

64. Satokata I, Benson G, Maas R. Sexually dimorphic sterility phenotypes in Hoxa10-deficient mice. Nature 1995;374:460–63.

65. Good DJ, Porter FD, Mahon KA, Parlow AF, Westphal H, Kirsch IR. Hypogonadism and obesity in mice with a targeted deletion of the Nhlh2 gene. Nat Gen 1997;15:397–401.

66. Luoh S-W, Bain PA, Polakiewicz RD, Goodheart ML, Gardner H, Jaenisch R, et al. Zfx mutation results in small animal size and reduced germ cell number in male and female mice. Development 1997;124:2275–84.

67. Fero ML, Rivkin M, Tasch M, Porter P, Carow CE, Firpo E, et al. A syndrome of multiorgan hyperplasia with features of gigantism, tumorigenesis, and female sterility in p27^{Kip1}-deficient mice. Cell 1996;85:733–44.

68. Kiyokawa H, Kineman RD, Manova-Todorova KO, Soares VC, Hoffman ES, Ono M, et al. Enhanced growth of mice lacking the cyclin-dependent kinase inhibitor function of p27^{Kip1}. Cell 1996;85:721–32.

69. Baker SM, Plug AW, Prolla TA, Bronner CE, Harris AC, Yao X, et al. Involvement of mouse Mlh1 in DNA mismatch repair and meiotic crossing over. Nat Gen 1996;13: 336–41.
70. Edelmann W, Cohen PE, Kane M, Lau K, Morrow B, Bennett S, et al. Meitoic pachytene arrest in MLH1-deficient mice. Cell 1996;85:1125–34.
71. Barlow C, Hirotsune S, Paylor R, Liyanage M, Eckhaus M, Collins F, et al. Atm-deficient mice: a paradigm of ataxia telangiectasia. Cell 1996;86:159–71.
72. Xu Y, T TA, Brainerd EE, Bronson RT, Meyn MS, Baltimore D. Targeted disruption of ATM leads to growth retardation, chromosomal fragmentation during meiosis, immune defects, and thymic lymphoma. Genes Dev 1996;10:2411–22.
73. Sicinski P, Donaher JL, Gene Y, Parker SB, Gardner H, Park MY, et al. Cyclin D2 is an FSH-responsive gene involved in gonadal cell proliferation and oncogenesis. Nature 1996;384:470–74.
74. Ormandy CJ, Camus A, Barra J, Damotte D, Lucas B, Buteau H, et al. Null mutation of the prolactin receptor gene produces multiple reproductive defects in the mouse. Genes Dev 1997;11:167–78.
75. Ruggiu M, Speed R, Taggart M, McKay SJ, Kilanowski F, Saunders P, et al. The mouse *Dazla* gene encodes a cytoplasmic protein essential for gametogenesis. Nature 1997;389:73–76.
76. Lu Q, Hasty P, Shur BD. Targeted mutation in b1,4-galactosyltransferase leads to pituitary insufficiency and neonatal lethality. Dev Biol 1997;181:257–67.
77. Toscani A, Mettus RV, Coupland R, Simpkins H, Litvin J, Orth J, et al. Arrest of spermatogenesis and defective breast development in mice lacking A-*myb*. Nature 1997;386:713–17.
78. Miyamoto N, Yoshida M, Kuratani S, Matsuo I, Aizawa S. Defects of urogenital development in mice lacking *Emx2*. Development 1997;124:1653–64.
79. Hsieh-Li HM, Witte DP, Weinstein M, Branford W, Li H, Small K, et al. *Hoxa 11* structure, extensive antisense transcription, and function in male and female fertility. Development 1995;121:1373–85.
80. Beck ARP, Miller IJ, Anderson P, Streuli M. RNA-binding protein TIAR is essential for primordial germ cell development. Proc Natl Acad Sci USA 1998;95: 2331–36.
81. Pittman DL, Cobb J, Schimenti KJ, Wilson LA, Cooper DM, Brignull E, et al. Meiotic prophase arrest with failure of chromosome synapsis in mice deficient for *Dmc1*, a germline-specific RecA homolog. Mol Cell 1998;1:697–705.
82. Yoshida K, Kondoh G, Matsuda Y, Habu T, Nishimune Y, Morita T. The mouse RecA-like gene Dmc1 is required for homologous chromosome synapsis during meiosis. Mol Cell 1998;1:707–18.
83. Lee H-W, Blasco MA, Gottlieb GJ, Horner JW, Greider CW, DePinho RA. Essential role of mouse telomerase in highly proliferative organs. Nature 1998;392: 569–77.

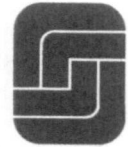

22

Studying the Genetics of Complex But Common Human Diseases Using Mice

Nobuyo Maeda and Oliver Smithies

Cardiovascular and cerebrovascular diseases that result from atherosclerosis and hypertension account for a large proportion of morbidity and mortality in the United States. Females are generally less susceptible to these conditions until after menopause, when they rapidly become as susceptible as males. In addition to gender, the genetic makeup of an individual is clearly important in the etiology of these diseases, and genetic analyses based on human population and family studies have successfully identified various factors associated with atherosclerosis and with hypertension thanks to the advancement of molecular techniques. The genetic heterogeneity of humans, however, makes it difficult to dissect the roles of individual genetic factors and to determine fundamental cause and effect relationships. In addition, environmental factors that significantly influence the development of these diseases are difficult to control in humans.

For these reasons, genetically modified mice, generated via gene targeting in embryonic stem cells or microinjection of DNA into the pronuclei of embryos, have been particularly useful for studying the phenotypes that result from specific changes in single genes in vivo for detecting synergistic or antagonistic interactions of mutations by combining genetic changes in a single animal. In contrast to the genetics of relatively uncommon diseases caused by single defects, the genetics of the common diseases are complex because the traits are often quantitative, and are probably a cue to unlucky combinations of allelic differences, often subtle, in multiple genes that may or may not be identical in any two individuals. In this chapter we will discuss some of the strategies that we are employing using mice to gain a better understanding of the genetics of atherosclerosis and hypertension in humans.

Quantitative Traits—Gene Titration

Phenotypes such as hypertension and plasma lipid levels are quantitative traits. Null mutations causing these quantitative variations are rare in human populations, and by and large the variation results from allelic changes that alter the levels of expression of the gene in a range of at most 50% below or above the population mean. To study the causative link between the levels of a specific gene product and a quantitative phenotype, Smithies and Kim (1) devised a strategy—gene titration—by which one of the two copies of a specific gene is either disrupted or duplicated at its own natural locus. Disruption of the gene is achieved by what is now a quite conventional knockout procedure, but the best animals are the heterozygotes for the disrupted allele in which there is a decrease (not an absence) of the gene product. Duplication of the gene, together with elements needed in controlling its expression, leads to an increase in the levels of gene product. If the gene is autosomal, animals can be generated that have one copy (heterozygous for the disrupted allele), two copies (wild type), three copies (heterozygous for the duplicated allele) or four copies (homozygous for the duplicated allele) of the gene. Because the spacial and temporal regulation of the gene expression is maintained, the physiological consequence in the resulting mice directly reflects the gene dosage effect.

As an example, we will discuss the application of this strategy to the *Npr1* gene coding for the natriuretic peptide receptor A (NPRA), which we had hypothesized was a candidate gene for controlling blood pressures by the following means. Natriuresis and diuresis, particularly in response to blood volume expansion following an increase in dietary salt, are important components of blood pressure maintenance. Atrial natriuretic peptide (ANP) is produced mainly in the right atrium of the heart, where its precursor (proANP) is stored in dense granules. The active peptide, the COOH-terminal 28 amino acids of the precursor, is cleaved and released into the circulation in response to an increase in atrial distension (2). ANP and its close relative BNP, which is produced primarily in the left ventricle, lower blood pressure and promote natriuresis via their interaction with the NPRA receptor in the vasculature, kidney, and adrenal grands. NPRA is a membrane-bound guanylyl cyclase (3) with an extracellular ligand-binding domain and two intracellular domains. One of the intracellular domains is homologous to a guanylyl cyclase catalytic domain; the other is homologous to a protein kinase catalytic domain, although kinase activity cannot be detected.

Figure 22.1 (left panel) shows the strategy used to duplicate the entire locus by a homologous recombination in the mouse embryonic stem cells. The targeting construct contains two regions of homology to the mouse gene: a 6.5 kb fragment of DNA homologous to a region 5' to the *Npr1* gene and a 1.3 kb fragment homologous to a region 3' to the gene. The two regions of homology are separated by a gap of 24 kb, but this is was efficiently filled

within the cell during the recombination event. The same 6.5 kb of 5' DNA
and another 1.5 kb DNA corresponding to exons 3 through 5 of the gene
were used to make a targeting construct to delete DNA corresponding to two
thirds of the ligand-binding domain of the NPRA (Fig. 22.1, right panel).
Both recombinational events took place efficiently and allowed us to generate mice with cells carrying either a duplicated or a disrupted *Npr1* gene. As
shown in Figure 22.2, the ANP-stimulated cGMP production of the lung
membrane preparation of the animals increases in direct proportion to the
number of copies of the *Npr1* gene, whereas the blood pressures decrease as
Npr1 gene copy number and NPRA expression increase. Thus, this experi-

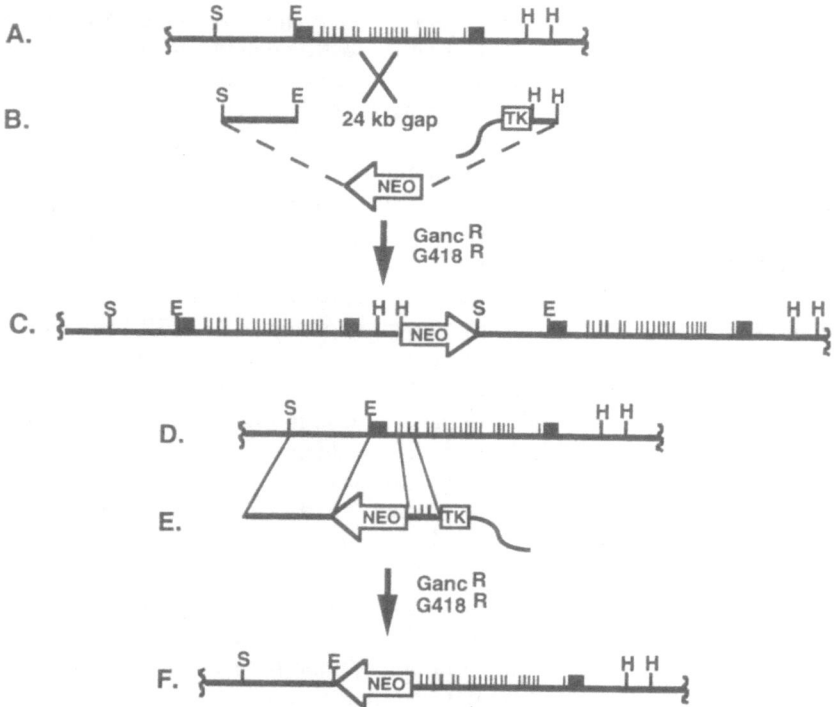

FIGURE 22.1. Targeted duplication (A,B,C) and disruption (D,E,F) of the mouse *Npr1*
gene. The gene is composed of 22 exons. A gap repair synthesis in conjunction with
a single crossover event between the endogenous locus (A) and the targeting construct (B) leads to a 32 kb duplication that includes the complete *Npr1* gene with
approximately 7 kb each of upstream and downstream flanking sequences (C). Homologous crossover between the locus (D) and the deletion targeting construct (E)
leads to deletion of exons 1–3 and their replacement by the neomycin resistant gene,
NEO, (F). TK indicates Herpes simplex thymidine kinase gene. The targeting plasmids were introduced into cells by electroporation and G418 and ganciclovir (Ganc)–
resistant cells were screened for the correct modification.

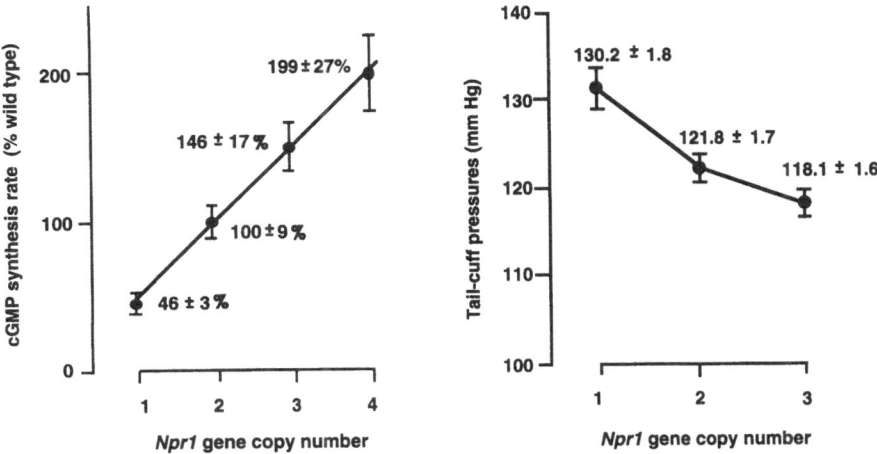

FIGURE 22.2. Guanylyl cyclase activity (left panel) and blood pressures (right panel) as a function of *Npr1* gene copy number. Mean ANP-stimulated cGMP synthesis in lung membrane preparations are expressed as a percentage of the rate of wild-type (two-copy) animals. Blood pressures of the F1 mice were measured by a computerized tail-cuff method.

ment clearly demonstrates that genetically determined quantitative changes in the expression of NPRA cause inverse changes in the blood pressures of mice that have all their normal homeostatic mechanisms intact. These causative effects shown in mice strongly suggest that any genetic polymorphisms affecting the levels of expression of the human gene coding for NPRA would also be of importance in determining the inheritance of higher or lower than normal blood pressures (4).

Although this is not a part of the gene titration experiment, zero-copy mice, which lack NPRA completely, have a chronic elevation of blood pressures about 18 mmHg above normal. It is striking that these mutants develop cardiac hypertrophy with extensive perivascular and interstitial fibrotic depositions. This pathological change is considerably more severe in males than it is in females in both extent and incidence. In addition, sudden death occurred in males before 6 months of age, some with pathology indicative of congestive heart failure and some with dissection of the thoracic aorta, or pulmonary artery embolism (5).

Subtle Allelic Changes—Gene Replacement

The second point we note in the genetics of common but complex diseases is that many polymorphic differences in humans that subtly but distinctively affect the phenotype in humans are due to small changes in protein sequence.

The isoforms of human apolipoprotein E(apoE), which is important for the transport of cholesterol and triglyceride, form a good example: there are three major alleles, *APOE*2*, *APOE*3*, and *APOE*4*, that occur at frequencies of 7.3, 78.3, and 14.3%, respectively. They differ at codons 112 and 158: The most common allele *APOE*3* codes for a cysteine at position 112 and an arginine at 158; the *APOE*2* allele codes for a cysteine at both positions, whereas the *APOE*4* allele codes for an arginine at both positions. Although apoE2 protein has only about 1% of binding affinity to the low-density lipoprotein receptor compared with 100% in apoE3 and apoE4, most individuals homozygous for apoE2 have lower than normal plasma cholesterol. A fraction of apoE2 homozygotes, however, develop type III hyperlipoproteinemia and premature atherosclerosis (6). Individuals homozygous for apoE4 have higher than normal plasma cholesterol and LDL cholesterol compared with E3 homozygotes and are at increased risk for developing coronary artery diseases (7). ApoE4 is also associated with the development of Alzheimer's disease (8).

To test the causative effects of human apoE isoforms on atherosclerosis and other diseases, we have generated mice expressing the three different human alleles by replacing the endogenous mouse gene with the human counterpart (Fig. 22.3). In the physiological range the mice generated by this "targeted gene replacement" approach express only the human apoE proteins corresponding to their respective allele. They differ from conventional transgenic mice made by pronuclear injection of human DNA in which levels of transgene expression are variable due to differences in chromosomal

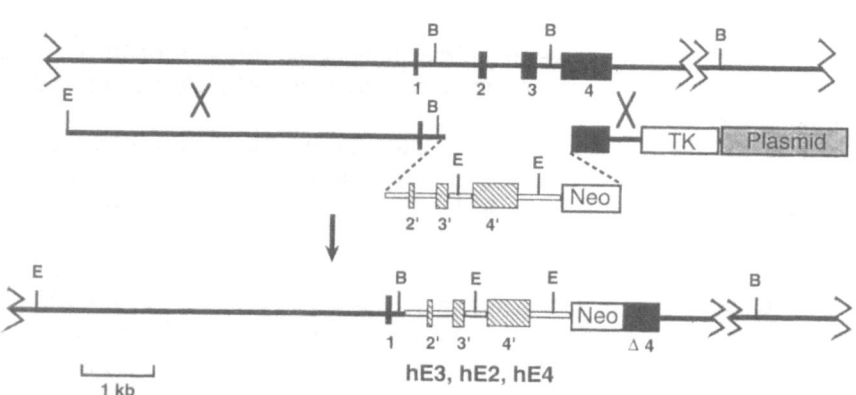

FIGURE 22.3. Replacement of the mouse Apoe gene with the human APOE alleles. The mouse Apoe gene containing exons 1–4 (black boxes) is shown in the top line. The targeting construct (middle line) contains a human DNA fragment that includes exons 2–4 (hatched boxes) of either the *APOE*2*, *APOE*3*, or *APOE*4* allele. Because exon 1 does not contain the coding sequence of the transcript, the targeted locus (bottom line) produces either hE2, hE3, or hE4 protein. EcoR1 (E) and BamH1 (B) sites are indicated.

location and copy number, and in which the mouse protein is still expressed. The physiological differences in the gene-replacement mice are thus the direct consequence of the allelic gene product. Figure 22.4 illustrates the plasma total cholesterol and triglycerides seen in the various animals. The plasma lipid profiles in the mice expressing human apoE3 (hE3/hE3) are quite similar to those seen in wild-type mice maintained on a normal chow. When the hE3/hE3 mice are fed a high fat and high cholesterol diet, however, they respond with a considerably greater (fivefold) increase in total cholesterol compared with the 1.5-fold increase in wild-type mice; and after 12 weeks on this diet they develop significantly larger atherosclerotic plaques in the aortic sinus area compared with their wild-type litter mates (9).

Mice that express apoE4 (hE4/hE4) have a slight increase in VLDL cholesterol and develop about twofold larger mean size plaques in response to the atherogenic diet compared with hE3/hE3 mice, although these differences are not statistically significant (Knouff C, et al., unpublished observation). Mice that express the human apoE2 isoform (hE2/hE2) have a more striking phenotype, which is abnormal even on normal chow diet: They have elevated cholesterol and triglycerides (both about three times normal), similar to that seen in human type III hyperlipoproteinemia. The hE2/hE2 mice develop atherosclerosis without dietary cholesterol supplement, although the

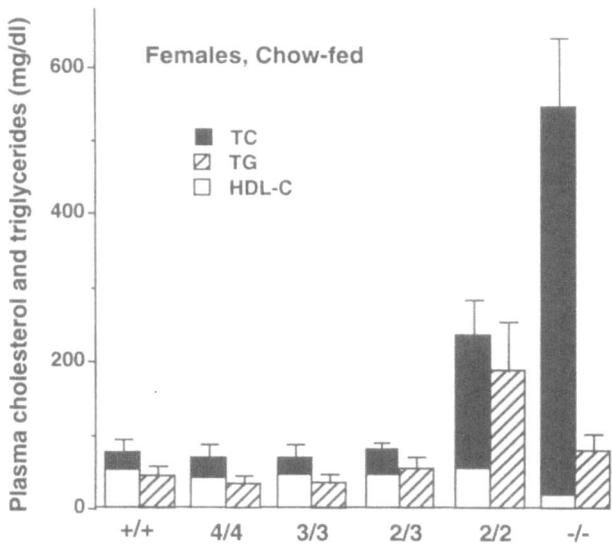

FIGURE 22.4. Plasma lipid levels in mice with different apoE alleles. Female mice on a chow diet were fasted overnight, and their plasma cholesterol (black bars), HDL cholesterol (white bars), and triglycerides (hatched bars) were measured. +/+, wild type mice; 4/4, mice homozygous for hE4; 3/3, mice homozygous for hE3; 2/3, mice heterozygous for hE3 and hE2; 2/2, mice homozygous for hE2; and -/-, mice lacking apoE.

plaques progress at a much slower rate than they do in the apoE null mice (10). All the hE2/hE2 mice, regardless of their sex, age, and diet, develop type III phenotype. This contrasts with human E2/E2 individuals in which only 5–10% develop the phenotype. We infer that there are some secondary differences in the susceptible humans, either genetic and/or environmental, that are already present in all the mice. A rigorous investigation of the isoform effects using these mice should help better understand the human disease.

Multigenic System—Genetic Synthesis

The third complexity in the genetics of common diseases is the involvement of multiple genes so that the disease in many patients may be the consequence of combinations of genetic alterations that individually have only modest effects. Studies of synergistic or antagonistic interactions between genes are clearly needed, and mice are excellent for carrying out such studies because genetic changes can be combined in them relatively easily. In other words, a phenotype can be "genetically synthesized" in mice. For example, apoE-null mice are useful models for the study of atherosclerosis because they are hypercholesterolemic and spontaneously develop atherosclerotic lesions very much like the human plaques (11). ApoE-null mice are also useful to investigate how mutations in a second or third gene modify the incidence and progression of the plaque development.

   We are currently testing how hypertension and atherosclerosis interact by using apoE-null mice into which we introduce an additional mutation in a gene known to affect the vascular system (Fig. 22.5). Thus, various population-based studies have shown that coronary vascular diseases that result

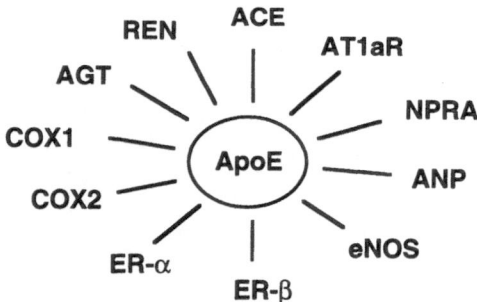

FIGURE 22.5. Scheme to study the interrelationships between atherosclerosis and hypertension. ApoE-/- mice crossed with mice having one of the modifications shown in the outer circles, respectively, will be evaluated for the development of atherosclerotic plaques. AGT, angiotensinogen; REN, renin; ACE, angiotensin converting enzyme; AT1aR, angiotensin type 1a receptor; NPRA, natriuretic peptide receptor A; ANP, atrial natriuretic peptide; eNOS, endothelial nitric oxide synthase; ER-α or ER-β, estrogen receptor type α or β; and COX1 or COX2, cyclo-oxygenase 1 or 2.

from atherosclerosis are associated with hypertension more frequently than would be expected by chance (12). A considerable body of evidence in laboratory animals has also established that the severity of atherosclerosis is enhanced when hypertension is experimentally induced (e.g., for example by the one-kidney one clip Goldblatt procedure) (13). Nevertheless, current data do not allow an unequivocal determination of whether the increase in risk associated with hypertension is a direct effect on atherogenesis of the increased blood pressure or is an effect on atherosclerosis of the agents that led to the hypertension. For example, lack of endothelial nitric oxide (14), lack of natriuretic peptide receptor A (10), and an increase in the angiotensinogen (15) all lead to about the same degrees of blood pressure elevation in mice—each by a different mechanism. By crossing these mutations with the apoE-null mutation, we hope to be able to determine how plaque development is affected by elevated blood pressures as such, and how it is affected by the different genetic factors that have altered the pressure.

Conclusion

With our aging population, a better understanding of the pathogenesis and the development of treatments for cardio- and cerebrovascular diseases is becoming acutely important. The incidence of these diseases in women after menopause is high, and understanding the underlying genetic causes should help the design of preventative strategies that are tailored to the individual. The value of mice as a tool for understanding the genetics underlying these complex but common diseases is obvious. The strategies we have discussed have the power to determine the causative effects of allelic differences that have been shown to be associated with different risks in humans. In the reverse direction, mutations generated in candidate genes in mice that prove to have important physiological effects point to the value of looking for variations in the same genes in human populations. This going back and forth between the human and mouse systems is expected to lead to our gaining a deeper understanding of the common debilitating diseases and to a disentanglement of the various genetic risk factors that affect their incidence.

Acknowledgment. We thank the members of our laboratories for their hard work in carrying out the experiments discussed here. This work was supported by NIH grants HL42630 and HL49277.

References

1. Smithies O, Kim H-S. Targeted gene duplication and disruption for analyzing quantitative genetic traits in mice. Proc Natl Acad Sci USA 1994; 91:3612–15.
2. Flynn TG, de Bold ML, de Bold AJ. The amino acid sequence of an arterial peptide

with potent diuretic and natriuretic properties. Biochem Biophys Res Commun 1983;90:859–65.

3. Garbers DL, Lowe DG. Guanylyl cyclase receptors. J Bio Chem 1994;269: 30741–44.

4. Oliver PM, John SWM, Purdy KE, Kim R, Maeda N, Goy MF, et al. Natriuretic peptide receptor 1 expression influences blood pressures of mice in a dose-dependent manner. Proc Natl Acad Sci USA 1998;95:2547–51.

5. Oliver PM, Fox JE, Kim R, Rockman HA, Kim H-S, Smithies O, et al. Hypertension, cardiac hypertrophy and sudden death in mice lacking natriuretic peptide receptor A. Proc Natl Acad Sci USA 1997;94:14730–35.

6. Mahley RW, Rall SC Jr. Type III hyperlipoproteinemia (dysbetalipoproteinemia): the role of apolipoprotein E in normal and abnormal lipoprotein metabolism. In: Beaudet AL, Sly WS, Valle D, eds. The metabolic and molecular basis of inherited diseases, seventh ed. New York: McGraw-Hill, 1995:1953–80.

7. Hixon JE. Apolipoprotein E polymorphisms affect atherosclerosis in young males. Pathological determination of atherosclerosis in youth (PDAY) research group. Arterioscler Thromb 1991;11:1237–44.

8. Corder EH, Saunders AM, Strittmatter WJ, Schmechel DE, Gaskell PC, Small GW, et al. Gene dose of apolipoprotein E type 4 allele and the risk of Alzheimer's disease in late onset families. Science 1993;261:921–23.

9. Sullivan PM, Mezdour H, Aratani Y, Knouff C, Najib J, Reddick R, et al. Targeted replacement of the mouse apolipoprotein E gene with the common human *APOE3* allele enhances diet-induced hypercholesterolemia and atherosclerosis. J Biol Chem 1997;272:17971–80.

10. Sullivan PM, Mezdour H, Quarfordt SH, Maeda N. Type III hyperlipoproteinemia and spontaneous atherosclerosis in mice resulting from gene replacement of mouse Apoe with human *APOE*2*. J Clin Invest 1998;102:130–35.

11. Reddick RL, Zhang SH, Maeda N. Atherosclerosis in mice lacking apoE. Arterioscler Thromb 1994;14:141–47.

12. McGill HC Jr, Strong JP, Tracy RE, McMahan CA, Oalmann MC. Relation of a postmortem renal index of hypertension to atherosclerosis in youth. The pathological determinants of atherosclerosis in youth (PDAY) research group. Arterioscler Thromb Vasc Biol 1995;15:2222–28.

13. Chobanian AV, Lichtenstein AH, Nilakhe V, Haudenschild CC, Drago R, Nickerson C. Influence of hypertension on aortic atherosclerosis in the Watanabe rabbit. Hypertension 1989;14:203–9.

14. Shesely EG, Maeda N, Kim S-S, Desai KM, Krege JH, Lauback VE, et al. Elevated blood pressures in mice lacking endothelial nitric oxide synthase. Proc Natl Acad Sci USA 1996;93:13176–81.

15. Kim H-S, Krege JH, Kluckman KD, Hagaman JR, Hodgin JB, Best CF, et al. Genetic control of blood pressure and the angiotensinogen locus. Proc Natl Acad Sci USA 1995;92:2735–39.

23

The Classic Steroid Hormone Receptors and ERβ, the Novel Estrogen Receptor

R. Rex Denton, Samir K. Ghosh,
Roland Baron, and Anuradha Ray

Steroid hormones control diverse biological functions, such as growth and development, regulation of immune functions, and cell death (1). The regulation of gene transcription by steroid hormones is mediated by intracellular receptors that belong to a complex superfamily of ligand-binding transcription factors (2,3). Members of this superfamily include the glucocorticoid receptor (GR), the mineralocorticoid receptor (MR), the estrogen receptor (ER), the progesterone receptor (PR), and the thyroid hormone receptor (TR). These receptors may be found in both the cytosolic and nuclear fractions of target tissues. Immunocytochemical studies with monoclonal antibodies against the various receptors indicate that most steroid hormone receptors (e.g., ER and PR) are nuclear even in the absence of hormone (4,5). GR and MR appear to be exceptions to this general rule. In the absence of hormone, these latter receptors are predominantly found in the cytoplasm complexed with other proteins (e.g., hsp 90) and concentrate in the nucleus only after hormone addition (6,7).

Domain Structure of the Glucocorticoid Receptor and the Estrogen Receptor

Although their ligands have diverse physiological functions, the nuclear hormone receptors including GR and ER share a common domain organization. The ligand-binding domains (LBD) are located in the C-terminal part of the proteins and share limited sequence similarity in the ligand-binding regions. Genetic analysis of the mouse ER LBD has allowed mapping of the hormone-binding region in LBD and another distinct region within the LBD comprising 22 amino acids that is responsible for hormone-induced receptor dimerization (8). Based on binding studies of agonists (estrogens) and antagonists (tamoxifen) to ER, Jensen and colleagues have proposed that there are two ligand-binding sites in ER, one that binds agonists and another that binds antagonists (9). They have also proposed that the agonistic behavior of some antagonists is due to binding to the agonist site, whereas the antagonist effect can be explained by binding to the antagonist site (9). Analysis of crystal structures of the LBD of ER complexed with the antagonist raloxifene, however, revealed that both agonists and antagonists bind to the same site but induce distinct conformations in the transactivation domain (AF-2) of the LBD (10). The AF-2 function is conserved among different members of the receptor family. It is dependent on binding of the ligand and is thought to undergo a dramatic change in conformation once the receptor binds ligand (11).

The amino terminal part of the nuclear hormone receptors is most diverse among the different members of the family. In GR, ER, and PR the transactivation function AF-1, which is independent of hormone binding, has been mapped to this region. The agonistic activity of some antihormones can be partly explained by induction of AF-1 activity (12). This region is also crucial for transcriptional synergism between DNA-bound receptor dimers.

The DNA-binding domain (DBD) is centrally located in the receptors and is most conserved between the different members. It has two zinc fingers and also contains a dimerization interface (the D loop). While GR-/- mice die shortly after birth, demonstrating an essential role for GR in survival (13), mice containing mutations in the D loop of GR (GR^{dim}) are viable and thus transactivational functions of GR which are dependent on DNA-binding of GR dimers are not essential for survival (14). It is interesting that even though the transactivational functions of GR were abolished by the D loop mutations, transrepression functions toward AP-1–driven genes were unaffected (14).

Transcripotional Activation: Interactions with Basal Transcription Factors and Coactivators

Steroid hormone receptors have been shown to interact with proteins of the basal transcriptional machinery. Transcription of eukaryotic genes by RNA polymerase II requires assembly of RNA polymerase and six general transcription factors—TFIIA, TFIIB, TFIID, TFIIE, TFIIF, and TFIIH—on the promoter to form a preinitiation complex. One of these factors, TFIID, is believed to be responsible for the recognition of two elements. One subunit of TFIID0, the TATA-binding protein (TBP), is responsible for the recognition of the TATA element. One or more of the remaining subunits of TFIID—the TBP-associated factors (TAFIIs)—are responsible for recognition of the downstream promoter element and are now regarded as highly promoter selective. Interactions of ER with a particular TAF, TAF$_{II}$30, has been demonstrated (15,16). Structure-function studies with different domains of ER revealed that only AF-2 interacts with TAF$_{II}$30 (15).

A search for coactivators that interact with hormone receptors was initiated when different classes of receptors were found to interfere with one another's transcriptional activity by squelching limiting factors that were not components of the basal transcription machinery. For example, several proteins were discovered that were found to interact with the LBD in ER in a ligand-dependent fashion. It is interesting that agonists, but not antagonists, promoted specific binding to these proteins. Two such proteins, called ERAP-160 and RIP140, interact with the AF-2 region in ER (17,18). Point mutations in AF-2 that abolish transactivation also eliminate interactions with RIP140, and RIP140 enhances transcriptional activity of the AF-2 domain of ER (18). It appears, therefore, that RIP140 is potentially an important regulator of ER function. ERAP-160 appears to be related to a family of proteins that includes SRC-1, which interacts with and enhances the ligand-dependent transactivation functions of several members of the nuclear receptor superfamily (19–21). The relationship of RIP140 to ERAP-160 is unclear. As suggested by Katzenellenbogen and colleagues, the effect of SRC-1 could be indirect and involve additional proteins (22). Of note, both nuclear receptors and SRC-1/ERAP-160 interact with another coactivator protein, CBP, to form a ternary complex (22,23). Most of the receptor-interacting proteins identified to date are ubiquitous and interact with multiple members of the nuclear hormone receptor superfamily, albeit with different affinities and specificities. Their major role is probably to link the receptors with the basal transcriptional machinery. Other roles, however, as in nuclear transport or DNA-binding, cannot be excluded.

Transcriptional Repression: Antiinflammatory Actions of Glucocorticoids

Just as transactivation by steroid hormones is crucial for growth and development, transcriptional repression of specific genes is critical for maintaining physiological homeostasis and inhibiting inflammation. To a large degree, the success of glucocorticoids as antiinflammatory agents stems from their ability to inhibit the cellular release of inflammatory mediators and cytokines. Cytokines whose expression is repressed by glucocorticoids include IL-1, IL-2, IL-5, IL-6, IL-8, TNF-α, and interferon-γ. Glucocorticoids also downregulate the production of chemoattractant cytokines (chemokines) and cell adhesion molecules that, along with cytokines, play important roles in inflammation. The target cells for the inhibitory actions of glucocorticoids include monocytes/macrophages, T cells, epithelial cells, endothelial cells, and fibroblasts.

The Role of nGREs in Transrepression by Glucocorticoids

Most genes that are negatively regulated by glucocorticoids lack consensus GREs in their regulatory regions. The term *negative GRE* (nGRE) has been used to describe GR-binding sequences in many such genes, including the human glycoprotein α-subunit gene, the bovine prolactin gene, the rat pro-opiomelanocortin gene, and the IL-6 gene (24). These nGREs are typically located close to or overlapping with binding sites for other transcription factors. It is unclear why the receptor does not mediate a positive effect when bound to nGREs as it does when it is bound to GREs. It is likely that the sequence of the nGREs, which are heterogeneous and often resemble only a half-site of the 15 bp GRE consensus, play an important role. Many genes that are negatively regulated by GR, however, lack either a GRE or an nGRE. Protein–protein interactions have been postulated to play a major role in repression of expression of these genes by glucocorticoids.

Interactions with Transcription Factors AP-1 and NF-κB in Transrepression by Glucocorticoids

A few years ago, a number of investigators described protein–protein interactions between GR and the transcription factor AP-1 in the inhibition of AP-1–dependent transcriptional activation by GR (25,26). The two AP-1–regulated target genes that have been most extensively studied in this regard are the collagenase gene and the proliferin gene. Although the collagenase gene has no GR-binding sites in its promoter, the proliferin gene contains a "composite GRE" that contains partially overlapping sites for the binding of the GR and AP-1. At this composite GRE, GR–AP-1 interaction was shown to result either in activation or inhibition of gene expression depending on the composition of the AP-1 dimer. Although induction of the proliferin gene by the AP-1 complex (fos/jun heterodimer) was repressed by a combination of dexamethasone (dex) and GR, induction by the jun/jun or jun/fra hetero-dimer was actually augmented by the receptor. It has been suggested that the repressive effect is probably mediated by GR monomers and not dimers be-cause point mutations in the D loop of GR that destroyed dimerization and transactivation by the receptor preserved repression of AP-1–dependent col-lagenase promoter activity. An altogether different mechanism has also been proposed in inhibition of AP-1 activity by GR. The coactivator CBP, which interacts with multiple nuclear receptors, also interacts with a variety of other transcription factors, including AP-1. It has been suggested that anatagonism of the AP-1 pathway by GR is a result of competition for a limiting pool of CBP. In support of this hypothesis, ectopically expressed CBP was shown to relieve GR inhibition of AP-1 activity (23). Thus, more information regarding structural requirements of GR/AP-1 versus GR/CBP interactions is needed for a better understanding of GR/AP-1 functional antagonism.

Interactions Between the GR and NF-κB: Cytokine Genes as Targets

An important target of glucocorticoids in their antiinflammatory actions is proinflammatory cytokine gene expression. The cytokine IL-6 is a key player in inflammation. It is at the core of a bidirectional relationship between the central neuroendocrine system and the immune system, being a crucial mediator of the acute phase response in the host in response to infection or trauma. In several studies we had demonstrated that IL-6 gene expression is profoundly inhibited by glucocorticoids in different cell types (see references in 27). Although a few nGREs were identified in the IL-6 gene, neither the nGREs nor an AP-1–dependent mechanism could explain GR inhibition of IL-6 gene transcription. Because NF-κB was shown by us and other investigators to play a crucial role in IL-6 gene expression, and was also known to be a key regulator of expression of multiple genes involved in inflammatory processes, we investigated whether NF-κB was another target in the antiinflammatory actions of glucocorticoids. Indeed, our studies have demonstrated mutual functional antagonism between the GR and NF-κB (28) and has also been reported by other investigators (29–31). The GR was also shown to interact with the other transcription factor NF-IL6 (C/EBPβ) that is also important for IL-6 gene expression (32). Dex has been shown to increase expression of IκB (i.e., the molecule that sequesters NF-κB in the cytoplasm in unstimulated cells) in certain cell types and that this mechanism was proposed as an additional mechanism for repression of NF-κB by glucocorticoids (33,34).

Negative Regulation of IL-6 Gene Expression by Estrogens

It appears that negative feedback by both glucocorticoids and estrogens (35) on endogenous IL-6 levels is an important regulator of physiological homeostasis. Disruption of this control, which leads to uninhibited IL-6 production in an estrogen-depleted state in postmenopausal women, has been proposed to be a contributing factor in the onset of postmenopausal osteoporosis (36,37).

Our initial studies on ER repression of IL-6 gene expression failed to reveal any ER-binding sites in the regulatory region of the IL-6 gene (Fig. 23.1) (38). Because GR inhibition of IL-6 gene expression was not predomi-

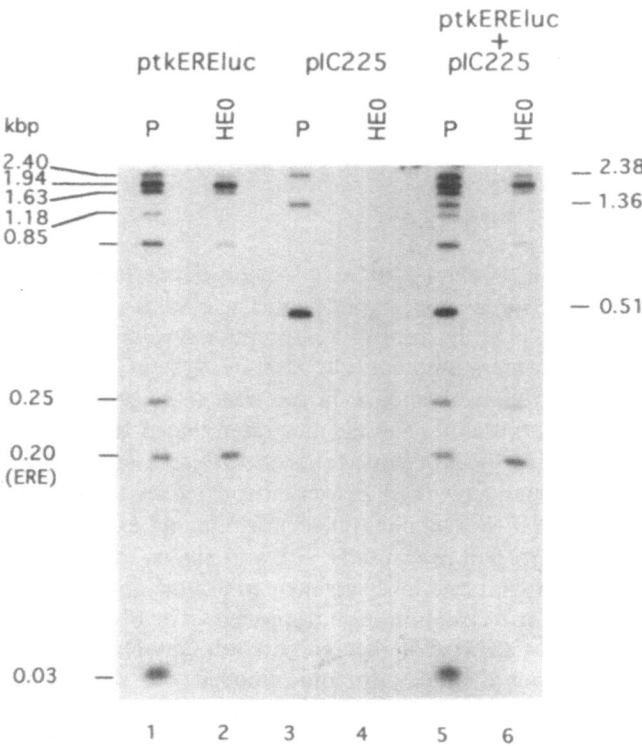

FIGURE 23.1. Lack of high-affinity ER-binding sites in the IL-6 promoter in a sequential DNA-binding immunoprecipitation assay. HeLa cell extract prepared from cells transfected with the expression vector for the estrogen receptor, HE0, was used to bind to and immuno-precipitate ER-binding fragments from either the vitellogenin or the IL-6 promoter. The promoter constructs used were pIC225, containing IL-6 promoter sequences between–225 and +13, and ptkEREluc, containing the estrogen response element (ERE) that is present in the vitellogenin promoter linked to the thymidine kinase (tk) promoter. The probes (P) were prepared by restriction enzyme digestion of the plasmids pIC225 and ptkEREluc followed by radiolabeling the resulting fragments. As illustrated, the probe used in each reaction was labeled fragments derived from either pIC225 or from ptkEREluc or was a mixture of the fragments from both plasmids. Samples in all lanes were immunoprecipitated with the anti-ERα monoclonal antibody H222. Immunoprecipitated DNA was analyzed by electrophoresis on a low-melting-point agarose gel comprising 3% NuSieve agarose and 1% SeaPlaque agarose. The gel was dried under vacuum and subjected to autoradiography. Numbers on the left denote molecular size of fragments (kbp) derived from ptkEREluc, whereas those on the right correspond to fragments derived from pIC225. Reprinted with permission from Ray et al., J Biol Chem (Ref. 38).

nantly driven by GR–DNA interactions (28,39), and because the two receptors shared structural similarity, protein–protein interactions between ER and transcription factors regulating IL-6 gene expression were a plausible explanation for ER inhibition of IL-6 gene expression. Using electrophoretic mobility shift assays and immunoprecipitation experiments, we showed that

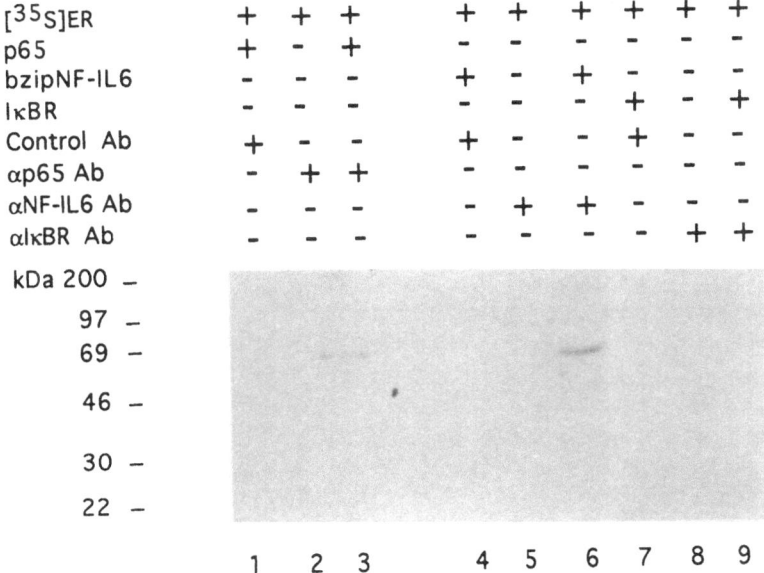

[³⁵S]ER	+	+	+		+	+	+	+	+	+
p65	+	-	+		-	-	-	-	-	-
bzipNF-IL6	-	-	-		+	-	+	-	-	-
IκBR	-	-	-		-	-	-	+	-	+
Control Ab	+	-	-		+	-	-	+	-	-
αp65 Ab	-	+	+		-	-	-	-	-	-
αNF-IL6 Ab	-	-	-		-	+	+	-	-	-
αIκBR Ab	-	-	-		-	-	-	-	+	+

kDa 200 –

97 –

69 –

46 –

30 –

22 –

1   2   3   4   5   6   7   8   9

FIGURE 23.2. Coimmunoprecipitation of ER-NF-IL6 and ER-p65 complexes. [35S]Labeled ER and unlabeled p65, bzipNF-IL6, or IκBR (an unrelated protein used as a control) were translated in wheat germ extracts. Ten microliter aliquots of the programmed extract containing labeled ERα were incubated in the presence or absence of p65 (10 μl), NF-IL6 (10 μl), IκBR (10 μl), anti-p65 antibody, anti-NF-IL6 antiserum, anti-IκBR antiserum, or control immunoglobulin. The proteins were immunoprecipitated as shown, and the resulting complexes were analyzed by fluorography following separation on a 10% SDS-polyacrylamide gel. The mobilities of the protein standards are indicated. Reprinted from Ray et al., FEBS Lett (Ref. 40).

ER interacts with both NF-κB and NF-IL6 (Fig. 23.2) (40). Thus, protein–protein interactions appear to play an important role in the regulation of immune functions by both corticosteroids and estrogens.

A Novel ER-ERβ

The cloning of a second estrogen receptor, ERβ, from both humans and rodents has been reported (41–43). Our deduction of the putative amino acid sequence of ERβ has revealed differences with the published sequence at two locations. First, in both humans and rodents the translational start site is located further upstream of the previously identified site in humans and rats. In addition, in the murine sequence, we have identified an additional 54 bp exon, exon 5B (Fig. 23.3A). Because this insertion is in frame, this would result in the addition of 18 amino acids in the LBD of mERβ. When compared with the crystal structure of ERα, these 18 amino acids should be

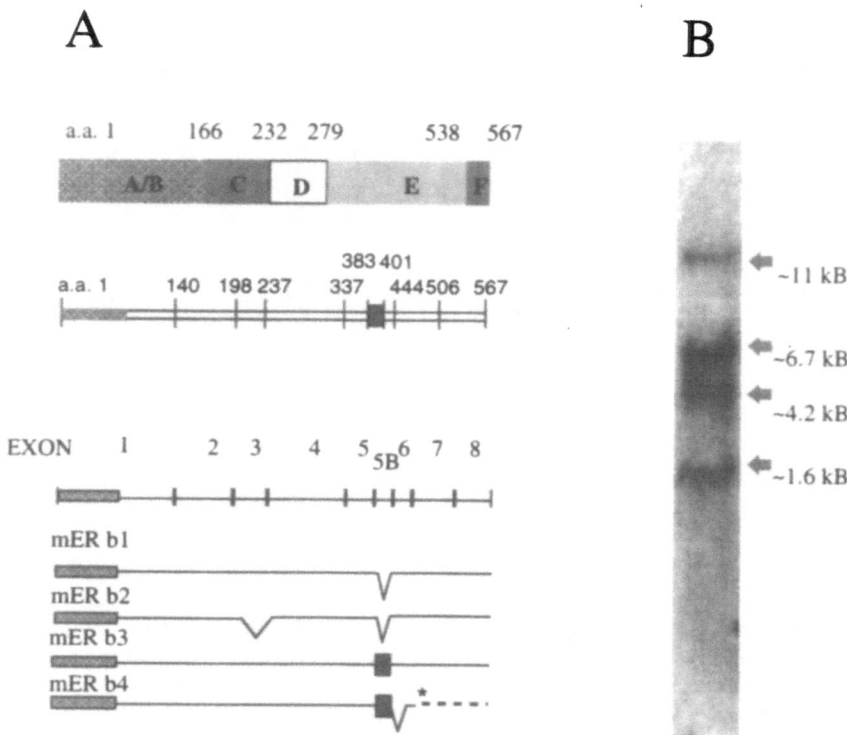

FIGURE 23.3. Schematic representation of mERβ. (A) The domains A–F of mERβ based on analogy with ERα are shown schematically. The stippled box within the LBD (domain E) denotes amino acids that correspond to exon 5B. The four isoforms (β1–β4) that are co-expressed in the mouse ovary are shown. The full-length isoform is mERβ-3. (B) Northern blot analysis of ERβ expression in the mouse ovary. The lengths of the different transcripts are shown on the right.

located before helix 7 of the LBD. The full-length isoform has been designated mERβ-3. We have also identified two other isoforms of mERβ that result from alternate splicing. One, mERβ-1 (1647 bp in length), does not contain exon 5B, but the other, mERβ-2 (1530 bp in length), is devoid of both exons 3 and 5B (Fig. 23.3A). As a consequence of the exon 3 deletion, mERβ-2 does not contain the second zinc finger and would therefore be unable to bind DNA; thus, it would most likely be devoid of transactivational functions. A fourth isoform, mERβ-4, which corresponds to a rarer message, was also identified. In this isoform, the region corresponding to exon 6 was spliced out and an in-frame stop codon that results from this is predicted to truncate the protein 12 amino acids after the beginning of exon 7 splice donor/acceptor junction (Fig. 23.3A).

Similar to observations by others, we have identified multiple species of mERβ mRNA that range in size from 1.1 to 11 kb by Northern blot analysis of total RNA isolated from murine ovaries (Fig. 23.3B). The large variation in the size of the transcripts indicates the presence of variable lengths of 5' and 3' UTRs in the different mRNA species.

Transcriptional Activation Functions of mERβ

To test whether all isoforms of mERβ could function as transactivators, transient trasfection assays were performed in COS-7 cells using a vit-tk-CAT construct as the reporter plasmid and different ligands. As shown in Figure 23.4, both mERβ-1 and mERβ-3 are capable of estradiol-stimulated transcription. mERβ-1 was found to be almost as active as mERα over a range of E2 concentrations. mERβ-3, however, displayed a much-reduced ability to transactivate (~40% of ERα activity) when compared with mERβ-1. Furthermore, this isoform was unable to activate at concentrations of E2 lower than 10^{-9} M. This was observed irrespective of the amount of expression vector used (30, 100, or 300 ng) in the assays, arguing against the difference in affinity that results from differences in levels of protein expression of the different isoforms in the cell. The mERβ-1 and mERβ-3 isoforms are co-expressed in murine ovaries as determined by RNase protection and quantita-

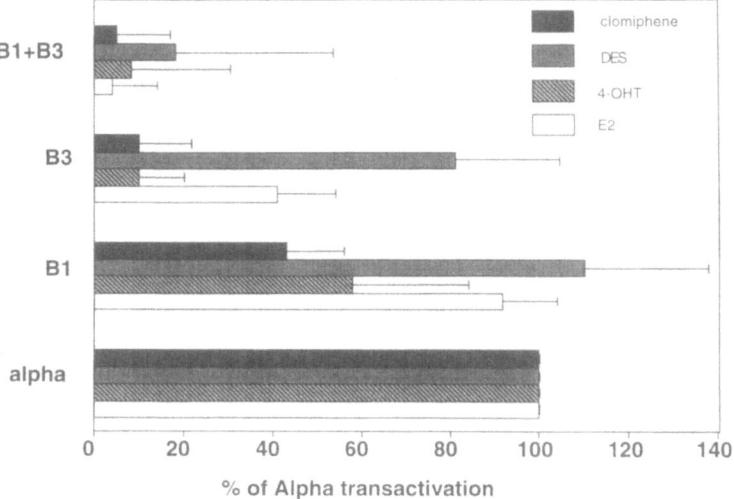

FIGURE 23.4. mERβ3 behaves as a dominant negative inhibitor of mERβ1 activity. COS-7 cells were transfected with the reporter construct vit-tk-CAT and the indicated expression constructs and an expression vector for β-galactosidase (the latter used as a control for transfection efficiency). Cells were either left unstimulated or stimulated with the indicated ligands (100 nM). CAT activity was measured in cell extracts and normalized for β-gal activity. The transactivation function of the β1 and β3 isoforms is shown relative to that of ERα considered as 100%.

tive RT-PCR experiments. To determine the biological significance of this co-expression, COS cells were transfected with equivalent quantities of the two expression plasmids. Cotransfection of both isoforms resulted in ablation of E_2-stimulated transcriptional activity of both mERβ-1 (Fig. 23.4) and ERα (data not shown). The suppression of transcription by mERβ-3 was observed at high concentrations of E_2 (100 nM). These experiments suggest that mERβ-3 has a potential to function as a dominant-negative modulator of mERβ-1 activity.

The discovery of ERβ has raised new questions about the action of estrogens on different tissues. Can some of the actions of estrogens be ascribed to ERβ activity alone (cardiovascular benefits of estrogen, Ref. 44)? The existence of different isoforms of mERβ adds to the complexity of estrogen action in target tissues in response to agonists and antagonists.

References

1. Blalock JE. The syntax of immune-neuroendocrine communication. Immunol Today 1994;15:504–11.
2. Lucas PC, Granner DK. Hormone response domains in gene transcription. Annu Rev Biochem 1992;61:1131–73.
3. Beato M, Chavez S, Truss M. Transcriptional regulation by steroid hormones. Steroids 1996;61:240–51.
4. King WJ, Greene GL. Monoclonal antibodies localize oestrogen receptor in the nuclei of target cells. Nature (Lond) 1984;307:745–47.
5. Perrot-Applanat M, Logeat F, Groyer-Picard MT, Milgrom E. Immunocytochemical study of mammalian progesterone receptor using monoclonal antibodies. Endocrinology 1985;116:1473–84.
6. Fuxe K, Wikstrom AC, Okret S, Agnati LF, Harfstrand A, Yu ZY, et al. Mapping of glucocorticoid receptor immunoreactive neurons in the rat tel- and diencephalon using a monoclonal antibody against rat liver glucocorticoid receptor. Endocrinology 1985;117:1803–12.
7. Lombes M, Farman N, Oblin ME, Baulieu EE, Bonvalet JP, Erlanger BF, et al. Immunohistochemical localization of renal mineralocorticoid receptor by using an anti-idiotypic antibody that is an internal image of aldosterone. Proc Natl Acad Sci USA 1990;87:1086–88.
8. Lees J, Fawell S, White R, Parker M. A 22-amino acid peptide restores DNA-binding activity to dimerization defective mutants of the estrogen receptor. Mol Cell Biol 1990;10:5529–31.
9. Hedden A, Muller V, Jensen EV. A new interpretation of antiestrogen action. Ann NY Acad Sci 1995;761:109–20.
10. Brzozowski AM, Pike AC, Dauter Z, et al. Molecular basis of agonism and antagonism in the oestrogen receptor. Nature 1997;389:753–58.
11. Wagner RL, Apriletti JW, McGrath ME, West BL, Baxter JD, Fletterick RJ. A structural role for hormone in the thyroid hormone receptor. Nature 1995;378:690–97.
12. Gronemeyer H, Meyer ME, Bocquel MT, Kastner P, Turcotte B, Chambon P. Progestin receptors: isoforms and antihormone action. J Steroid Biochem Mol Biol 1991;40:271–78.

13. Cole TJ, Blendy JA, Monaghan AP, et al. Targeted disruption of the glucocorticoid receptor gene blocks adrenergic chromaffin cell development and severely retards lung maturation. Genes Dev 1995;9:1608–21.
14. Reichardt HM, Kaestner KH, Tuckermann J, et al. DNA binding of the glucocorticoid receptor is not essential for survival [see comments]. Cell 1998;93:531–41.
15. Jacq X, Brou C, Lutz Y, Davidson I, Chambon P, Tora L. Human TAFII30 is present in a distinct TFIID complex and is required for transcriptional activation by the estrogen receptor. Cell 1994;79:107–17.
16. Brou C, Wu J, Ali S, et al. Different TBP-associated factors are required for mediating the stimulation of transcription in vitro by the acidic transactivator GAL-VP16 and the two nonacidic activation functions of the estrogen receptor. Nucleic Acids Res 1993;21:5–12.
17. Halachmi S, Marden E, Martin G, Mackay H, Abbondanza C, Brown M. Estrogen-receptor associated proteins: possible mediators of hormone-induced transcription. Science 1994;264:1455–58.
18. Cavailles V, Dauvois S, Danielian PS, Parker MG. Interaction of proteins with transcriptionally active estrogen receptors. Proc Natl Acad Sci USA 1994:91:10009–13.
19. Onate SA, Tsai-S-Y, Tsai M-J, O'Malley BW. Sequence and characterisation of a coactivator for the steroid hormone receptor superfamily. Science 1995;270:1354–57.
20. Hong H, Kohli K, Trivedi A, Johnson DL, Stallcup MR. GRIP1, a novel mouse protein that serves as a transcriptional coactivator for the hormone-binding domains of steroid receptors. Proc Natl Acad Sci USA 1996;93:4948–52.
21. Voegel JJ, Heine MJS, Zechal C, Chambon P, Gronemeyer H. TIF2, a 160 kDa transcriptional mediator for the ligand-dependent activation function AF-2 of nuclear receptors. EMBO J 1996;15:3667–75.
22. McInerney EM, Tsai M-J, O'Malley BW, Katzenellenbogen BS. Analysis of estrogen receptor transcriptional enhancement by a nuclear hormone receptor coactivator. Proc Natl Acad Sci USA 1996;93:10069–73.
23. Kamei Y, Xu L, Heinzel T, et al. A CBP integrator complex mediates transcriptional activation and AP-1 inhibition by nuclear receptors. Cell 1996;85:403–14.
24. Ray A, Zhang DH, Siegel MD, Ray P. Regulation of Interleukin-6 gene expression by steroids. Ann NY Acad Sci 1995;762:79–88.
25. Cato ACB, Wade E. Molecular mechanisms of antiinflammatory actions of glucocorticoids. Bioessays 1996;18:371–78.
26. Pfahl M. Nuclear receptor/AP-1 interaction. Endocrinol Rev 1993;14:651–58.
27. Ray A, LaForge KS, Sehgal PB. On the mechanism for efficient repression of the interleukin-6 promoter by glucocorticoids: enhancer, TATA box and RNA start site (Inr motif) occlusion. Mol Cell Biol 1990;10:5736–46.
28. Ray A, Prefontaine KE. Physical association and functional antagonism between the p65 subunit of transcription factor NF-κB and the glucocorticoid receptor. Proc Natl Acad Sci USA 1994;91:752–56.
29. Caldenhoven E, Liden J, Wissink S, et al. Negative cross-talk between RelA and the glucocorticoid receptor: a possible mechanism for the antiinflammatory actions of glucocorticoids. Mol Endocrinol 1995;9:401–12.
30. Scheinman RI, Gualberto A, Jewell CM, Cidlowski JA, Baldwin AS Jr. Characterization of mechanisms involved in transrepression of NF-κB by activated glucocorticoid receptors. Mol Cell Biol 1995;15:943–53.

31. Bosscher KD, Schmitz ML, Berghie WV, Plaisance S, Fiers W, Haegeman G. Gluco-corticoid-mediated repression of nuclear factor-κB-dependent transcription involves direct interference with transactivation. Proc Natl Acad Sci USA 1997;94:13504–9.

32. Nishio Y, Isshiki H, Kishimoto T, Akira S. A nuclear factor for interleukin-6 (IL-6) expression (NF-IL6) and the glucocorticoid receptor synergistically activate tran-scription of the rat α1-acid glycoprotein gene via direct protein–protein interaction. Mol Cell Biol 1993;13:1854–62.

33. Scheinman RI, Cogswell PC, Lofquist AK, Baldwin AS Jr. Role of transcriptional activation of I kappa B alpha in mediation of immunosuppression by glucocorti-coids. Science 1995;270:283–86.

34. Auphan N, DiDonato JA, Rosette C, Helmberg A, Karin M. Immunosuppression by glucocorticoids: inhibition of NF-κB activity through induction of IκB synthesis. Science 1995;270:286–90.

35. Tabibzadeh SS, Santhanam U, Sehgal PB, May LT. Cytokine induced production of IFN-β2/IL-6 by freshly explanted human endometrial stromal cells. J Immunol 1989;142:3134–39.

36. Girasole G, Jilka RL, Passeri G, et al. 17β-Estradiol inhibits interleukin-6 production by bone marrow cells and osteoblasts in vitro: a potential mechanism for the antiosteoporotic effect of estrogen. J Clin Invest 1992;89:883–91.

37. Jilka RL, Hangoc G, Girasole G, et al. Increased osteoclast development after es-trogen loss: mediation by interleukin-6. Science 1992;257:88–91.

38. Ray A, Prefontaine K, Ray P. Down-modulation of interleukin-6 gene expression by 17b-estradiol in the absence of high affinity DNA binding by the estrogen re-ceptor. J Biol Chem 1994;269:12940–46.

39. Ray A, LaForge KS, Sehgal PB. Repressor to activator switch by mutations in the first Zn finger of the glucocorticoid receptor: is direct DNA-binding necessary? Proc Natl Acad Sci USA 1991;88:7086–90.

40. Ray P, Ghosh SK, Zhang D-H, Ray A. Repression of Interleukin-6 gene expression by 17β-estradiol: inhibition of the DNA-binding activity of the transcription factors NF-IL6 and NF-κB by the estrogen receptor. FEBS Lett 1997;409:79–85.

41. Kuiper GGMJ, Enmark E, Pelto-Huikko M, Nilsson S, Gustafsson J-A. Cloning of a novel estrogen receptor expressed in rat prostate and ovary. Proc Natl Acad Sci USA 1996;93:5925–30.

42. Mosselman S, Polman J, Dijkema R. ERβ: identification and characterization of a novel human estrogen receptor. FEBS Lett 1996;392:49–53.

43. Tremblay GB, Tremblay A, Copeland NG, et al. Cloning, chromosomal localiza-tion, and functional analysis of the murine estrogen receptor beta. Mol Endocrinol 1997;11:353–65.

44. Iafrati MD, Karas RH, Aronovitz M, et al. Estrogen inhibits the vascular injury response in estrogen receptor alpha-deficient mice. Nat Med 1997;3:545–48.

24

Estrogen Receptor Alpha versus Beta: New Estrogen Responsive Tissues and New Potentials for HRT

ISTVAN MERCHENTHALER AND PAUL J. SHUGHRUE

Introduction

The steroid hormone estrogen induces the growth, differentiation, and function of many target tissues, including the female and male reproductive organs (e.g., ovary, uterus, vagina, mammary gland, testis, prostate, and epididymis) and certain neuronal populations in the brain. In the developing brain, estrogen is thought to establish a sex dimorphism in the "wiring" of certain neuronal pathways and thereby determine the phenotype of adult brain function, including the pattern of gonadotropin release, proceptive and receptive behavior, and other aspects of procreation. In the adult brain, biochemical, physiological, and histochemical techniques have revealed that estrogen regulates neurotransmitter production and release, enzyme activity, membrane potentials, dendritic arborization, synaptogenesis, and other events involved in neuronal function. Evidence suggests that estrogen action in the brain may also mediate nonreproductive events that include cognitive functions such as learning and memory. Estrogen also plays an important role in bone maintenance and in the cardiovascular system, where estrogen exerts cardioprotective effects. Estrogen is thought to regulate these physiological events by binding to its nuclear estrogen receptor (ER), interacting with a specific DNA response element and thereby modulating the expression of certain genes. Because some of the actions of estrogen occur within minutes or in regions that lack the classical estrogen receptor (ERα), it was thought that these actions were mediated through a putative membrane receptor, activation of second messenger systems, or by way of interneuronal connections (1–4). The discovery of a new nuclear ER (ERβ) by Gustafsson's group (5) provides a

259

novel mechanism for estrogen action and may explain how estrogen mediates its effects in some tissues that lack ERα.

Comparison of the Two Estrogen Receptors (ERα and ERβ)

The classical estrogen receptor (now designated ERα) that was cloned in 1986 (6) consists of several functional domains (Fig. 24.1). The N-terminal A/B or AF1 domain contains a transactivation function that appears to be important for the receptor interacting with the transcriptional machinery. The C domain contains the specific DNA-binding region (DNA-binding domain) and is also involved in receptor dimerization. The E/F or ligand-binding domain (also called AF2) contains the region that binds to estrogen and modulates dimerization, nuclear localization, and interaction with transcriptional co-activators and co-repressors (1,2). Estrogen receptor beta (ERβ) also has these same domains, which are highly homologous (Fig. 24.1) to ERα, particularly in the DNA-binding domain (95% amino acid homology) and ligand-binding domain (almost 60% amino acid homology). Since the initial discovery of ERβ in the rat prostate and ovary (5), the mouse and human homologues of ERβ have also been cloned and shown to have similar homologies to their respective ERαs (7–9). Ligand-binding experiments have shown that ERβ specifically binds [³H]-17β-estradiol with an affinity ($K_D = 0.1$ nM) similar to ERα (2). Moreover, the evaluation of binding affinities for a variety of synthetic and naturally occurring estrogenic compounds (including phytoestrogens and environmental estrogenic compounds) revealed that the relative binding affinities for ERα versus ERβ were similar (1,2). It is interesting that some phytoestrogens (e.g., genistein and coumestrol) had a somewhat higher biding affinity for ERβ when compared with ERα (Table 24.1) (10,11).

Additional studies have also shown that both ERα and ERβ are able to stimulate the transcription of an estrogen-response element (ERE)-driven reporter gene in a dose-dependent manner (2,5,7,9,12–14). In transient transfection experiments, the estrogen-induced transcriptional activity of ERβ was attenuated or enhanced when compared with ERα (9,12–14), depending on the cell lines used in these studies, which perhaps reflects the differential interactions of ERα and ERβ subtypes with certain cell-type specific transcriptional co-activator and/or co-suppressor proteins. ERβ has been shown to form a homodimer complex in vitro with consensus ERE-oligonucleotides, apparently with a similar affinity as seen for ERα (13). In addition, ERα and ERβ form heterodimeric complexes with ERE-oligonucleotides in vitro, as well as within intact cells (12,13,15). The discovery of ERβ and the putative formation of heterodimers suggests that estrogen may signal a cell that expresses both ERs via an ERα or ERβ homodimer or ERα/ERβ heterodimer. It is not known, however, whether target genes that exclusively interact with either the homodimers or heterodimers exist, but this could provide additional selectivity in a cell response to estrogen.

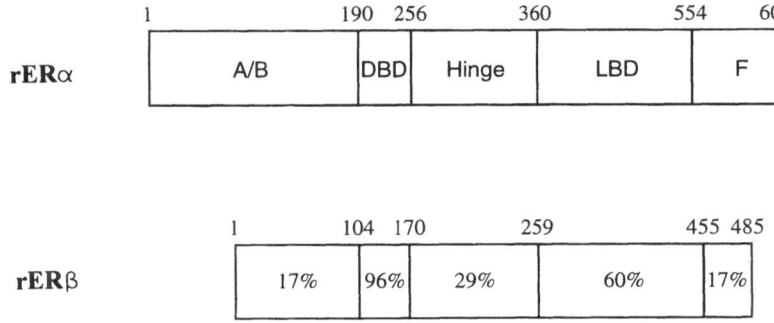

FIGURE 24.1. Comparison between ERα and ERβ protein. Percentage amino acid identities in the A/B, C, D, E, and F regions. See text for details (1,2).

Tissue-Specific Expression of ERβ mRNA

Based on RT-PCR experiments, ERβ expression was detected in the brain, spinal cord, pituitary, prostate, testis, epididymis, bladder, ovary, uterus, thymus, and the gastrointestinal tract, including the stomach and small and large intestines (Table 24.2). Concurrent in situ hybridization experiments also evaluated the regional and cellular distribution of ERβ mRNA in these

TABLE 24.1. Relative binding affinity of various compounds for ERα and ERβ.

Compound	RBA+		Ki (nM)*	
	ERα	ERβ	ERα	ERβ
17β-estradiol	100	100	0.13	0.12
17α-estradiol	58	11	0.2	1.2
Diethylstilbestrol	468	295	0.04	0.05
Estrone	60	37	0.3	0.4
Estriol	14	21	1.4	0.7
Moxestrol	43	5	0.5	2.6
4-OH-tamoxifen	178	339	0.1	0.04
ICI-164384	85	166	0.2	0.08
Nafoxidine	44	16	0.3	0.8
Clomifene	25	12	0.9	1.2
Tamoxifen	7	6	3.4	2.5
5-androstanediol	6	17	3.6	0.9
3β-androstanediol	3	7	6.0	2.0
5α-DHT	0.05	0.17	221.0	73.0
Coumestrol	94	185	0.14	0.07
Genistein	5	36	2.6	0.3

Modified from Ref. 10.

TABLE 24.2. ER expression in different organs by RT-PCR.

	ERα	ERβ
Pituitary	++++	+
Testis	+	++
Prostate	+	+++++
Epididymis	+++	+
Ovary	+++	+++++
Uterus	+++++	+
Mammary gland	++	+
Stomach	−	+
Small intestine	−	+
Large intestine	−	+
Liver	+	−
Thymus	−	+
Kidney	++	−
Adrenal	++	+
Bladder	+	+++
Lung	−	++
Heart	+	−
Vessels	+	+
Bone (osteoblast)	+	++

Based on Refs. 10, 11, and 52.

organs (16–18). The results of these studies are in good agreement with RT-PCR studies, although some differences were seen. Information on the presence of ERβ protein obtained with immunocytochemistry or Western blot analysis has been published, although the results of these studies are still controversial (19–20) and need to be repeated by other laboratories and with other antisera.

ERα and ERβ mRNA in the Brain

Through the use of in situ hybridization, the distribution of ERα and ERβ mRNA have been evaluated in adjacent sections from the rat brain (Fig. 24.2). The results of these studies revealed that the distribution of ERα and ERβ mRNA-containing perikarya was similar throughout the rostral-caudal extent of the rat brain, but some differences were noted (17). In the telencephalon, both ERs have been detected in the hippocampus (CAI–CA3 regions), the medial and lateral septum, diagonal band of Broca, basal nucleus of Maynert and in the organum vasculosum of the lamina terminalis, bed nucleus of the stria terminalis, and medial amygdala. In the diencephalon, the majority of the ERα and ERβ mRNA-containing cells are localized in a few nuclei of the hypothalamus, including the medial preoptic area, periventricular nucleus, and arcuate nucleus. In the brainstem, scattered labeled cells are present in the periaqueductal gray, parabrachial nuclei, locus ceruleus, dorsal tegmental nucleus, lateral reticular nucleus, and the nucleus

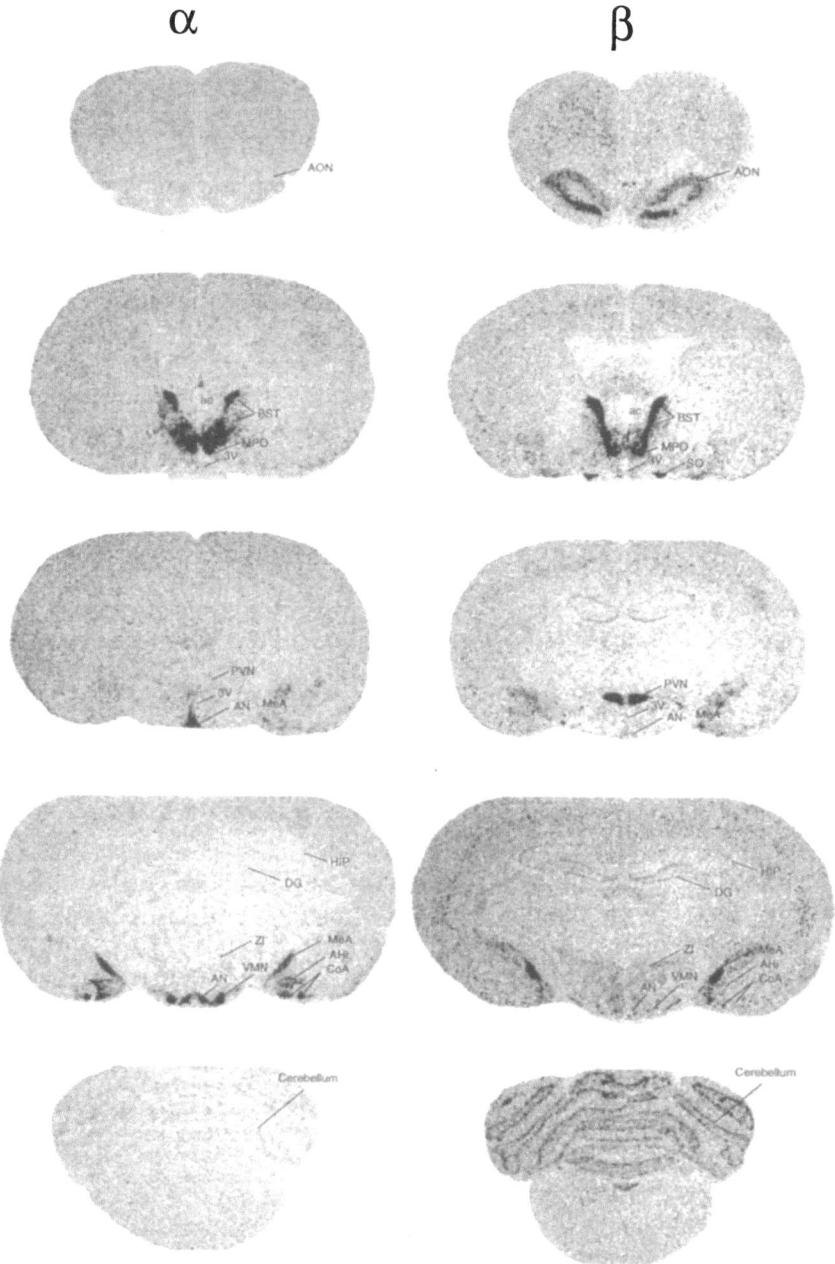

FIGURE 24.2. Representative autoradiograms of ERα and ERβ mRNAs at different levels of the rat brain after in situ hybridization with antisense riboprobes. Ac, anterior commissure; AHi, amygdalohippocampal area; AN, arcuate nucleus; AON, anterior olfactory nucleus; BST, bed nucleus of stria terminalis; CoA, cortical nucleus of the amygdala; DG, dentate gyrus of hippocampus; HIP, hippocampus; MeA, medial nucleus of amygdala; PVN, paraventricular hypothalamic nucleus; VMN, ventromedial hypothalamic nucleus; ZI: zona incerta. (Modified after Refs. 10 and 11.)

of the solitary tract. In addition, certain brain regions were found to express either ERα or ERβ exclusively (17). For example, the cerebral cortex, paraventricular nucleus, and medial tuberal hypothalamic nucleus contained only ERβ mRNA-expressing cells, whereas the ventromedial hypothalamic nucleus contains only ERα (17). The presence of ERβ mRNA in the brain has been confirmed by others (21).

ERβ mRNA Expression in ERαKO Mice

Based on initial observations in the rat brain (16,17), we were interested to see if the expression of ER-β mRNA matched the pattern of residual binding seen in the ERαKO mouse brain (22). The results of in situ hybridization studies revealed that ERβ mRNA was expressed in specific regions of the ERαKO mouse forebrain (23), with a pattern of distribution that was similar to the rat brain (see earlier and Ref. 17), although some differences were noted. In the ERαKO brain, ERβ mRNA-expressing cells were concentrated in the suprachiasmatic nucleus, an area with only sparse signal in the rat, and more abundant in the entorhinal cortex and raphe nuclei. In contrast, very few labeled cells were seen in the ERαKO supraoptic nucleus. The differences between the expression of ERβ mRNA in the rat and ERαKO mouse brain appear to be due to species differences, because a comparable distribution of ERβ mRNA was detected in the wild-type mouse brain (23).

The Presence of Functionally Active ERβ Protein in the Brain

Studies in the late 1990s have described the distribution, regulation, and pharmacology of ERβ based on its mRNA or in vitro studies. In order to ascertain if ERβ mRNA is translated into a functional protein, we used in vivo steroid autoradiography with a ligand that binds equally to ERα and ERβ. In an earlier study (22), before the discovery of ERβ (5), we detected residual binding in the preoptic area, bed nucleus of the stria terminalis, and amygdala of the ERαKO mouse brain, and showed that estrogen was capable of inducing progesterone receptor (PR) mRNA in the preoptic area of ovariectomized ERαKO mice. The presence of neurons that accumulated radiolabeled estrogen and the modulation of PR gene expression by estrogen in ERαKO mice suggested the presence of another estrogen receptor. Since the discovery of ERβ (5), we have learned that the radiolabeled ligand (MIE2) used in our earlier in vivo binding studies has an approximately 30-fold higher binding affinity for ERα versus ERβ (unpublished observations), which perhaps explains the weak binding seen in the original ERαKO studies (22). The in vivo binding studies in ERαKO mice, therefore, were repeated with a radiolabeled ligand (17α- ^{125}Iodovinyl 11β- methoxy estradiol [$17\alpha IE_2$]) that binds equally to ERα and ERβ (24). The distribution of $17\alpha IE_2$ binding seen in ERαKO mouse brain was comparable to the known distribution of ERβ mRNA. In addition, competition studies in wild-type animals with an ERα selective agonist (16α-iodo 17β-estradiol) revealed a pattern of binding that was comparable to the binding seen in the ERαKO brain and the distribution of ERβ mRNA (Fig. 24.3). These data provide the first evidence that the ERβ mRNA is translated into a protein that binds estrogen. In addition, our earlier studies

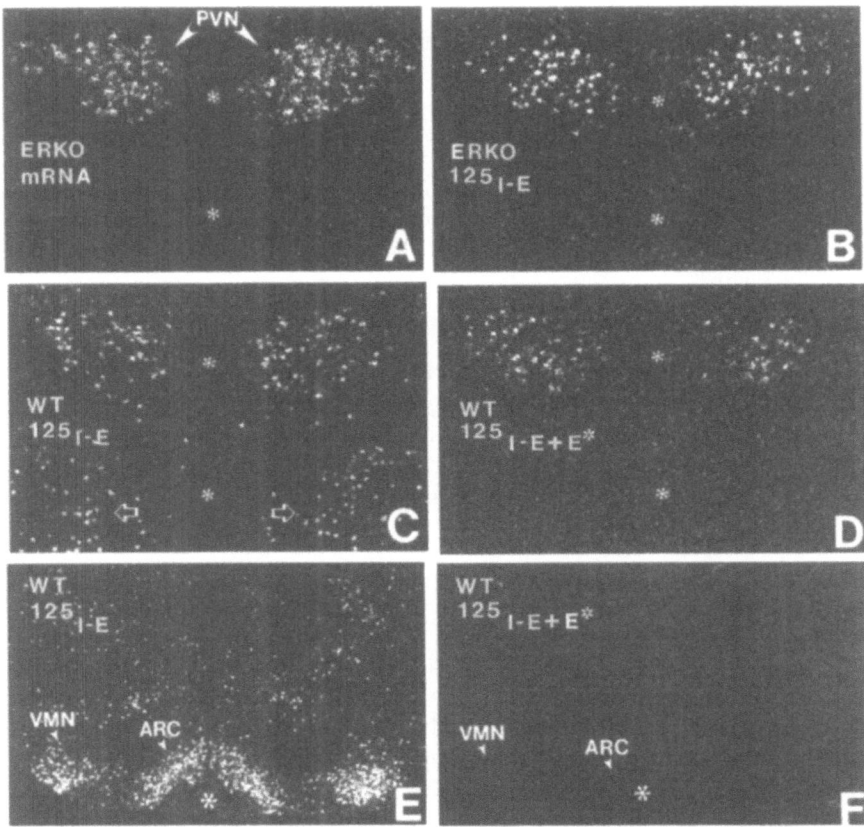

FIGURE 24.3. Autoradiographic images of ERβ mRNA (A) or ^{125}I-estrogen (^{125}I-E) binding (B–D) in the paraventricular nucleus (A–D) and the medial basal hypothalamus [arcuate nucleus (ARC) and ventromedial nucleus (VMN)] of ERαKO (A,B) and wild-type (C–F) female mice by in vivo binding and in situ hybridization histochemistry. A comparison of the distribution of ERβ mRNA (A) with ^{125}I-estrogen binding (B) in the paraventricular nucleus (PVN) of ERαKO mice suggests that the radiolabeled ligand is binding to ERβ. In the wild-type brain, ^{125}I-estrogen binding is seen in the PVN (C), the dorsal hypothalamus (open arrows in C) below the PVN, and in the medial basal hypothalamus, including the ARC and the VMN, where only ERα is expressed (E). Note that the treatment of wild-type animals with an ERα-selective agonist (E*), prior to the injection of ^{125}I-estrogen, eliminated specific binding in areas where ERα is expressed (dorsal hypothalamus in D; ARC, VMN in F) but had little or no effect on binding in the PVN (D), where ERβ is exclusively expressed. Asterisks indicate the third ventricle. Reproduced with permission from Shughrue PJ, Lane MV, Merchenthaler I. Biologically active estrogen receptor-β. Endocrinology 1999;140:2613–20. © The Endocrine Society.

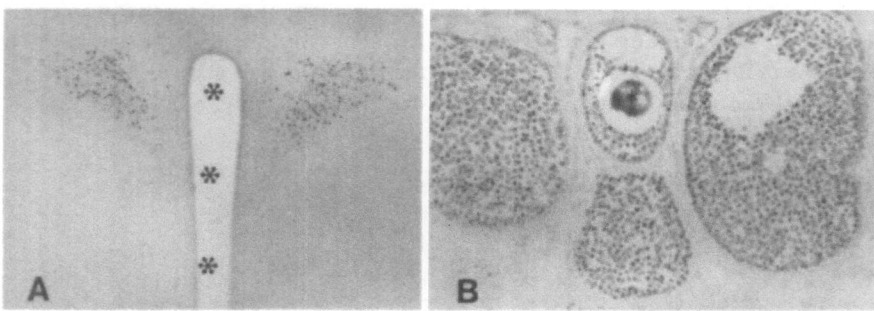

FIGURE 24.4. Immunocytochemical detection of ERβ protein in the paraventricular nucleus of the rat hypothalamus (A) and the mouse ovary (B) with an antiserum raised against the C-terminus of ERβ (FMS#20). The star labels the ovum and the asterisks indicate the third ventricle.

(22) showed that estrogen increases PR expression in the ERαKO preoptic area. Together, these observations suggest that ERβ both binds estrogen and is capable of modulating gene expression. The presence of immunoreactive ERβ protein has been confirmed with immunocytochemistry using antisera against the N- and C-terminus of the ERβ protein (Fig. 24.4A).

Distribution of ERβ mRNA and Protein in Peripheral Tissues

Expression of ERβ in Prostate and Testis

In the rat prostate, ERβ has been localized by in situ hybridization and immunohistochemistry to the secretory epithelial cells (5,18), and its expression has been shown to be upregulated by estrogen (unpublished observations). The presence of ERα mRNA and protein in stromal cells is controversial. Although RT-PCR studies suggest that ERα mRNA is present in the prostate (5) and immunocytochemical observations indicate that stromal cells are immunopositive for ERα (25), we have been unable to detect ERα mRNA with a cRNA probe that provides an excellent signal in brain and uterus (18). In situ hybridization analysis of the testis revealed that ERβ mRNA is present in the germinal cells (spermatocytes and spermatids) and Sertoli cells of the rat testis (18). The presence of ERβ protein in these cells has also been confirmed with immunocytochemistry (26). In the efferent ductules of the testis and in the epididymis, both ERα and ERβ are expressed (27). It is interesting that estrogens have been found in relatively high concentrations in semen. The function of estrogens in the testis and epididymis is largely unknown. The emerging picture, however, is that ERs, either ERα or ERβ, are co-expressed with androgen receptor in the male urogenital tract.

Because most androgens can be converted to estrogens, some of the functions in the male urogenital tract previously ascribed to androgens may result from estrogen action (28).

ERβ in Ovary and Uterus

Estrogens critically affect the growth and development of ovarian follicles during the female reproductive cycle. In the rat, both ERα and ERβ are expressed in the ovary, although their distribution is quite different (18,29). ERβ mRNA (18) and protein (Fig. 24.4B and Ref. 25) are present in the granulosa cells of small, growing, and preovulatory follicles and the adjacent thecal cells, whereas additional weak labeling can also be seen in a subset of corpora lutea (5,16,18,29). Although ERα mRNA was clearly detected by RT-PCR in ovary (5,29), examination by in situ hybridization showed that ERα mRNA is expressed at a low level throughout the ovary with no particular cellular localization (18). The presence of ERα protein in the ovary remains debatable, varying among laboratories, antibodies, and methodology.

The rat and mouse uterus contain both ERα and ERβ mRNA measured by RT-PCR, although the amount of ERα mRNA is much higher (5,30). We were unable to detect ERβ mRNA in the uterus of ovariectomized rats and only low, almost undetectable levels in intact rats or ovariectomized rats treated with estrogen (18). Although others reported low levels of ERβ immunoreactivity in the uterus (26), by using another, highly specific antiserum against the C-terminal portion of ERβ we could not detect any signal even in the intact uterus (Merchenthaler, unpublished). The lack of ERβ in the uterus is further supported by the observations that, although plasma estrogen levels are extremely high in ERαKO mice, and estrogen seems to upregulate the expression of ERβ in the uterus, the uterus of ERαKO mice does not contain immunoreactive ERβ (Merchenthaler, unpublished observations).

Expression of ERβ in Bone

Bone is an important target tissue for gonadal hormones. Ovariectomy leads to a deficit in bone ash weight and bone mineral density in adult rats and monkeys, which are changes that can be entirely prevented with exogenous estrogens (31). In osteoblastic cells, as well as osteosarcoma cell lines, ER has been detected, although since these studies were conducted prior to the discovery of ERβ it is uncertain whether these cells contained ERα, ERβ, or both receptors. One study using a transformed, human fetal osteoblastic cell line (SV-HFO) detected both ERα and ERβ mRNA (32). In primary osteoblastic cells isolated from the bone of neonatal rats, both ERα and ERβ mRNA were found, although the level of ERβ mRNA was much higher than was that of ERα mRNA (33).

ERβ mRNA in the Cardiovascular System

The beneficial effect of estrogen on the cardiovascular system may involve several endpoints, including decreasing the vascular tone, elevating high-density lipoprotein cholesterol (HDL) and decreasing low-density lipoprotein cholesterol (LDL), inhibiting LDL oxidation, decreasing plasminogen and fibrinogen concentrations, inhibiting smooth muscle cell proliferation and migration (myointimal proliferation), and decreasing the expression of different adhesion molecules (34,35). Although the level of expression of both ERs in the cardiovascular system is extremely low, RT-PCR data collected from in vitro studies utilizing different cell types indicate that at least the mRNAs of both ERs are expressed in endothelial cells and/or smooth muscle cells (36,37).

Estrogen inhibits myointimal hyperplasia in ERαKO mice after vascular injury (37). This suggests that (1) this action of estrogen is specific for ERβ, (2) ERβ can substitute for ERα, or (3) this action is mediated via a mechanism that does not involve either ERα or ERβ, such as a membrane effect. The ERβKO mouse, under development in several laboratories, and soon available for evaluation, should help answer these questions.

Tissue- and Cell-Specific Action of Estrogens, Partial Agonists, and Antagonists: The Potential Role of ERβ

Most recent drugs targeted to the ER, such as tamoxifen, ICI-182780, and raloxifene (also known as selective estrogen receptor modulators or SERMs) exhibit a curious pharmacological profile, acting as either ER antagonist or agonist depending on the animal species, target tissue, and the dose administered (38). For instance, the benzothiophene analog raloxifene has been reported to act as an antiestrogen in breast tumor tissue and the brain, even though it has potentially beneficial estrogenlike effects in nonreproductive tissues such as bone and in lipid profile (39–42). Another example is tamoxifen, which was developed as an antiestrogen for the treatment of breast cancer and subsequently shown to have estrogenlike effects on bone and the cardiovascular system (38). The potentially beneficial effects of tamoxifen in reducing the risk of osteoporotic fractures and coronary heart disease in postmenopausal women, however, are at least partially offset by the estrogenic effects in the uterus, increased risk of endometrial cancer development, and antagonist activity in the brain, including an increased incidence of hot flushes (38,42,43). Estrogen replacement therapy in postmenopausal women is used increasingly with the aim of relieving hot flashes and mood changes, reducing skin, urinary, and reproductive tract atrophy, delaying of atherosclerosis, and loss of bone mass. It is interesting that estrogen replacement therapy also appears enhancing to cognitive function (44), with the onset of Alzheimer's disease in women treated with conjugated estrogens being significantly delayed when compared with age-matched controls (45). Although the mechanism by which estrogen improves cognitive function remains unclear, the discovery of ERβ and its localization in brain regions associated with learning and memory (i.e., cerebral cortex, hippocampus, amygdala, etc.) may provide new insight to these questions. The most compelling problem of estrogen replacement therapy is the putative increase in breast cancer and inconvenience due to irregular spotting and bleeding. Although the precise mechanism by which SERMs exert their tissue-specific action is not known, one possible mechanism is that they induce conformational changes in the ER, thus altering the affinity of the receptor/ligand complex to the known ERE or other response elements in a tissue-specific fashion (46–49). The genes for transforming growth factor β3 and quinone reductase are examples of ERα-regulated genes controlled by promoters that contain response elements distinct from the "classical" EREs that are activated by certain SERMs, but are inhibited by or much less efficiently regulated by other estrogens (49). The action of ERβ at either of these promoters has not been reported.

One must also consider that ERα as well as ERβ can mediate gene transcription from an AP1-enhancer element that requires ligand and the AP1 transcription factors Fos and Jun for transcriptional activation (50,51). In transient transfection systems, estrogen bound to ERα or ERβ was shown to signal in opposite directions from an AP1 site. Estrogen bound to ERα estrogen activated transcription, but bound to ERβ it inhibited transcription. The ER ligands tamoxifen, raloxifene, and ICI-182780 were activators with ERβ as well as ERα, although the degree of agonism differed between cell types (50). Thus, the role of estrogen bound to ERβ would be to turn off the transcription of these genes, whereas the tissue-selective estrogens could override this blockade and activate gene transcription. The finding that ERα and ERβ respond differently to certain ligands at AP1 sites adds another possible mechanism for the tissue-selective effects of estrogens and other partial agonists and antagonists. Moreover, the potential tissue- and cell-specific expression of co-activators and co-suppressors with either ERα, ERβ, or both, and their different binding affinities to ERs makes the tissue-, cell-, and, perhaps, gene-specific action of ER ligands even more complicated (2). The challenge for the future will be to ascertain the relative importance of each of these proposed mechanisms with the aim to develop the next generation of tissue-, cell-, and gene-specific estrogen receptor ligands. The discovery of ERβ and its localization in the brain and peripheral tissues allow one to hypothesize that an ERβ-selective ligand could be a well-tolerated hormone replacement therapy for the treatment of declining cognitive functions, osteoporosis, atherosclerosis, and so on, without the unwanted effects in the uterus and breast.

References

1. Kuiper GG, Gustafsson J-A. The novel estrogen receptor-β type: potential role in the cell- and promoter-specific actions of estrogens and anti-estrogens. FEBS Lett 1997;410:87–90.

270   I. Merchenthaler and P.J. Shughrue

2. Kuiper G, Shughrue PJ, Pelto-Huikko M, Merchenthaler I, Gustafsson J-A. The estrogen receptor β subtype: a novel mediator of estrogen action in neuroendocrine systems. Front Neuroendocrinol 1998;19:253–86.
3. Joels M. Steroid hormones and excitability in the mammalian brain. Front Neuroendocrinol 1997;18:2–48.
4. Wehling M. Specific, nongenomic actions of steroid hormones. Annu Rev Physiol 1997;59:365–93.
5. Kuiper GG, Enmark E, Pelto-Huikko M, Nilsson S, Gustafsson J-A. Cloning of a novel estrogen receptor expressed in rat prostate and ovary. Proc Natl Acad Sci USA 1996;93:5925–30.
6. Greene GL, Gilna P, Waterfield M, Baker A, Hort Y, Shine J. Sequence and expression of human estrogen receptor complementary DNA. Science 1986;231:11505–54.
7. Mosselman S, Polman J, Dijkema R. ERβ: identification and characterization of a novel human estrogen receptor. FEBS Lett 1996;392:49–53.
8. Enmark E, Pelto-Huikko M, Grandien K, Lagercrantz S, Lagercrantz J, Fried G, et al. Human estrogen receptor β—gene structure, chromosomal localization and expression pattern. J Clin Endocrinol Metab 1997;82:4258–65.
9. Tremblay GBA, Tremblay A, Copeland NG, Gilbert DJ, Jenkins NA, Labrie F, et al. Cloning, chromosomal localization and functional analysis of the murine estrogen receptor β. Mol Endocrinol 11:353–65.
10. Kuiper GG, Lemmen JG, Carlsson B, Corton JC, Safe SH, van der Saag P, et al. Interaction of estrogenic chemicals and phytoestrogens with estrogen receptor β. Endocrinology 1998;139:4252–63.
11. Kuiper GG, Carlsson B, Grandien KAJ, Enmark E, Haggblad J, Nilsson S, et al. Comparison of the ligand binding specificity and transcript tissue distribution of estrogen receptor α and β. Endocrinology 1997;138:863–70.
12. Cowley SM, Hoare S, Mosselman S, Parker MG. Estrogen receptors α and β form heteromdimers on DNA. J Biol Chem 1997;272:19858–62.
13. Pettersson K, Grandien K, Kuiper GG, Gustafsson J-Å. Mouse estrogen receptor β-forms estrogen response element-binding heterodimers with estrogen receptor α. Mol Endocrinol 1997;11:1486–98.
14. Watanabe T, Inoue S, Ogawa S, Ishii Y, Hiroi H, Ikeda K, et al. Agonistic effect of tamoxifen is dependent on cell type, ERE-promoter context, and estrogen receptor subtype: functional difference between estrogen receptors α and β. Biochem Biophys Res Commun 1997;236:140–45.
15. Pace P, Taylor J, Suntharalingam S, Coombes C, Ali S. Human estrogen receptor β binds DNA in a manner similar to and dimerizes with estrogen receptor α. J Biol Chem 1997;272:25832–38.
16. Shughrue PJ, Komm B, Merchenthaler I. The distribution of estrogen receptor-β mRNA in the rat hypothalamus. Steroids 1996;61:678–81.
17. Shughrue PJ, Lane MV, Merchenthaler I. Comparative distribution of estrogen receptor-α and -β mRNA in the rat central nervous system. J Comp Neurol 1997;388:507–25.
18. Shughrue PJ, Lane MV, Scrimo PJ, Merchenthaler I. Comparative distribution of estrogen receptor-α (ER-α) and b (ER-β) mRNA in the rat pituitary, gonad, and reproductive tract. Steroids 1998; 63:498–504.
19. Li X, Schwartz PE, Rissman EF. Distribution of estrogen receptor-β-like immunoreactivity in rat forebrain. Neuroendocrinology 1997;66:63–67.

20. Rosenfeld CS, Ganjam VK, Taylor JA, Yuan X, Stiehr JR, Hardy MP, et al. Transcription and translation of estrogen receptor-β in the male reproductive tract of estrogen receptor-α knock-out and wild-type mice. Endocrinology 1998;139: 2982–87.

21. Laflamme N, Nappi RE, Drolet G, Labrie C, Rivest S. Expression and neuropeptidergic characterization of estrogen receptors (ERα and ERβ) throughout the rat brain: anatomical evidence of distinct roles of each subtype. J Neurobiol 1998;36:357–78.

22. Shughrue PJ, Lubahn DB, Negro-Vilar A, Korach KS, Merchenthaler I. Responses in the brain of estrogen receptor- α-disrupted mice. Proc Natl Acad Sci USA 1997;94:11008–12.

23. Shughrue P, Scrimo P, Lane M, Askew R, Merchenthaler I. The distribution of estrogen receptor-β mRNA in forebrain regions of the estrogen receptor-α knockout mouse. Endocrinology 1997;138:5649–52.

24. Shughrue PJ, Lane MV, Merchenthaler I. Biologically active estrogen receptor-β: evidence from in vivo autoradiographic studies with estrogen receptor α-knockout mice. Endocrinology 1999;140:2613–20.

25. Ehara H, Koji T, Deguchi T, Yoshii A, Nakano M, Nakane PK, et al. Expression of estrogen receptor in diseased human prostate assessed by non-radioactive in situ hybridization and immunohistochemistry. Prostate 1995;27:304–13.

26. Saunders PTK, Maguire SM, Gaughan J, Millar MR. Expression of oestrogen receptor beta (ERβ) in multiple rat tissues visualised by immunohistochemistry. J Endocrinol 1997;154:R13–16.

27. Hess RA, Gist DH, Bunick D, Lubahn DB, Farrell A, Bahr J, et al. Estrogen receptor (α and β) expression in the excurrent ducts of the adult male rat reproductive tract. J Androl 1997;18:602–11.

28. Sharpe RM. Do males rely on female hormones? Nature 1997;390:447–48.

29. Byers M, Kuiper GG, Gustafsson J-A, Park-Sarge OK. Estrogen receptor-β mRNA expression in rat ovary: downregulation by gonadotropins. Mol Endocrinol 1997;10:119–31.

30. Couse JF, Lindzey J, Grandien K, Gustafsson J-Å, Korach KS. Tissue distribution and quantitative analysis of estrogen receptor-α (ER-α) and estrogen receptor-β (ER-β) messenger RNA in the wild-type and ERα-knockout mouse. Endocrinology 1997;138:4613–48.

31. Turner RT, Riggs BL, Spelsberg TC. Skeletal effects of estrogen. Endocrinol Rev 1994;15:275–96.

32. Arts J, Kuiper GG, Janssen JMMF, Gustafsson J-Å, Löwik CWGM, Pols HAP, et al. Differential expression of estrogen receptors α and βmRNA during differentiation of human osteoblast SV-HFO cells. Endocrinology 1997;138:5067–70.

33. Onoe Y, Miyaura C, Ohta H, Nozawa S, Suda T. Expression of estrogen receptor β in rat bone. Endocrinology 1997;138:4509–12.

34. Farhat MY, Lavigne MC, Ramwell PW. The vascular protective effects of estrogen. FASEB J 1996;10:615–24.

35. Foegh ML, Ramwell PW. Cardiovascular effects of estrogen: implications of the discovery of the estrogen receptor subtype β. Curr Opin Nephrol Hypertension 1998;7:83–89.

36. Karas RH, Patterson BL, Mendelson ME. Human vascular smooth muscle cells contain functional estrogen receptor. Circulation 1994;89:1943–50.

37. Iafrati MD, Karas RH, Aronovitz M, et al. Estrogen inhibits the vascular injury response in estrogen receptor a-deficient mice. Nat Med 1997;3:545–48.
38. Katzenellenbogen JA, O'Malley BW, Katzenellenbogen BS. Tripartite steroid hormone receptor pharmacology: interaction with multiple effector sites as a basis for the cell- and promoter-specific action of these hormones. Mol Endocrinol 1996;10:119–31.
39. Delmas PD, Bjarnasan NH, Mitlak BH, Ravoux AC, Shah AS, Huster WJ, et al. Effects of raloxifene on bone mineral density, serum cholesterol concentrations, and uterine endometrium in post-menopausal women. N Engl J Med 1997;337: 1641–47.
40. Gustafsson J-Å. Raloxifene: magic bullet for heart and bone? Nat Med 1998;4: 152–53.
41. Sato M, Rippy MK, Bryant HU. Raloxifene, tamoxifen, nafoxidene, or estrogen effects on reproductive and nonreproductive tissues in ovariectomized rats. FASEB J 1996;10:905–12.
42. Shughrue PJ, Lane MV, Merchenthaler I. Regulation of progesterone receptor messenger ribonucleic acid in the rat medial preoptic nucleus by estrogenic and antiestrogenic compounds: an in situ hybridization study. Endocrinology 1997;138:5476–84.
43. Webb P, Lopez GN, Uht RM, Kushner PJ. Tamoxifen activation of the estrogen receptor/AP-1 pathway: potential origin for the cell-specific estrogen-like effects of antiestrogens. Mol Endocrinol 1995;9:443–46.
44. Birge SJ. Is there a role for oestrogen replacement therapy in the prevention and treatment of dementia? J Am Geriatr Soc 1996;44:865–70.
45. Tang M-X, Jacobs D, Stern Y, Marder K, Schofield P, Gurland B, et al. Effect of oestrogen during menopause on risk and age at onset of Alzheimer's disease. Lancet 1996;348:429–32.
46. Brzozowski AM, Pike ACW, Dauter Z, Hubbard RE, Bonn T, Engström O, et al. Molecular basis of agonism and antagonism in the oestrogen receptor. Nature 1997;389:753–58.
47. Grese TA, Sluka JP, Bryant HU, Cullinan GJ, Glasebrook AL, Jones CD, et al. Molecular determinants of tissue selectivity in estrogen receptor modulators. Proc Natl Acad Sci USA 1997;94:14105–10.
48. McDonnell DP, Clemm DL, Hermann T, Goldman ME, Pike JW. Analysis of estrogen receptor function in vitro reveals three distinct classes of antiestrogens. Mol Endocrinol 1995;9:659–69.
49. Yang NN, Venugopalan M, Hardikar S, Glasebrook A. Identification of an estrogen response element activated by metabolites of 17β-estradiol and raloxifene. Science 1996;273:1222–25.
50. Paech K, Webb P, Kuiper GG, Nilsson S, Gustafsson J-Å, Kushner P, et al. Differential ligand activation of estrogen receptors ERα and ERβ at AP1 sites. Science 1997;277:1508–10.
51. Weisz A, Rosales R. Identification of an estrogen response element upstream of the human c-fos gene that binds the estrogen receptor and the AP1 transcription factor. Nucl Acid Res 1990;18:5097–106.
52. Campbell-Thompson ML. Estrogen receptor α and β expression in upper gastrointestinal tract with regulation of trefoil factor family 2 mRNA levels in ovariectomized rats. Biochem Biophys Res Commun 1997;240:478–83.

25

Putting the Menopause into Perspective: Lessons Learned from the *Biology of Menopause*

FREDERICK NAFTOLIN

As the millenium draws to a close, it is particularly gratifying to look back upon the emergence of women as a force throughout our entire society. During the past century, women's life spans doubled and their influence in all walks of life far exceeded that growth. This book is about the physiologic adjustments that face these newly long-lived women as they elect to continue or change their activities with aging. The chapters have stressed the critical importance of ovarian aging in determining the timing of the menopausal transition, menopause, and the postmenopause (combined to form the climacteric). Work done by many laboratories on the aging rodent and nonhuman primate has been discussed. It is important to note from the limitations of models that during their normal life span they do not undergo ovarian follicle depletion, instead undergoing hypothalamic failure with aging, as has been discussed. Even though estrogen production and secretion by the ovaries and nonovarian tissues was featured, the underlying metabolism also underlines the key roles of androgens and progestins as both prohormones and in their oft-overlooked roles in the modification of estrogen action. The vagaries of perimenopausal ovarian function have been unveiled. In addition to the recent and wonderful complication of the presence of a second estrogen receptor and its isoforms, and the possibility of heterodimeric control of "estrogen action" at the receptor-DNA level, nonreceptor-mediated activities and secondary regulatory compound perturbations have been brought to center stage. Compounds other than the natural steroidal estrogens have entered the picture.

In respect of all these disclosures, what is the mission of the National Institute on Aging (NIA) in menopause research? The NIA has studies on aging as its mission. In considering the climacteric, research must distinguish between the effects of aging and the effects of hormonal deficiency. Because both are found in the aging menopausal woman (and the aging male) mechanisms must be developed that allow both factors to be studied as independent variables or as covariables. Although these often include studies on the perimenopause and/or menopause, the distinction and root interest in aging must be clearly represented.

The work presented in this volume is of direct importance to womens' health. In considering the many remarkable processes that were elucidated, it is important to see the interrelationship and continuity of the climacteric woman with her younger sisters and daughters. Women may thusly be encouraged to understand the origins of their success in reproduction and in evolutionary fitness, as well as how this can play a deleterious role during the climacteric, unless extraevolutionary steps such as lifestyle changes, dietary changes, and, perhaps, hormone replacement therapy are employed. Within this volume a veritable cornucopia of new information was disclosed/reviewed that defines women's responses during the climacteric; however, it also became clear that more attention must be paid to the bases of climacteric responses. In this way accounting for the precise contributions of aging and prevention can be accomplished. On first inspection the responses to the climacteric are largely the same responses that are seen in all women who become hypogonadal, rather than being specific to the climacteric. This is generally a true observation, varying according to the woman's age, the completeness of ovarian loss (ablation vs. follicle failure), the abruptness of the change, and the degree of morbidity already present. Why, then, should elderly women respond to hypogonadism in the same manner as younger, reproductive-age women? What is the role of age in the menopausal response? Is there an evolutionary basis to the response? If so, should the response to hypogonadism not differ between the reproductive and nonreproductive years?

The menopausal cessation of reproduction forecloses Darwinian evolution toward "fitness" for the menopause, and, in the absence of a specific "menopausal response," women show responses to the climacteric that are the same ones they show during hypogonadism during the reproductive years (1). These are adaptive responses that are the basis of successful passage of individuals' gene complements during reproduction; therefore, they have been highly honed during evolution to favor reproductive success; other responses have been weeded out by reproductive failure. Thus, during the perimenopausal period in response to increasing hypogonadism, climacteric women continue to express responses to the changing endocrine milieu that during evolutionary selection favored successful reproduction and maintenance of the offspring until the F2 generation became pubertal. A good exposition of such responses is seen during the puerperium. These include elevated central arterial pressure to maintain blood flow during the postpartum period despite massive circulatory changes and blood loss, elevation of blood lipids and calcium to produce lactational milk, disintegrated sleep patterns to foster vigilance and avoid inadvertent damage to the newborn, radiation of heat in the form of peripheral vasodilatation or hot flushes to warm the newborn, and so on. When they arise during the climacteric, however, these previously adaptive responses often have adverse effects (e.g., cardiovascular disease, osteoporosis, poor mentation, and hot flushes). Understanding the basis of these responses should color research on the climacteric and on the experimental climacteric surrogates, as well as raise questions about their effect on young "normal" women, such as: Are arteriosclerotic precursors present in reproductive-age women; What does pregnancy do to peak bone mass?

Several general areas exposed in this symposium are noteworthy in their relationship to future research goals.

Models

Whereas humans are unique, all animals have value in the study of aging menopausal symptoms. Even though rodents do not undergo ovarian failure, there are many aspects of hormonal deprivation and aging that can be more effectively and ethically addressed. This will become of increasing importance as the use of genetically altered animals increases.

Primates are closer to humans, but they should not automatically be considered as identical or complete surrogates for aging humans. This is especially important because it has become acceptable to use castrate monkeys as menopausal surrogates, a paradigm that does allow separation of aging from hypogonadism.

Wherever feasible, human subjects should be studied. If not, the interpretation of the studies must be circumspect and should follow from clear, hypothesis-driven questions.

What Should Be Studied in Aging Women Undergoing Menopause?

Endocrinology is just one of the factors that come into play in the aging individual. The United States is multicultural and pluralistic. There are many antecedent factors that contribute to the response to the loss of ovarian function at any age. Differences in genetic composition and secular changes must be considered lest our information be devalued. The field is widely open, from motivational studies that test women's knowledge and interest to quite esoteric dissection of changes due to aging versus changes due to hypogonadism, per se.

Factors Underlying Physiological Ovarian Failure

The function of the hypothalamus and gonads or their deterioration must be pursued beginning during the reproductive years (i.e., prior to age 35 when ovarian function declines) (2). A number of leads have already been raised in this volume, indicating that the human central nervous system and pituitary may fail with age.

Ovarian studies that have been lengthy and difficult are now coming to fruition and should be completed. Two important questions coming to resolution are the precise timing and course of the beginning of physiological ovarian deficiency, and the effects of ovarian deficiency, per se, versus the effects of age itself.

Steroid Hormone-Related Issues

The discovery of the second estrogen receptor and its subtypes, and the lack of direct investigation of different estrogen receptor–bearing areas of the body, particularly the brain, are of obvious importance. Along these lines, little is known of the basis of androgen action in the aging individual and the effects of androgen depletion. The same may be said for the role of progestins in the postmenopausal period. Clinically driven questions are also important (e.g., clinical pharmacology and the role of local versus systemic administration of hormones in the aging female).

Xenoestrogens

Time did not permit entry in depth into this increasingly important area of daily life. Evidence indicates that we are knowingly and unknowingly bombarded by phytoestrogens and other xenoestrogens that, to varying degrees, obey some of the newly evolving rules that steroid receptorology is teaching us. Our ignorance of this subject and of the actual manufacturing process of individual preparations is soon to be addressed by regulatory agencies of the federal and state governments. It behooves us as scientists to have some of the answers or to be planning to get them.

Issues Related to Specific Systems

Although it is tempting to give a progress report on studies of specific systems, this would generally be repetitive of the content of the preceding chapters. Space does not permit such an approach. It is clear, however, that there are specific areas that deserve special attention. In the brain, cerebrovascular insufficiency in association with hypogonadism, better definition of cognitive elements, and studies on cerebral dystrophies like Alzheimer's syndrome and Parkinson's syndrome are of the greatest importance. In the cardiovascular area, much progress is being made, all of which moves our understanding of the basis of disease in the aging toward poor diet and prevention in the young. Because cardiovascular death is the fate of most of us, we must ignore research in this area at our own peril. The immune system remains little known in the aged individual. If we are now finishing the "decade of the brain," then it seems that the next will be the "decade of nonimmune regulation by the immune system." This is nowhere as important as it is in the aging individual. The skeletal system has spawned a new and important wave of studies to understand bone cell kinetics better. Finally, cancer is increasingly unraveled and deserves to be seen as a separatable concomitant of aging.

The Value of This Volume

Even though it is easy to lay out a laundry list of areas of important interests, it is a harder task to lay out the guidelines for attacking those interests and prioritizing them. Volumes such as this one are the kind of "update and consensus" meetings that will lead to rapid and appropriate progress toward understanding both the biology of aging and the biology of menopause. It is, therefore, particularly important to prioritize future studies in light of difficulties imposed by limited time and resources. The organizing committee had to make similar, difficult choices and priorities in order to develop a cohesive theme for this volume. Let us hope that more such meetings will follow and that many of the research goals expressed here will come to fruition.

Acknowledgment. I would like to thank Dr. Frank Bellino for his helpful discussions.

References

1. Naftolin F, Whitten P, Keefe D. An evolutionary perspective on the climacteric and menopause. Menopause 1994;4:223–25.
2. Seifer D, Naftolin F. Moving toward an earlier and better understanding of the perimenopause. Fertil Steril 1998;69:387–88.

Author Index

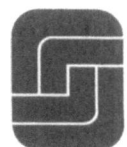

Subject Index

PROCEEDINGS IN THE SERONO SYMPOSIA USA SERIES

(Continued)